Die Grundlehren der mathematischen Wissenschaften

in Einzeldarstellungen
mit besonderer Berücksichtigung
der Anwendungsgebiete

Band 156

B. L. van der Waerden

Mathematical Statistics

With 39 Figures

Springer-Verlag New York · Heidelberg · Berlin 1969

Translation of

Mathematische Statistik

(Grundlehren der mathematischen Wissenschaften,

Bd. 87, 2. Auflage, 1965)

Printed in Germany. Title No. 5139

Foreword to the English Edition

Chapters 1 – 7 were translated by Mrs. Virginia Thompson, and Chapters 8 – 14 by Miss Ellen Sherman, at the University of California, Berkeley, and the University of Hull, England. In my opinion, both translators have done a superb job. In particular, they have taken care to find the correct translations of technical terms in widely divergent fields like biology and economics. They also have corrected several errors in formulae. They asked me every time they did not fully understand the reasoning. Several points are now explained more clearly than in the original German text.

Thanks are due to E. L. Lehmann (Berkeley), who initiated the translation project and whose help was very valuable. Following his advice, I have added to § 4 a subsection on testing the mean of a normal distribution with given variance.

Zürich, July 1968 B. L. van der Waerden

Foreword

Ever since my days as a student, economists, doctors, physiologists, biologists, and engineers have come to me with queries of a statistical nature. This book is the product of my long interest in practical solutions to such problems. Study of the literature and my own ideas have repeatedly led me to improved methods, which will be established here and applied to instructive examples taken from the natural and social sciences. Thus, I hope to help the reader avoid the many fruitless directions in which I worked at first. The examples are not artificially constructed from theoretical considerations but are instead taken from real situations; consequently, many of the examples require detailed explanation.

The presentation of the basic mathematical concepts is, I hope, as brief as possible without becoming incomprehensible. Some rather long theoretical arguments have been necessary, but, whenever possible, references for the more difficult proofs have been made to good textbooks already in existence. There would be no point in developing again the mathematical theories which have been presented clearly and in detail by Kolmogorov, Carathéodory, and Cramér.

A knowledge of the elements of the theory of functions and of the Lebesgue theory of integration has been assumed. This does not imply that a reader without such knowledge will not be able to understand the book: he will have either to accept certain theorems without proof or to confine himself to the more elementary sections, in which only analytic geometry and calculus are used (Chapters 1 through 4, 10, and 12).

The book is only meant to be an introduction. No attempt has been made at completeness, and such important topics as sequential analysis, decision theory, and stochastic processes have had to be omitted. However, distinguished experts have written specialized books devoted to such topics:

A. Wald, *Sequential Analysis.* New York: John Wiley & Sons 1947;

A. Wald, *Statistical Decision Functions.* New York: John Wiley & Sons 1950;

J. L. Doob, *Stochastic Processes.* New York: John Wiley & Sons 1953.

In many places references for further reading have been given. These have been placed where they are readily available, in the text or as footnotes. The new style of putting footnotes at the end of the book or, even worse, at the ends of the chapters causes a horrible amount of flipping back and forth through the pages. I also consider it preferable to write p. 5 rather than just 5. No attempt has been made at uniformity in quotations and extensive abbreviations have been avoided.

The first draft of this book was written in 1945 and served as the basis for a course on error theory and statistics at the Shell Laboratory in Amsterdam. A later version was read critically by Dr. E. Batschelet (Basel). I wish to express my thanks to him and also to Professor E. L. Lehmann (Berkeley) for their most valuable criticism. Also I thank Mr. H. R. Fischer and Mr. E. Nievergelt (Zürich) for drawing the figures and for their help with the proofreading.

September 1956 B. L. van der Waerden

Foreword to the Second Edition

In the new edition, Fig. 28, which was in error, has been redrawn by Mr. H. Studer.

April 1965 B. L. van der Waerden

Contents

Introduction 1

Chapter One: General Foundations

§ 1. Foundations of Probability Theory 3
§ 2. Random Variables and Distribution Functions 8
§ 3. Expectation and Standard Deviation 12
§ 4. Integral Representation for Expectations and Probabilities 17

Chapter Two: Probabilities and Frequencies

§ 5. Binomial Distribution 24
§ 6. Deviation of the Frequency h from the Probability p 27
§ 7. Confidence Bounds for Unknown Probabilities 32
§ 8. Sampling 37
§ 9. Comparison of Two Probabilities 40
§ 10. Frequency of Rare Events 47

Chapter Three: Mathematical Tools

§ 11. Multiple Integrals. Transformations to Polar Coordinates 52
§ 12. Beta and Gamma Functions 55
§ 13. Orthogonal Transformations 60
§ 14. Quadratic Forms and Their Invariants 61

Chapter Four: Empirical Determination of Distribution Functions, Expectations, and Standard Deviations

§ 15. The Quetelet Curve 67
§ 16. Empirical Determination of Distribution Functions 69
§ 17. Order Statistics 75
§ 18. Sample Mean and Sample Variance 79
§ 19. Sheppard's Correction 82
§ 20. Other Mean and Dispersion Measures 85

Chapter Five: Fourier Integrals and Limit Theorems

§ 21. Characteristic Functions 89
§ 22. Examples 93
§ 23. The χ^2 Distribution 95
§ 24. Limit Theorems 97
§ 25. Rectangular Distribution. Rounding Errors 104

Chapter Six: Gauss Theory of Errors and Student's Test

§ 26. Gauss Theory of Errors 108
§ 27. The Distribution of s^2 113
§ 28. Student's Test 118
§ 29. Comparison of Two Means 121

Chapter Seven: Method of Least Squares

§ 30. Smoothing Observational Errors 127
§ 31. Expectations and Standard Deviations of the Estimates $\tilde{\vartheta}$ 133
§ 32. Estimation of the Variance σ^2 139
§ 33. Linear Regression 144
§ 34. Causal Interpretation of Dependence between Economic Variables 149

Chapter Eight: Estimation of Unknown Constants

§ 35. R. A. Fisher's Method of Maximum Likelihood 151
§ 36. Determination of the Maximum 155
§ 37. An Inequality Due to Fréchet 160
§ 38. Sufficiency and Minimum Variance 162
§ 39. Examples 165
§ 40. Conditional Expectation 168
§ 41. Sufficient Statistics 170
§ 42. Application to the Problem of Unbiased Estimation 173
§ 43. Applications 175
§ 44. Estimation of the Variance of a Normal Distribution 179
§ 45. Asymptotic Properties 182

Chapter Nine: Inferences Based on Observed Frequencies

§ 46. The Maximum Likelihood Method 185
§ 47. Consistency of the Maximum Likelihood Estimate 189
§ 48. Maximum Likelihood, Minimum χ^2, and Least Squares 192
§ 49. Asymptotic Distributions of χ^2 and $\tilde{\vartheta}$ 197
§ 50. Efficiency 203
§ 51. The χ^2-Test 207

Chapter Ten: Bio-Assay

§ 52. Response Curves and Logarithmic Response Curves 212
§ 53. Integral-Approximation Method of Behrens and Kärber 214
§ 54. Methods Based on the Normality Assumption 217
§ 55. "Up and Down" Methods 221

Chapter Eleven: Hypothesis Testing

§ 56. Applications of the χ^2-Test 225
§ 57. The Variance-Ratio Test (F-Test) 242
§ 58. The Analysis of Variance 246
§ 59. General Principles. Most Powerful Tests 256
§ 60. Composite Hypotheses 263

Chapter Twelve: Order Tests

§ 61. The Sign Test 267
§ 62. The Two-Sample Problem 271
§ 63. Wilcoxon's Test 273
§ 64. The Power of the Wilcoxon Test 282
§ 65. The X-Test 290

Chapter Thirteen: Correlation

§ 66. Covariance and the Correlation Coefficient 301
§ 67. The Correlation Coefficient as a Characteristic of Dependence 305
§ 68. Partial Correlation Coefficients 310
§ 69. Distribution of the Coefficient r for Dependent Variables 316
§ 70. Spearman's Rank Correlation R 323
§ 71. Kendall's Rank Correlation T 333

Chapter Fourteen: Tables

Tables 1 – 13 339

Examples, Arranged According to Subject Matter 359
Author and Subject Index 361

Introduction

In earlier works on statistics, and especially in German books on *Kollektivmaßlehre*, the concepts of frequency, mean, standard deviation, etc., were developed for the case of fixed finite collections only, and attention was restricted entirely to that case. On the other hand, English and American statisticians considered any set of statistical data to be a random sample from an infinite collection, or *population*, of possibilities. From this point of view, the frequency of an event is only an estimate of the corresponding *probability* of the event, and the sample mean is only an estimate of the *population mean*, or *expectation*. The central question of statistics then becomes: *by how much can the quantities calculated for the random sample differ from the corresponding quantities for the population?* Thus, today, we have come to base mathematical statistics on the theory of probability.

The theory of probability as an exact mathematical discipline was first developed to the extent required here by Kolmogorov. We shall start with his axioms as a basis for our theory, without investigating the origins of the concept of probability. The purely mathematical theory constructed in this way proves as great a success in applications as Euclidean geometry or Newtonian mechanics do in their respective fields. The philosophical discussion of the concept of probability is certainly interesting and important, but it has no place in a book like this.

The logical structure of the book is set out schematically in the Table of Contents. Chapters 1 through 6 cover Kolmogorov's axiomatic theory and several direct applications, including the theory of confidence intervals for an unknown probability and confidence bands for an unknown distribution function, various simple cases of the χ^2-test, the Gaussian theory of errors, and Student's test. The statistical applications in Chapters 2, 4, 6 are based upon the mathematical methods developed in Chapters 1, 3, 5.

Two closely related sections make up the central part of the book: the theory of estimation (Chapters 7 through 9) and the theory of hypothesis testing (Chapters 11 and 12).

The theory of estimation has its origin in the least squares method developed by Gauss. Gauss advanced two arguments in favor of this method. The first is that the most plausible values of the unknown parameters are those which make the probability of the observed result as large as possible. The second argument, which Gauss himself preferred, arises from the requirement of making the mean square errors of the estimates as small as possible.

R. A. Fisher extended both criteria to a number of more general estimation problems. The condition that the estimates make the probability of the observed values as large as possible leads to maximum likelihood estimation. The requirement that the mean square error be as small as possible leads to the notion of efficient estimation. In a wide class of problems the maximum likelihood principle leads to an efficient estimate. The precise definitions of these concepts and the exact proofs given by Fréchet, Rao, and Lehmann and Scheffé will be found in Chapter 8. The concepts are then applied to the case of observed frequencies in Chapter 9.

The development of the modern theory of hypothesis testing began with Pearson's χ^2-test and Student's t-test. R. A. Fisher greatly expanded the scope of application of these methods, introduced the concept of "degrees of freedom", and revealed the connection with estimation theory by pointing out the necessity for using only efficient estimates in the χ^2-test. J. Neyman and E. S. Pearson furnished exact proofs of his assertions. They also formulated the general principles on which the modern theory of testing is based. These topics are discussed and illustrated in Chapter 11.

These principles are also valuable in the theory of rank tests (Chapter 12). The mathematical background needed for this chapter is not extensive: only the main points of Chapters 1 and 2 and occasional use of the limit theorems from Chapter 5 are required.

Bio-assay is treated in Chapter 10. Although it is an estimation problem in the sense of Chapter 8, the exposition here is wholly based upon Chapters 1 and 2.

The final chapter treats correlation coefficients and rank correlation. It requires only Chapters 1 through 6.

Chapter One

General Foundations

§ 1. Foundations of Probability Theory

A. Preliminary Explanation and Examples

In probability theory *events* are considered whose occurrence depend on chance and whose *probabilities* are expressible as numbers.

The concept of probability is a *statistical* concept. The numerical value of the probability of a chance event can be determined statistically in so far as the conditions under which the event occurs can be replicated and the *frequency* with which the event occurs can be observed. To say that the probability of an event is p, is to say that in repeated sequences of n replications the event occurs on average pn times per sequence. The number of occurrences will vary about the mean pn, by an amount which we shall estimate more accurately later.

Events will be designated by capital letters $A, B, \ldots$. We use the following notation:

AB (read: A and B) is the event that A and B both occur.

$\bar{A}$ (read: not A) is the event that A does not occur.

E is the event which always occurs.

$A \dotplus B$ (read: A or B) is the event that A or B, or both occur, or equivalently, at least one of the events A or B occurs.

If A and B are mutually exclusive in the sense that the two can never occur simultaneously, we write $A+B$ (again read: A or B) instead of $A \dotplus B$. Similarly, with a finite number of mutually exclusive events $A_1, A_2, \ldots, A_n$, or with countably many mutually exclusive events $A_1, A_2, \ldots$, the event "at least one of the events $A_1, \ldots, A_n$ occurs" and the event "at least one of the events $A_1, A_2, \ldots$, occurs" will be denoted by

$$\sum_1^n A_i = A_1 + A_2 + \cdots + A_n$$

and

$$\sum_1^\infty A_i = A_1 + A_2 + \cdots,$$

respectively.

The probability of an event A will be denoted by $\mathscr{P}(A)$. The following examples should help to illustrate the use of these words.

Example 1. A die is thrown three times. Events are all possible throws such as the triple (6, 1, 1) denoting a six followed by two ones, and all combinations of such outcomes formed by using the word "or"; *e.g.* "(6,1,1) or (4,5,6)" is an event which occurs if either the triple (6,1,1) or the triple (4,5,6) occurs. The probability of getting a six on the first throw need not be $\frac{1}{6}$, for the die can be biased or possess chance irregularities. If the die is approximately symmetrical and homogeneous, then it is reasonable to assume that the probability would be about $\frac{1}{6}$. Otherwise, we can determine the probability approximately only after a long sequence of throws in which we establish how frequently 6 is thrown.

Example 2. A person shoots at a target, aiming at the center. What happens is idealized in that we replace the position of impact by a point and assume that the shot always hits the target. A hit in any well defined region of the target constitutes an event. Therefore, to every portion of the target there corresponds an event, in particular the whole target corresponds to the event E. The larger the area of the portion and nearer the center it lies, the greater will be the probability of such an event (since one is aiming at the center). To hit one single point is also an event, but the probability of this event is zero, since a point has no area. If two events A and B correspond to certain portions of the target, then $A+B$ corresponds to the union of the two portions, the product AB to the intersection.

B. Events

If we want to interpret the operations AB, $\bar{A}$, $A \dotplus B$, and $A+B$, mathematically, there are two ways open. We can interpret the field of "events" as a *Boolean algebra* or as a *field of sets*. In the first interpretation, the "events" are undefined objects and the operations need only satisfy certain axioms (see C. Carathéodory, Maß und Integral, Basel 1956). According to the second interpretation, the "events" are subsets of a set E and AB is the intersection, $\bar{A}$ the complement, $A \dotplus B$ the union.

The two interpretations are equivalent, since by Stone's[1] well-known theorem every Boolean algebra is isomorphic to a suitable field of sets. The first interpretation is perhaps more natural (see D. A. Kappos, Zur mathematischen Begründung der Wahrscheinlichkeitstheorie, Sitzungsber. Bayer. Akad. München, 1948), but the second is more simple mathematically. Therefore, we shall follow Kolmogorov[2] and interpret all "events" as sets of "elementary events".

With this interpretation, E is the set of all elementary events which may be considered possible in a given situation.

---

[1] See M. H. Stone. Trans. Amer. math. Soc. 40 (1936) p. 37 or G. Birkhoff, Lattice Theory.

[2] A. Kolmogorov. Grundbegriffe der Wahrscheinlichkeitsrechnung, Ergebn. der Math. II 3, Berlin 1933, or H. Reichenbach's book Wahrscheinlichkeitslehre (translated by E. Hutten and M. Reichenbach, The Theory of Probability, Univ. of California Press 1949).

C. Probabilities

According to Kolmogorov, we can construct a theory of probability from the following axioms:

1. If A and B are events, then $\bar{A}$, AB and $A \dotplus B$ are also events. All events are subsets of E.

2. To each event A there corresponds a real number $\mathscr{P}(A) \geqq 0$. E is an event with $\mathscr{P}(E)=1$.

3. If A and B are mutually exclusive events, then $\mathscr{P}(A+B)=\mathscr{P}(A)+\mathscr{P}(B)$.

4. If $A_1, A_2, \ldots$ are events which never all occur at the same time, then

$$\lim_{n\to\infty} \mathscr{P}(A_1 A_2 \cdots A_n)=0.$$

From axioms 2 and 3, it follows that

$$\mathscr{P}(\bar{A})=1-\mathscr{P}(A) \tag{1}$$

and further, that the highest value of $\mathscr{P}(A)$ is 1:

$$0 \leqq \mathscr{P}(A) \leqq 1. \tag{2}$$

If $A_1, A_2, \ldots, A_n$ are mutually exclusive, the *Addition Rule* follows:

$$\mathscr{P}(A_1+\cdots+A_n)=\mathscr{P}(A_1)+\cdots+\mathscr{P}(A_n). \tag{3}$$

From the continuity axiom 4, it follows that the Addition Rule holds for countably many (mutually exclusive) events if $A=A_1+A_2+\cdots$ is an event:

$$\mathscr{P}(A)=\mathscr{P}(A_1)+\mathscr{P}(A_2)+\cdots. \tag{4}$$

For a simple proof see Kolmogorov's work cited above.

D. Conditional Probabilities

Suppose $\mathscr{P}(A) \neq 0$. The *conditional probability* of B under the assumption that A has occurred is defined by

$$\mathscr{P}_A(B)=\frac{\mathscr{P}(AB)}{\mathscr{P}(A)}. \tag{5}$$

It follows that the *Product Rule* holds:

$$\mathscr{P}(AB)=\mathscr{P}(A)\,\mathscr{P}_A(B). \tag{6}$$

This formula is also true for $\mathscr{P}(A)=0$, no matter what value we assign to the factor $\mathscr{P}_A(B)$.

In applications, as H. Richter[3] has quite correctly remarked, the conditional probability $\mathscr{P}_A(B)$ is almost never calculated from definition (5). Instead certain assumptions are made about $\mathscr{P}_A(B)$, and $\mathscr{P}_A(B)$ is calculated on the basis of these assumptions according to (6). We shall see this in Example 3 below. Hence, it would be better not to define the conditional probability $\mathscr{P}_A(B)$ by (5), but rather to accept it as an undefined concept. Then (6) could be taken as an additional axiom. However, following Kolmogorov, we shall take (5) as the basic definition and not discuss these axiomatic questions here.

Example 3. From an urn with r white and $N-r$ black balls, two balls are drawn one after the other without replacement. What is the probability of drawing (a) a white ball on the first draw, (b) a white ball on both the first and second draw, (c) a white ball on the second draw? We shall assume that the balls are well mixed, so that any ball is equally likely to be drawn. Given that one ball has already been drawn, the conditional probabilities of any of the $N-1$ remaining balls being drawn are also assumed to be equal.

Let A_j be the event that on the first draw ball number j is drawn. Similarly, let B_k be the event that on the second draw ball number k is drawn. From the given assumptions, it follows that

$$\mathscr{P}(A_j)=\frac{1}{N}$$

and

$$\mathscr{P}_{A_j}(B_k)=\frac{1}{N-1} \qquad (j \neq k).$$

According to the *Product Rule*, the probability of drawing ball number j on the first draw and ball number k on the second is equal for all pairs with $j \neq k$, namely,

$$\mathscr{P}(A_j B_k)=\frac{1}{N}\cdot\frac{1}{N-1}.$$

There are $r(N-1)$ pairs with the first ball white, so the probability of drawing a white ball on the first draw is

$$\frac{r(N-1)}{N(N-1)}=\frac{r}{N}.$$

Likewise, the number of pairs in which the second ball is white is equal to $r(N-1)$; therefore, the probability of drawing a white ball on the second draw is also equal to r/N. Finally, the number of white pairs (j,k) is equal to $r(r-1)$; therefore, the probability of drawing white balls both times is

$$\frac{r(r-1)}{N(N-1)}.$$

[3] H. Richter, Grundlegung der Wahrscheinlichkeitsrechnung. Math. Annalen 125 and 126.

E. Rule of Total Probability

By an *experiment* or *trial* we mean a partition

$$E = A_1 + A_2 + \cdots + A_n,$$

where the possible elements A_k are events. From (3) and (6), the *Rule of Total Probability*

$$\mathscr{P}(B) = \sum_k \mathscr{P}(A_k)\, \mathscr{P}_{A_k}(B) \tag{7}$$

holds for each such decomposition.

F. Independence

Two or more decompositions

$$E = A_1 + A_2 + \cdots + A_n, \qquad E = B_1 + \cdots + B_m, \ldots$$

are said to be *independent*, if for all $h, i, \ldots, k$

$$\mathscr{P}(A_h\, B_i \cdots D_k) = \mathscr{P}(A_h)\, \mathscr{P}(B_i) \cdots \mathscr{P}(D_k) \tag{8}$$

holds. A number of events $A, B, \ldots, D$ are said to be *independent* if the decompositions $E = A + \bar{A}$, $E = B + \bar{B}$, $\ldots$, $E = D + \bar{D}$ are independent.

This implies

$$\mathscr{P}(AB \cdots D) = \mathscr{P}(A)\, \mathscr{P}(B) \cdots \mathscr{P}(D),$$
$$\mathscr{P}(\bar{A}B \cdots D) = \mathscr{P}(\bar{A})\, \mathscr{P}(B) \cdots \mathscr{P}(D), \quad \text{and so on}.$$

However, most often in applications independence is not defined by (8), but postulated. We shall assume two experiments independent if the outcome of one has no influence on the outcome of the other.

G. Infinite Sums

A countably infinite sum $A_1 + A_2 + \cdots$ of mutually exclusive events need not be an event nor have a probability. However, using the methods of Lebesgue measure theory, we can extend the field of "events" to a σ-field of "measurable sets" A^*, and a measure $\mathscr{P}^*$ can be defined on this σ-field of sets so that in this extention axioms 1 to 5 are satisfied and the measure $\mathscr{P}^*$ agrees with the probability $\mathscr{P}$ in the sense that for all original events A

$$\mathscr{P}^*(A) = \mathscr{P}(A).$$

If, in addition, the probability $\mathscr{P}$ depends on an unknown parameter ϑ, then we confine ourselves to the σ-fields of sets A^* which are measurable for all ϑ.

Every countable union of sets A^* is again a set in the σ-field and the *Countable Additivity Property* holds:

$$\mathscr{P}^*(A_1^* + A_2^* + \cdots) = \mathscr{P}^*(A_1^*) + \mathscr{P}^*(A_2^*) + \cdots. \tag{9}$$

For a proof, see C. Carathéodory, Reelle Funktionen, or Halmos, Measure Theory.

Subsequently, we shall consider this extension completed whenever necessary, without distinguishing the new sets and their measures from the original events and probabilities. Hence, we shall assume that a sum of events $A_1 + A_2 + \cdots$ is also an event and that the probability $\mathscr{P}$ is countably additive.

§ 2. Random Variables and Distribution Functions

A. Random Variables

A *random variable* or *stochastic variable* is popularly expressed as a function whose value depends upon chance. More exactly, suppose a real function $\mathbf{x}$ is defined on a set E so that for each elementary event ξ, the value $\mathbf{x}(\xi)$ is a real number. Let the function be measurable in the sense that for each real number t, the set of ξ for which $[\mathbf{x} < t]$ is a measurable set. We have agreed to designate each measurable set of E as an event (in the extended sense). Therefore, the assumption of measurability means that for each t, $[\mathbf{x} < t]$ is an event.

The simplest case is the one in which the set E is partitioned into a finite number of parts

$$E = A_1 + A_2 + \cdots A_n$$

such that on each part A_k the function $\mathbf{x}$ assumes a constant value x_k. If the A_k are events, then the assumption of measurability is fulfilled.

In Example 1 (§ 1), the sum of the face numbers appearing in the first three throws of the die is a random variable which takes only a finite number of values (from 3 to 18). In Example 2, the two coordinates $\mathbf{x}$, $\mathbf{y}$ of the point of impact are random variables.

B. Distribution Functions

If $\mathbf{x}$ is a random variable and we allow t to vary from $-\infty < t < \infty$, then the probability of the event $[\mathbf{x} < t]$ is a nondecreasing function of t which is continuous from the left. We shall call it, as Kolmogorov[4] does, the *distribution function* $F(t)$ of the random variable $\mathbf{x}$,

$$F(t) = \mathscr{P}[\mathbf{x} < t]. \tag{1}$$

[4] Other authors define $F(t)$ as the probability of the event $[\mathbf{x} \leqq t]$. Then, this $F(t)$ is continuous from the right.

As t tends to $-\infty$, $F(t)$ tends to zero; as $t \to +\infty$, $F(t) \to 1$. This follows easily from the continuity axiom (5). If t tends to a from the left, then $F(t)$ tends to $F(a)$, but if t tends to a from the right, then $F(t)$ tends to $\mathscr{P}[\mathbf{x} \leqq a]$. The difference of these two limiting values

$$\Delta F(a) = F(a+0) - F(a-0)$$

is the probability that $\mathbf{x}$ is exactly equal to a. Further, for $a<b$,

$$F(b) - F(a) = \mathscr{P}[a \leqq \mathbf{x} < b]. \tag{2}$$

Two extreme cases are particularly important for applications. If the random variable $\mathbf{x}$ assumes only a finite number of values $t_1, \ldots, t_n$ with probabilities $p_1, \ldots, p_n$, then $F(t)$ is a *step function* which jumps in height p_i at the point $t = t_i$.

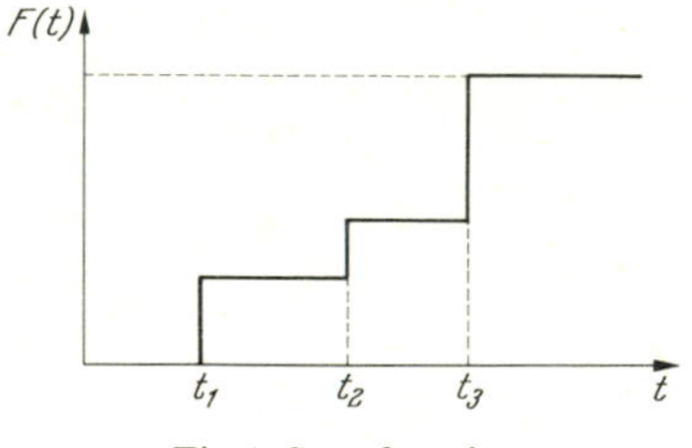

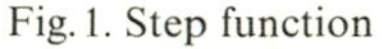
Fig. 1. Step function

C. Probability Densities

The other extreme case occurs when $F(t)$ is continuously differentiable:

$$F'(t) = f(t).$$

Then, from (2)

$$\mathscr{P}[a \leqq \mathbf{x} < b] = F(b) - F(a) = \int_a^b f(t)\,dt. \tag{3}$$

By taking the limit as $a \to -\infty$, it follows from (3) that

$$\mathscr{P}[\mathbf{x} < b] = F(b) = \int_{-\infty}^{b} f(t)\,dt,$$

and by taking the limit as $b \to +\infty$,

$$\int_{-\infty}^{\infty} f(t)\,dt = 1. \tag{4}$$

The function $f(t)$ is called the *probability density* of the random variable $\mathbf{x}$. Roughly speaking, $f(t)\,dt$ is the probability that $\mathbf{x}$ lies between t and $t + dt$.

D. Normal Distribution

The Gaussian error function

$$f(t)=\frac{1}{c\sqrt{2\pi}}\,e^{-\frac{1}{2}\frac{(t-a)^2}{c^2}} \tag{5}$$

is a famous example of a probability density.

If we consider the random variable $(\mathbf{x}-a)/c$, instead of $\mathbf{x}$, the probability density assumes the simpler form

$$f(t)=\frac{1}{\sqrt{2\pi}}\,e^{-\frac{1}{2}t^2} \tag{6}$$

(see the upper part of Fig. 2). In § 12, we shall show that

$$\int_{-\infty}^{\infty} e^{-\frac{1}{2}t^2}\,dt=\sqrt{2\pi}; \tag{7}$$

from this it follows that condition (4) is satisfied. The corresponding distribution function is

$$\Phi(t)=\frac{1}{\sqrt{2\pi}}\int_{-\infty}^{t} e^{-\frac{1}{2}\tau^2}\,d\tau \tag{8}$$

(see the lower part of Fig. 2 and Table 1 at the end of the book).

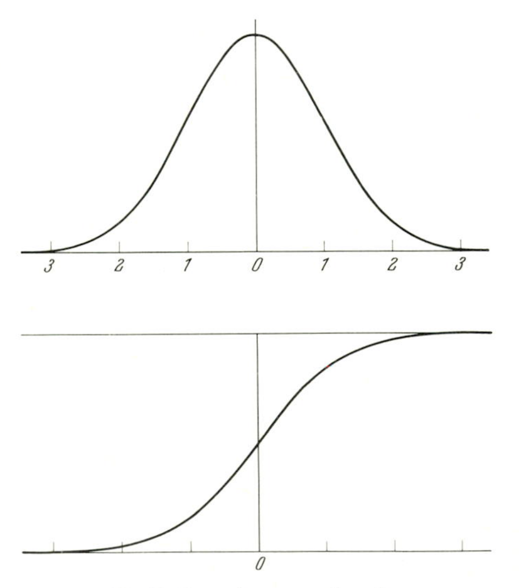

Fig. 2. Gaussian error function

The expression "Gaussian error function" originates from Gauss' assumption that the random error in astronomical observations has a probability density of the form (5). Many other variables in nature and mathematical statistics also have probability densities of exactly or approximately this form. Thus, it is worthwhile to study functions (6) and (8) more closely.

In the numerical calculation of the error integral for moderate sized t, we expand $f(t)$ in an infinite series

$$f(t)=\frac{1}{\sqrt{2\pi}}\left(1-\frac{t^2}{2}+\frac{t^4}{2!\,2^2}-\frac{t^6}{3!\,2^3}+\cdots\right)$$

and integrate

$$\Phi(t)=\frac{1}{2}+\frac{1}{\sqrt{2\pi}}\left(t-\frac{t^3}{2\cdot 3}+\frac{t^5}{2!\,2^2\cdot 5}-\frac{t^7}{3!\,2^3\cdot 7}+\cdots\right). \tag{9}$$

For large t, there is an asymptotic expansion which we obtain in the following way. We have

$$1-\Phi(t)=\frac{1}{\sqrt{2\pi}}\int_t^\infty e^{-\frac{1}{2}\tau^2}\,d\tau.$$

If we set $\frac{1}{2}\tau^2=x$ and $\frac{1}{2}t^2=u$, we have

$$1-\Phi(t)=\frac{1}{2\sqrt{\pi}}\int_u^\infty e^{x}\,x^{-\frac{1}{2}}\,dx. \tag{10}$$

Integration by parts gives

$$\int_u^\infty e^{-x}\,x^{-\frac{1}{2}}\,dx=e^{-u}\,u^{-\frac{1}{2}}-\frac{1}{2}\int_u^\infty e^{-x}\,x^{-\frac{3}{2}}\,dx=e^{-u}\,u^{-\frac{1}{2}}-R_1, \tag{11}$$

where $-R_1$ is a negative remainder. In order to evaluate R_1, we use integration by parts once more:

$$R_1=\frac{1}{2}\int_u^\infty e^{-x}\,x^{-\frac{3}{2}}\,dx=\frac{1}{2}e^{-u}\,u^{-\frac{3}{2}}-\frac{1}{2}\cdot\frac{3}{2}\int_u^\infty e^{-x}\,x^{-\frac{5}{2}}\,dx<\frac{1}{2}e^{-u}\,u^{-\frac{3}{2}}.$$

If we substitute (11) in (10), we obtain the desired asymptotic formula

$$\Phi(t)=1-\frac{1}{\sqrt{2\pi}}e^{-\frac{1}{2}t^2}(t^{-1}-S_1), \qquad 0<S_1<t^{-3}. \tag{12}$$

If we want the remainder to be of order of magnitude t^{-5}, we need only use integration by parts once more

$$\Phi(t)=1-\frac{1}{\sqrt{2\pi}}e^{-\frac{1}{2}t^2}(t^{-1}-t^{-3}+S_2), \qquad 0<S_2<1\cdot 3\cdot t^{-5}. \tag{13}$$

Except for the factor $\exp(-\frac{1}{2}t^2)$, we always get a segment of the alternating series

$$t^{-1}-t^{-3}+1\cdot 3\cdot t^{-5}-1\cdot 3\cdot 5\cdot t^{-7}+\cdots \tag{14}$$

and the remainder is always smaller than the first term neglected.

The asymptotic series (14) is not very useful for calculations requiring extreme accuracy because, although the series terms decrease at first, they increase again. Van Wijngaarden[5] has given a transformation that converts (14) into a convergent series which is very useful for numerical calculations for moderate and large t.

Table 1 at the end of the book gives values for the function $u=\Phi(t)$, Table 2 for the inverse function $t=\Psi(u)$. The notations Φ and Ψ will be retained throughout the book.

The points of inflection of the curve $y=f(t)$ lie at ± 1 since the second derivative

$$y''=\frac{t^2-1}{\sqrt{2\pi}}e^{-\frac{1}{2}t^2}$$

is negative from -1 to $+1$ and positive outside. The function decreases very sharply from the points of inflection outwards. More that 95% of the area under the curve lies within the interval $(-2, +2)$, and 99.7% lies within the interval $(-3, +3)$. For all practical purposes, we can confine ourselves to the interval $(-4, +4)$ since the probability that the random variable $(\mathbf{x}-a)/c$ falls outside this region is almost zero.

All distribution functions with probability densities having the form of (5) are called *normal distributions*. The constant a, indicates the location of the maximum of $f(t)$, whereas c determines the distance of the two points of inflection from the line of symmetry and is thus a measure of the dispersion.

§ 3. Expectation and Standard Deviation

A. Expectation

The *expectation* of a random variable $\mathbf{x}$ which takes only a finite number of values $t_1, \ldots, t_n$, is defined as the sum of these values multiplied by their probabilities

$$\hat{\mathbf{x}}=\mathscr{E}\mathbf{x}=\sum t_k p_k. \tag{1}$$

[5] A. van Wijngaarden, A transformation of formal series. Proc. Kon. Ned. Akad. Amsterdam, Section of Science A 56 (1953) p. 537.

It is clear from the definition that $\hat{\mathbf{x}}$ lies between the largest and smallest possible value of $\mathbf{x}$. The expectation can also be defined as the center of gravity of the possible values t_k with weights p_k.

If the random variable $\mathbf{x}$ has a probability density $f(t)$, then instead of the sum we have an integral

$$\mathscr{E}\mathbf{x} = \int_{-\infty}^{\infty} t f(t)\, dt. \tag{2}$$

In the general case, for an arbitrary distribution function $F(t)$, we have to replace integral (2) by a Stieltjes integral

$$\mathscr{E}\mathbf{x} = \int_{-\infty}^{\infty} t\, dF(t). \tag{3}$$

We don't need the general concept of Stieltjes integration here, since we are only concerned with the special cases (1) and (2). Still we shall give, for the sake of completeness, Fréchet's definition of integral (3):

$$\mathscr{E}\mathbf{x} = \lim_{h \to 0} \sum_{k=-\infty}^{\infty} k\,h \{F(k\,h+h) - F(k\,h)\}. \tag{4}$$

The expectation is a special case of the Lebesgue integral. If E_k is the event $[k\,h \leqq \mathbf{x} < (k+1)\,h]$, then the Lebesgue integral of $\mathbf{x}$ over B with respect to the measure $\mathscr{P}(A)$ is defined as

$$\int_B \mathbf{x}\, \mathscr{P}(dE) = \lim_{h \to 0} \sum_{k=-\infty}^{\infty} k\,h\, \mathscr{P}(B E_k). \tag{5}$$

In particular, if we set $B = E$, then (5) becomes (4).

It can be shown from Lebesgue integration theory that the sum $\mathbf{x}+\mathbf{y}$ of two measurable functions $\mathbf{x}$ and $\mathbf{y}$ is also measurable and that the integral of the sum over any measurable set B is equal to the sum of the integrals of the summands:

$$\int_B (\mathbf{x}+\mathbf{y})\, \mathscr{P}(dE) = \int_B \mathbf{x}\, \mathscr{P}(dE) + \int_B \mathbf{y}\, \mathscr{P}(dE),$$

where it is assumed that the two integrals on the right are finite. For the proof, refer to C. Carathéodory, Vorlesungen über reelle Funktionen, or to Halmos, Measure Theory.

Again, if we set $B = E$, then it follows that the expectation of a sum is equal to the sum of the expectations of the summands, if these are finite:

$$\mathscr{E}(\mathbf{x}+\mathbf{y}) = \mathscr{E}\mathbf{x} + \mathscr{E}\mathbf{y}. \tag{6}$$

Similarly for more than two summands,

$$\mathscr{E}(\mathbf{x}_1 + \cdots + \mathbf{x}_n) = \mathscr{E}\mathbf{x}_1 + \cdots + \mathscr{E}\mathbf{x}_n. \tag{7}$$

If c is a constant, then obviously,

$$\mathscr{E}(c\,\mathbf{x}) = c\,\mathscr{E}\,\mathbf{x}. \tag{8}$$

For convenience, we shall sometimes use the notation $\hat{\mathbf{x}}$ instead of $\mathscr{E}\,\mathbf{x}$.

B. Independent Variables

Two or more random variables $\mathbf{x}, \mathbf{y}, \ldots$ are called *independent* if the events

$$[\mathbf{x} < t],\ [\mathbf{y} < u],\ \ldots$$

are independent for arbitrary real numbers $t, u, \ldots$. It follows that the events

$$[a \leqq \mathbf{x} < b],\ [c \leqq \mathbf{y} < d],\ \ldots$$

are also independent. Hence, if $\mathbf{x}$ and $\mathbf{y}$ are independent random variables, we have

$$\mathscr{P}[a \leqq \mathbf{x} < b,\ c \leqq \mathbf{y} < d] = \mathscr{P}[a \leqq \mathbf{x} < b]\,\mathscr{P}[c \leqq \mathbf{y} < d].$$

In this case, the expectation of the product is equal to the product of the expectations:

$$\mathscr{E}\,\mathbf{x}\,\mathbf{y} = \mathscr{E}\,\mathbf{x} \cdot \mathscr{E}\,\mathbf{y}. \tag{9}$$

Analogously for more than two independent variables,

$$\mathscr{E}\,\mathbf{x}_1\,\mathbf{x}_2 \cdots \mathbf{x}_n = \hat{\mathbf{x}}_1\,\hat{\mathbf{x}}_2 \cdots \hat{\mathbf{x}}_n. \tag{10}$$

For the proof of (9), see Kolmogorov, The Theory of Probability (translated by E. Hotten and H. Reichenbach).

C. Standard Deviation and Variance

The *standard deviation* $\sigma = \sigma_{\mathbf{x}}$ of a random variable $\mathbf{x}$ is defined as the positive square root of the variance

$$\sigma^2 = \mathscr{E}(\mathbf{x} - \hat{\mathbf{x}})^2 = \mathscr{E}(\mathbf{x}^2 - 2\,\mathbf{x}\,\hat{\mathbf{x}} + \hat{\mathbf{x}}^2). \tag{11}$$

If we use (7) and (8) to evaluate the right-hand side, we have

$$\sigma^2 = \mathscr{E}\,\mathbf{x}^2 - 2\,\hat{\mathbf{x}}^2 + \hat{\mathbf{x}}^2 = \mathscr{E}\,\mathbf{x}^2 - (\mathscr{E}\,\mathbf{x})^2. \tag{12}$$

Clearly, for c constant,

$$\sigma_{c\mathbf{x}} = |c| \cdot \sigma_{\mathbf{x}}. \tag{13}$$

The variance of a sum is

$$\sigma^2_{\mathbf{x}+\mathbf{y}} = \mathscr{E}(\mathbf{x}-\hat{\mathbf{x}}+\mathbf{y}-\hat{\mathbf{y}})^2$$
$$= \mathscr{E}(\mathbf{x}-\hat{\mathbf{x}})^2 - 2\mathscr{E}[(\mathbf{x}-\hat{\mathbf{x}})(\mathbf{y}-\hat{\mathbf{y}})] + \mathscr{E}(\mathbf{y}-\hat{\mathbf{y}})^2.$$

If $\mathbf{x}$ and $\mathbf{y}$ are independent, then the middle term is equal to zero

$$\mathscr{E}[(\mathbf{x}-\hat{\mathbf{x}})(\mathbf{y}-\hat{\mathbf{y}})] = \mathscr{E}(\mathbf{x}-\hat{\mathbf{x}}) \cdot \mathscr{E}(\mathbf{y}-\hat{\mathbf{y}}) = 0$$

and we obtain

$$\sigma^2_{\mathbf{x}+\mathbf{y}} = \sigma^2_{\mathbf{x}} + \sigma^2_{\mathbf{y}}. \tag{14}$$

The same formula holds for the difference of two independent random variables:

$$\sigma^2_{\mathbf{x}-\mathbf{y}} = \sigma^2_{\mathbf{x}} + \sigma^2_{\mathbf{y}}.$$

Similarly, for more than two pairwise independent random variables, the equation

$$\sigma^2 = \sigma_1^2 + \cdots + \sigma_n^2 \tag{15}$$

holds for the variance of the sum $\mathbf{x} = \mathbf{x}_1 + \mathbf{x}_2 + \cdots + \mathbf{x}_n$.

A generalization of formula (2) is useful for computing the expectation and standard deviation:

$$\mathscr{E} g(\mathbf{x}) = \int_{-\infty}^{\infty} g(t) f(t)\, dt. \tag{16}$$

In § 4, we shall give a more generalized form of this equation and a proof. We shall apply it now in an example.

Example 4. The random variable $\mathbf{x}$ has a Gaussian probability density

$$f(t) = \frac{1}{\sqrt{2\pi}} e^{-\frac{1}{2}t^2}.$$

What are the expectation and standard deviation of $\mathbf{x}$? From (2), the expectation is

$$\hat{\mathbf{x}} = \int_{-\infty}^{\infty} t f(t)\, dt = \frac{1}{\sqrt{2\pi}} \int_{-\infty}^{\infty} t\, e^{-\frac{1}{2}t^2} dt = 0.$$

The variance is the expectation of $(\mathbf{x}-0)^2 = \mathbf{x}^2$. Applying equation (16) with $g(t) = t^2$, we obtain

$$\sigma^2 = \mathscr{E}(\mathbf{x}^2) = \int_{-\infty}^{\infty} t^2 f(t)\, dt = \frac{1}{\sqrt{2\pi}} \int_{-\infty}^{\infty} t^2 e^{-\frac{1}{2}t^2} dt.$$

Integration by parts gives

$$\sigma^2 = \left[-\frac{1}{\sqrt{2\pi}}\, t\, e^{-\frac{1}{2}t^2} \right]_{-\infty}^{\infty} + \frac{1}{\sqrt{2\pi}} \int_{-\infty}^{\infty} e^{-\frac{1}{2}t^2} dt = 0 + 1 = 1;$$

hence $\sigma=1$. In the general case where

$$f(t)=\frac{1}{c\sqrt{2\pi}}\,e^{-\frac{1}{2}\frac{(t-a)^2}{c^2}},$$

we find $\mathscr{E}\mathbf{x}=a$ and $\sigma=c$. Therefore, from now on we shall always write σ instead of c:

$$f(t)=\frac{1}{\sigma\sqrt{2\pi}}\,e^{-\frac{1}{2}\frac{(t-a)^2}{\sigma^2}}.$$

D. Tchebychev's Inequality

The practical significance of the standard deviation lies in the fact that deviations of a random variable $\mathbf{x}$ from its mean which are large as compared with the standard deviation σ are highly improbable. More precisely, Tchebychev's Inequality states:

If g is any arbitrary positive number, then the probability of the event $[|\mathbf{x}-\hat{\mathbf{x}}|>g\,\sigma]$ *is less than* g^{-2}.

The proof is very simple. We proceed from the definition of the variance

$$\sigma^2=\mathscr{E}(\mathbf{x}-\hat{\mathbf{x}})^2.$$

The expectation is, by definition, either a sum or an integral. If it is a sum, we have to multiply the possible values of the variable

$$\mathbf{y}=(\mathbf{x}-\hat{\mathbf{x}})^2$$

by their probabilities and add. The possible values decompose into those $\leqq g^2\sigma^2$ and those $>g^2\sigma^2$. Correspondingly, the sum decomposes into two parts. The first part (for values $\leqq g^2\sigma^2$) is positive or zero everywhere, since all values of squares are positive or zero. The second part of the sum is larger than $g^2\sigma^2$ multiplied by the probability $\mathscr{P}$ of the event $[(\mathbf{x}-\mathbf{x})^2>g^2\sigma^2]$. Assume $\mathscr{P}\neq 0$; if $\mathscr{P}=0$, the proof is trivial. Thus, for the whole sum, we have the inequality

$$\sigma^2>g^2\sigma^2\mathscr{P}.$$

From this

$$\mathscr{P}<g^{-2}$$

follows.

The proof is exactly analogous if the expectation is an integral. Suppose $F(t)$ is the distribution function of the random variable $\mathbf{y}$. The integral

$$\sigma^2=\mathscr{E}\mathbf{y}=\int_{-\infty}^{\infty} t\,dF(t)$$

will decompose into an integral over $t\leqq g^2\sigma^2$ and an integral over $t>g^2\sigma^2$. The first part is $\geqq 0$, the second part $>g^2\sigma^2\mathscr{P}$. Thus, again, $\sigma^2>g^2\sigma^2\mathscr{P}$ and $\mathscr{P}<g^{-2}$.

In most cases, the probability $\mathscr{P}$ is considerably smaller than g^{-2}. In the case of the normal distribution, for example, the probability that $[|\mathbf{x}-a|>3]$ is only 0.0027, which is much smaller than $\frac{1}{9}$.

For $\sigma=0$, we have a special case of Tchebychev's inequality:

If the standard deviation is zero, then the probability that $\mathbf{x}$ *differs from the constant value* $\hat{\mathbf{x}}$ *is zero.*

§ 4. Integral Representation for Expectations and Probabilities

A. Rectangles and Open Sets

By a *rectangle* in the (u, v)-plane we mean a point set

$$a \leqq u < b, \qquad c \leqq v < d.$$

Every open point set M in the (u, v)-plane can be represented as the sum of countably many rectangles. In fact, if we partition the plane into equal rectangles by vertical and horizontal lines and pick out of these rectangles those which are included in M, and then partition the remaining rectangles into equal quarters and again pick out those which lie in M, and so forth, we see finally that each point of the set M is contained in one of the rectangles, so

$$M = R_1 + R_2 + \cdots.$$

Now, if $\mathbf{x}$ and $\mathbf{y}$ are random variables, then $[a \leqq \mathbf{x} < b]$ and $[c \leqq \mathbf{y} < d]$ are events. The intersection of these two events is the event that occurs when both inequalities are fulfilled, that is when the point $\mathbf{z}$ with coordinates $\mathbf{x}, \mathbf{y}$ belongs to the rectangle $R(a \leqq u < b, c \leqq v < d)$. This holds for every rectangle, thus the membership of $\mathbf{z}$ in $M = R_1 + R_2 + \cdots$ is also an event.

Such sets M, in the (u, v)-plane, for which the membership of point $\mathbf{z}$ in M is an event that has a probability, are called *measurable*, and the probability of the event is called the *measure* of the set M. The measure function is absolutely additive.

We have seen that every open set is measurable. Consequently, every closed set is also measurable, since its complement is open.

Suppose that $\mathbf{x}$ and $\mathbf{y}$ are random variables and that $g(u, v)$ is a continuous function of the real variables u and v. Then $g(\mathbf{x}, \mathbf{y})$ is also a random variable. In order to prove this, we need only to show that for arbitrary real t

$$[g(\mathbf{x}, \mathbf{y}) < t]$$

is an event. This is clear from the above, since the set of all points (u, v) for which $[g(u, v) < t]$ is an open set, because g is continuous.

Suppose that the plane E is decomposed into finitely many measurable sets and that the function $g(u, v)$ is continuous on each of these sets. Such a function is called *piecewise continuous*. It is easy to see that any piecewise continuous function is measurable. Hereafter this fact will only be used in simple cases, in which $g(u, v)$ is either piecewise constant or coincides with a given continuous function on M_1 and is zero on $E - M_1$.

B. Bivariate Probability Densities

We say the random variables $\mathbf{x}$ and $\mathbf{y}$ have a *joint probability density* $f(u, v)$ if the probability of the event

$$\{[a \leqq \mathbf{x} < b] \quad \text{and} \quad [c \leqq \mathbf{y} < d]\}$$

equals the integral of the function $f(u, v)$ over the rectangle

$$(a \leqq u < b, c \leqq v < d)$$

in the (u, v)-plane:

$$\mathscr{P}\{[a \leqq \mathbf{x} < b], [c \leqq \mathbf{y} < d]\} = \int_a^b \int_c^d f(u, v)\, dv\, du.$$

The right side of the above equation is an ordinary Riemann integral; $f(u, v)$ will be assumed integrable in the Riemann sense.

Now we can extend formula (16), § 3, to functions of two variables.

Theorem I. *If the random variables* $\mathbf{x}, \mathbf{y}$ *have the joint probability density* $f(u, v)$ *and the function* $g(u, v)$ *is integrable in the Riemann sense, then*

$$\mathscr{E} g(\mathbf{x}, \mathbf{y}) = \int_{-\infty}^{\infty} \int_{-\infty}^{\infty} g(u, v) f(u, v)\, du\, dv, \tag{1}$$

provided that the integral on the right converges. The analogous result holds for more than two variables $\mathbf{x}, \mathbf{y}, \ldots$.

Proof. First, we restrict ourselves to a rectangle $(-n \leqq u < n, -n \leqq v < n)$ in the (u, v)-plane and replace the function $f(u, v)$ by zero outside this rectangle. We subdivide the rectangle into m^2 equal subrectangles and replace the function $g(u, v)$ in each of these by a lower or upper bound. Thus, we obtain two piecewise constant functions g_1, a lower bound of g, and g_2, an upper bound of g. The functions $g_1(u, v)$ and $g_2(u, v)$ only take a finite number of values. For g_1 and g_2, formula (1) becomes

$$\mathscr{E} g_1(\mathbf{x}, \mathbf{y}) = \iint g_1(u, v) f(u, v)\, du\, dv, \tag{2}$$

$$\mathscr{E} g_2(\mathbf{x}, \mathbf{y}) = \iint g_2(u, v) f(u, v)\, du\, dv. \tag{3}$$

The left side of (2) is the sum of the possible values of g_1 multiplied by their probabilities and the right side is a sum of integrals over the subrectangles. If we factor the constant $g_1(u, v)$ out of each integral in the sum, then the right side becomes the same as the left. Analogously for (3).

Clearly, for any arbitrary values x, y of the random variables $\mathbf{x}, \mathbf{y}$,

$$g_1(x, y) \leqq g(x, y) \leqq g_2(x, y).$$

Thus, it follows that

$$\mathscr{E} g_1(\mathbf{x}, \mathbf{y}) \leqq \mathscr{E} g(\mathbf{x}, \mathbf{y}) \leqq \mathscr{E} g_2(\mathbf{x}, \mathbf{y}).$$

Now, if we let the sides of the subrectangles tend to zero, the right sides of (2) and (3) both tend to the right side of (1). Since the expectation of $g(\mathbf{x}, \mathbf{y})$ is squeezed between the two, (1) follows for the modified function $g^{(n)}(u, v)$ which is zero outside the original rectangle $(-n \leqq u < n, -n \leqq v < n)$.

In order to take the limit as $n \to \infty$, we use a theorem from Lebesgue integration theory which states that an integral over a sum $A_1 + A_2 + \cdots$ is equal to the sum of integrals over $A_1, A_2, \ldots$. In our case, A_1 is the event that occurs when the point $(\mathbf{x}, \mathbf{y})$ falls into the square $(-1 \leqq u < 1, -1 \leqq v < 1)$, similarly $A_1 + A_2$ is the event that the point falls into $(-2 \leqq u < 2, -2 \leqq v < 2)$, etc. The Lebesgue integral of the random variable $g(\mathbf{x}, \mathbf{y})$ over $A_1 + A_2 + \cdots + A_n$ is equal to the Riemann integral of the product $g(u, v) f(u, v)$ over the rectangle $(-n \leqq u < n, -n \leqq v < n)$. Passage to the limit yields (1).

C. Computation of Probabilities by Integration

The important application of Theorem I is the following. Suppose G is any open set in the (u, v)-plane. Suppose $\mathbf{z}$ is a random point with coordinates $(\mathbf{x}, \mathbf{y})$. We want to find the probability of the event [$\mathbf{z}$ lies in G].

In order to apply Theorem I, we use a function $g(u, v)$:

$$g(u, v) = 1 \quad \text{in } G,$$

$$g(u, v) = 0 \quad \text{outside of } G.$$

Now the expectation of the random variable $g(\mathbf{x}, \mathbf{y})$ equals the probability that $\mathbf{z}$ lies in G. From formula (1), we have

$$\mathscr{P}(\mathbf{z} \text{ in } G) = \int_G f(u, v)\, du\, dv. \tag{4}$$

The result can be generalized for more than two random variables:

Theorem II. *If the system of random variables* $\mathbf{x}, \mathbf{y}, \ldots$ *has the joint probability density* $f(x, y, \ldots)$, *then the probability that the point* $\mathbf{z}$ *with coordinates* $(\mathbf{x}, \mathbf{y}, \ldots)$ *falls into any region* G *is equal to the integral of function* f *over this region:*

$$\mathscr{P}(\mathbf{z} \text{ in } G) = \int_G f(u, v, \ldots)\, du\, dv \ldots. \tag{5}$$

D. Distribution Function of a Sum x + y

We want to apply Theorem II to the following problem. Two independent random variables have probability densities $f(u)$ and $g(u)$. What is the distribution function of their sum $\mathbf{x}+\mathbf{y}$?

The probability that $\mathbf{x}$ lies between a and b and $\mathbf{y}$ between c and d is equal to the product

$$\int_a^b f(u)\, du \cdot \int_c^d g(v)\, dv = \int_a^b \int_c^d f(u)\, g(v)\, du\, dv$$

because $\mathbf{x}$ and $\mathbf{y}$ are independent. Thus, the random variables $\mathbf{x}, \mathbf{y}$ have the joint probability density

$$f(u, v) = f(u)\, g(v).$$

The distribution function $H(t)$ of $\mathbf{x}+\mathbf{y}$ is the probability that $[\mathbf{x}+\mathbf{y}<t]$ occurs:

$$H(t) = \mathscr{P}[\mathbf{x}+\mathbf{y}<t].$$

According to Theorem II, this probability is equal to the double integral

$$H(t) = \iint_{(u+v<t)} f(u)\, g(v)\, du\, dv.$$

The integral can be evaluated by iteration:

$$H(t) = \int_{-\infty}^{\infty} du \int_{-\infty}^{t-u} f(u)\, g(v)\, dv.$$

If we introduce $w = u + v$ as a new variable in the inner integral, we obtain

$$H(t) = \int_{-\infty}^{\infty} du \int_{-\infty}^{t} f(u)\, g(w-u)\, dw.$$

We have non-negative functions so we can change the order of integration

$$H(t) = \int_{-\infty}^{t} dw \int_{-\infty}^{\infty} f(u)\, g(w-u)\, du. \tag{6}$$

This distribution has the probability density

$$h(t)=\int_{-\infty}^{\infty} f(u)\, g(t-u)\, du. \tag{7}$$

Hence, we have:

Theorem III. *The probability density of a sum of independent random variables with probability densities $f(t)$ and $g(t)$ is given by* (7).

E. The Sum of Two Normal Variables

Suppose the random variables **x** and **y** are normally distributed. What is the distribution function of their sum $\mathbf{x}+\mathbf{y}$?

Suppose the probability densities are

$$f(t)=\frac{1}{\sigma\sqrt{2\pi}}\, e^{-\frac{1}{2}\left(\frac{t-a}{\sigma}\right)^2}, \tag{8}$$

and

$$g(t)=\frac{1}{\tau\sqrt{2\pi}}\, e^{-\frac{1}{2}\left(\frac{t-b}{\tau}\right)^2}. \tag{9}$$

Since we can substitute $\mathbf{x}-a$ and $\mathbf{y}-b$ for **x** and **y**, we can assume $a=0$ and $b=0$. According to Theorem III, the sum $\mathbf{z}=\mathbf{x}+\mathbf{y}$ has probability density

$$\begin{aligned} h(t) &= \int_{-\infty}^{\infty} f(u)\, g(t-u)\, du \\ &= \frac{1}{2\pi\sigma\tau}\int_{-\infty}^{\infty} e^{-\frac{1}{2}\sigma^{-2}u^2-\frac{1}{2}\tau^{-2}(t-u)^2}\, du \\ &= \frac{1}{2\pi\sigma\tau}\int_{-\infty}^{\infty} e^{-\frac{1}{2}(\alpha u^2-2\beta u+\gamma)}\, du, \end{aligned} \tag{10}$$

where

$$\alpha=\sigma^{-2}+\tau^{-2}, \qquad \beta=t\,\tau^{-2}, \qquad \gamma=t^2\,\tau^{-2}.$$

If we introduce

$$u-\frac{\beta}{\alpha}=v \tag{11}$$

as a new variable, then we have

$$h(t)=\frac{1}{2\pi\sigma\tau}\int_{-\infty}^{\infty} e^{-\frac{1}{2}\alpha v^2-\frac{1}{2}\delta}\, dv, \tag{12}$$

where

$$\delta=\frac{\alpha\gamma-\beta^2}{\alpha}=\frac{(\sigma^{-2}+\tau^{-2})\,t^2\tau^{-2}-t^2\tau^{-4}}{\sigma^{-2}+\tau^{-2}}=\frac{t^2}{\sigma^2+\tau^2}$$

or

$$\begin{aligned} h(t)&=\frac{1}{2\pi\sigma\tau}\,e^{-\frac{1}{2}\delta}\int_{-\infty}^{\infty}e^{-\frac{1}{2}\alpha v^2}\,dv\\ &=\frac{1}{2\pi\sigma\tau}\,e^{-\frac{1}{2}\delta}\left(\frac{2\pi}{\alpha}\right)^{\frac{1}{2}}\\ &=(2\pi)^{-\frac{1}{2}}(\sigma^2+\tau^2)^{-\frac{1}{2}}\,e^{-\frac{1}{2}t^2(\sigma^2+\tau^2)^{-1}}. \end{aligned} \tag{13}$$

Thus, for **z**, we obtain a normal distribution with mean zero and standard deviation $(\sigma^2+\tau^2)^{\frac{1}{2}}$. The original sum is thus normally distributed with expectation $(a+b)$ and standard deviation $(\sigma^2+\tau^2)^{\frac{1}{2}}$.

The same holds for the difference $\mathbf{x}-\mathbf{y}$, except that the expectation is $a-b$.

By repeated application, it follows that a sum of more than two normally distributed independent random variables is also normally distributed.

Example 5. Confidence limits for the expectation of a normal distribution. First suppose we have made one observation x_1 of a normally distributed variable **x**. Suppose that the standard deviation σ of the distribution is known, but the expectation a unknown. Can we find limits for the deviation x_1-a such that the probability for x_1-a to lie outside the limits is small?

The answer is easy. The variable

$$\mathbf{y}=\frac{\mathbf{x}-a}{\sigma}$$

has a normal distribution with expectation 0 and standard deviation 1. The distribution function of **y** is the function $\Phi(t)$ considered in § 2D. The probability for **y** to exceed a given limit g is $1-\Phi(g)$.

Now let a small probability $\alpha<\frac{1}{2}$ be given, *e.g.* $\alpha=0.01$. From the table of the function Φ, we may determine a positive real number g such that

$$1-\Phi(g)=\alpha. \tag{14}$$

Then the probability of the event $[\mathbf{y}>g]$ is just α. For reasons of symmetry, the probability of the event $[\mathbf{y}<-g]$ is also α. Hence the probability for **y** to lie outside the limits $-g$ and $+g$ is 2α. In other words: the probability of the event

$$[|\mathbf{x}-a|>g\,\sigma] \tag{15}$$

is 2α, and the probability of the complementary event

$$[|\mathbf{x}-a|\leqq g\,\sigma] \tag{16}$$

is $1-2\alpha$.

The value of **x** found in the experiment is x_1. Hence event (16) happens if $|x_1-a|\leqq g\,\sigma$. This means x_1-a lies between the limits $-g$ and $+g$ with probability $1-2\alpha$, and outside these limits with probability 2α.

Instead of (16) we may also write

$$\mathbf{x} - g\,\sigma = a = \mathbf{x} + g\,\sigma. \tag{17}$$

Hence, if we determine, after the experiment, the limits $x_1 - g\,\sigma$ and $x_1 + g\,\sigma$, we may expect the true value a to lie between these limits. The error probability of the assertion that a lies between these limits is just 2α.

Note that a is fixed, although unknown, but that x_1 and hence the limits $x_1 - g$ and $x_1 + g$ depend on chance. If α is 0.01, the probability for the unknown true value a to lie between these chance limits is 0.98. The limits $x_1 \pm g$ are called *confidence limits* for a. They depend, of course, on the assigned error level 2α.

In Table 3 the values of g for several usual error levels are tabulated. The first column gives the *one-sided* error level α in percent, *i.e.*, the error probability of each one of the two limits $x_1 - g$ and $x_1 + g$. The second column gives the *two-sided* error level 2α. In the third column we find the required factor g.

Next suppose that we have made n observations $x_1, \ldots, x_n$ and form the expectation

$$m = \frac{x_1 + \cdots + x_n}{n}.$$

The sum $x_1 + \cdots + x_n$ is a particular value of a random variable $\mathbf{x}_1 + \cdots + \mathbf{x}_n$, which is normally distributed with expectation $n\,a$ and variance $n\,\sigma$. Hence, the expectation m is a particular value of a random variable $\mathbf{m}$ having a normal distribution with expectation a, variance σ^2/n, and standard deviation

$$\sigma_m = \frac{\sigma}{\sqrt{n}}.$$

Hence, we may assign to a the confidence limits $m - g\,\sigma_m$ and $m + g\,\sigma_m$. The probability for a to lie within these limits is $1 - 2\alpha$. The probability for a to be less than $m - g\,\sigma_m$ is α, and the probability to exceed $m + g\,\sigma_m$ is also α.

Chapter Two

Probabilities and Frequencies

The first three sections (§§ 5, 6, 7) of this chapter cover basic theory. The last three (§§ 8, 9, 10) discuss applications in demography, medicine, biology and physics. If you are primarily interested in the sample mean and sample standard deviation, skip from § 7 directly to § 18.

§ 5. Binomial Distribution

A. Bernoulli's Formula

We may be able to repeat an experiment, whose results depend on chance, under identical conditions. For example, a botanist might grow offspring from a certain parent stock by means of cross breeding and classify them according to flower color or other characteristics. Each offspring is a random product and the different flower colors are the possible results of the experiment. Or, a surgeon may perform the same operation several times and count the number of patients which recover and the number which die.

To simplify matters, we shall assume that in each experiment only two possibilities A and $\bar{A}$ are distinguished, for example, whether the patient dies or not. The death of the first patient will be event A_1, say, the death of the second A_2, and so on until A_n. We shall assume 1) that the experiments $E=A_1+\bar{A}_1$, $E=A_2+\bar{A}_2, \ldots, E=A_n+\bar{A}_n$ are all independent, and 2) that the probabilities $P(A_1), P(A_2), \ldots, P(A_n)$ all have the same value p.

It may seem as if condition 2) would seldom be fulfilled in real applied problems. The patients which a surgeon treats don't all have the same constitution: a weak heart, age, or sex may make a great deal of difference. Nevertheless, this doesn't interfere with the application of the theory provided that the patients are selected at random and not systematically. If our surgeon treats his patients independently of each other as they come to him operating on a random sequence of men *and* women (excluding those patients for which the operation would be dangerous or unnecessary, of course), we can consider the condition of uniform probability fulfilled. Let us say that $\mathscr{P}(F)$ is the probability that a woman comes to the operation and that $\mathscr{P}_F(D)$ is the probability of death for a woman; likewise let $\mathscr{P}(M)$ and $\mathscr{P}_M(D)$ be the corresponding probabilities for men. Then, according to the Rule of Total Probability (§ 1), the probability of death for a randomly selected patient is

$$\mathscr{P}(D)=\mathscr{P}(F)\,\mathscr{P}_F(D)+\mathscr{P}(M)\,\mathscr{P}_M(D). \tag{1}$$

This probability $\mathscr{P}(D)$ remains constant and thus condition 2) is fulfilled; this holds even if the conditional probability of death is different for men and women. We can argue in the same way when the patients are classified according to any other characteristics, with different death probabilities for different classes, rather than according to sex.

In order to make the conditions most nearly equal for all animals in pharmacological experiments processing specific animal poisons, sometimes equal doses per kg body weight are given rather than the same absolute dose. However, if the animals are chosen at random, irrespective of their weights, this is unnecessary; we can allot the same absolute dose and still assume that the probabilities of reaction to it are equal. The consideration is the same as for the sex of the patients above. So, in my opinion, we can neglect differences in weight as well as differences in susceptibility to the poison.

In probability theory, it is customary to set

$$\mathscr{P}(A_i)=p \quad \text{and} \quad \mathscr{P}(\bar{A}_i)=q, \qquad p+q=1,$$

and to call the occurrence of A_i a success and the occurrence of $\bar{A}_i$ a failure. The probability that in a sequence of n experiments, or trials, an event A_i occurs k times and event $\bar{A}_i$ occurs the remaining l times is

$$W_k=\binom{n}{k}p^k q^l=\frac{n!}{k!\,l!}\,p^k q^l \tag{2}$$

according to Bernoulli.

This fact can be proved as follows. The probability that A_i occurs in k definite trials and $\bar{A}_i$ occurs in the remaining trials is $p^k q^l$. The number of combinations of k from n is $\binom{n}{k}$. Thus, the probability of success in any k trials and failure in the remaining trials is $\binom{n}{k}p^k q^{n-k}$.

B. Expectation and Standard Deviation of the Number of Successes

If we define for each i a random variable $\mathbf{x}_i$ whose value is $+1$ or 0 according to whether a success occurs on the ith trial, then the sum

$$\mathbf{x}=\mathbf{x}_1+\mathbf{x}_2+\cdots+\mathbf{x}_n$$

is a random variable whose value in any particular case is equal to the number of successes. The individual $\mathbf{x}_i$ are independent because the trials $E = A_i + \bar{A}$ are assumed to be independent. The expectation of $\mathbf{x}_i$ is

$$\mathscr{E}\mathbf{x}_i = p \cdot 1 + 0 \cdot q = p.$$

Likewise, the expectation of $\mathbf{x}_i^2$ is

$$\mathscr{E}\mathbf{x}_i^2 = p \cdot 1 + q \cdot 0 = p.$$

Thus, the variance of $\mathbf{x}_i$ is

$$\sigma_i^2 = \mathscr{E}\mathbf{x}_i^2 - (\mathscr{E}\mathbf{x}_i)^2 = p - p^2 = p(1-p) = pq.$$

The expectation and variance of $\mathbf{x} = \mathbf{x}_1 + \cdots + \mathbf{x}_n$ are

$$\begin{aligned} \mathscr{E}\mathbf{x} &= np \\ \sigma^2 &= \sigma_1^2 + \cdots + \sigma_n^2 = npq. \end{aligned} \tag{3}$$

Thus, the standard deviation of $\mathbf{x}$ is

$$\sigma = \sqrt{npq}. \tag{4}$$

C. Law of Large Numbers

According to the Tchebychev inequality (§ 3), the values of $\mathbf{x} - pn$ which are large compared with σ have an extremely small probability. *However, we are accustomed to considering events which have a very small probability as almost impossible;* we don't expect their occurrence at a single realization of the conditions under which theoretically they are possible. Practical applications of probability theory are all based on this principle. Thus, in practice, we say the values $k - pn$ of $\mathbf{x} - pn$ have at most the order of magnitude $\sqrt{pqn}$. We can also omit the factor q and say for all practical purposes the values $k - pn$ have at the most the order of magnitude $\sqrt{pn}$. This last formulation can be used to advantage when a rare event is concerned, that is, when p is near zero and q is near one.

The *frequency* is the number k of successes divided by n:

$$h = \frac{k}{n}. \tag{5}$$

We now form a random variable

$$h = \frac{\mathbf{x}}{n}$$

that takes the values $h = k/n$ where k is the number of successes in n trials. Thus, using (3) and (4), the expectation and standard deviation of h are given by

$$\mathcal{E} h = p, \qquad \sigma_h = \left(\frac{pq}{n}\right)^{\frac{1}{2}}.$$

The conceptual difference between the random variable $\mathbf{h}$ and an individual value h of $\mathbf{h}$ is very troublesome. For this reason, it will be dropped henceforth. We shall simply speak of the frequency h, keeping in mind that this frequency is a random variable with an expectation and standard deviation. We shall proceed likewise in all analogous cases. Bold face type will only be used occasionally in order to recall this difference.

The expectation of h is p. Thus, the values of $|h-p|$ which we observe in practice have at most the order of magnitude

$$\left(\frac{pq}{n}\right)^{\frac{1}{2}}.$$

Since pq is at most $\frac{1}{4}$, we can also say that *for all practical purposes,* $|h-p|$ *has at most the order of magnitude* $n^{-\frac{1}{2}}$. By this we mean that the values of $|h-p|$ which are large as compared with $n^{-\frac{1}{2}}$ have all together only a very small probability.

If we let the number of trials increase, then $n^{-\frac{1}{2}}$ tends to zero. Consequently, the frequency forms a more and more exact estimate of the probability p. The possibility of estimating probabilities statistically rests on this *Law of Large Numbers.*

§ 6. Deviation of the Frequency h from the Probability p

A. Normal Approximation to the Binomial Distribution

De Moivre and Laplace studied the binomial distribution for large n and approximated it with a very suitable continuous distribution. The results are well known and will be briefly summarized here[1]. An approximation for the probability W_k for large np and nq is

$$W_k \sim (\sigma\sqrt{2\pi})^{-1} e^{-\frac{1}{2}(z/\sigma)^2} [1 + \tfrac{1}{2}\sigma^{-2}(p-q)z - \tfrac{1}{6}\sigma^{-4}(p-q)z^3], \qquad (1)$$

[1] For the derivation, which depends upon Stirling's formula

$$n! \sim n^n e^{-n} \sqrt{2\pi n}\left(1 + \frac{\vartheta}{12n}\right) \qquad (0 < \vartheta < 1),$$

see W. Feller, Introduction to Probability Theory, Vol. 1, Wiley (New York) 1957.

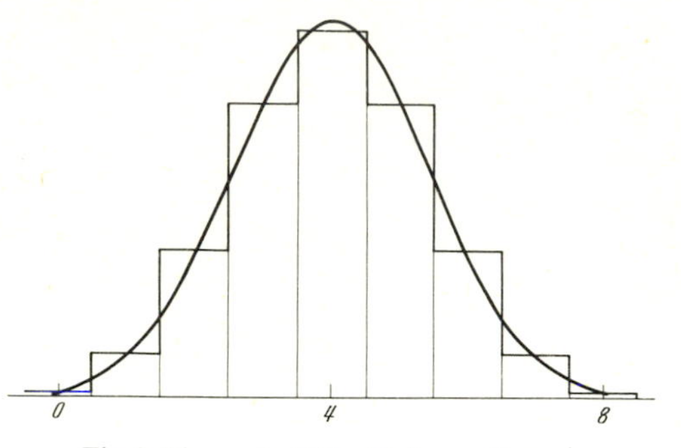

Fig. 3. The probability W_k for $n=8$, $p=\frac{1}{2}$

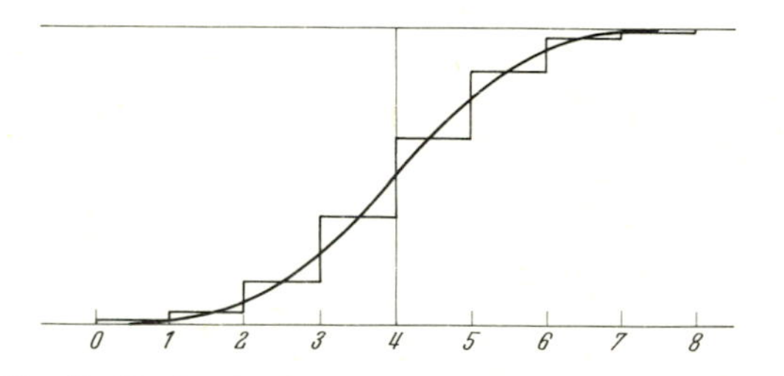

Fig. 4. The distribution function of the random variable **x** for $n=8$, $p=\frac{1}{2}$

where $z=k-np$. The derivation is valid for those values of z which are, at most, of order of magnitude $\sigma=(npq)^{\frac{1}{2}}$. However, we can also apply the formula for large values of z/σ, since then both sides of (1) are very small. Fig. 3 shows how good approximation (1) is for moderate n. The exact values of W_k for $n=8$, $p=q=\frac{1}{2}$ are represented as heights of rectangles and the approximation as a smooth curve. In this case, the curve is a Gaussian error curve because $p-q=0$; if $p-q\neq 0$, the additional terms cause a skewness in the curve which disappears, however, as $n\to\infty$.

From (1), it follows that the distribution function of the random variable x can be approximated by an integrated Gaussian error curve[2]. The distribution function of x is really a step function as shown in Fig. 4, since x can take only a finite number of values, $k=0, 1, \ldots, n$.

Now, we calculate by summation the probability of those values of k for which the absolute value of $k-np$ does not exceed a certain multiple $g\sigma$ of the standard deviation $\sigma=(pqn)^{\frac{1}{2}}$:

$$|k-np|\leqq g(pqn)^{\frac{1}{2}} \tag{2}$$

[2] There are even better continuous approximations for the binomial distribution with moderate n. See M. E. Wise, Proc. Kon. Ned. Akad. Amsterdam (Science section) A 57, p. 513.

or equivalently, for which

$$|h-p| \leqq g \left(\frac{pq}{n}\right)^{\frac{1}{2}} \tag{3}$$

holds. This probability is approximately equal to

$$2\Phi(g)-1=\frac{2}{\sqrt{2\pi}} \int_0^g e^{-\frac{1}{2}t^2} dt. \tag{4}$$

$\Phi(g)$ tends to one very quickly, so that for $g=2.58$, the probability is 0.99 and for $g=3$ it is 0.9973. That is to say, *the values of $k-np$ which are larger than three times the standard deviation in absolute value are so improbable that we scarcely need consider them.*

Numerical calculations show that this result is valid not only for very large n, but even for moderately large values of n. If pn and qn are both larger than six, we can use the above 3σ-rule without hesitation.

B. Estimation of σ

One difficulty in the application of the 3σ-rule is that although we may know k, l, and n and thus the frequency, in practice, we don't know p and q, and therefore the standard deviation σ is also unknown. We can overcome this difficulty in several ways.

If we replace pq by its largest possible value $\frac{1}{4}$, thus replacing $\sigma=(pqn)^{\frac{1}{2}}$ by $\sqrt{n}/2$, we are always on the safe side. This procedure is very good when the observed frequency lies near $\frac{1}{2}$.

A second possibility is to replace p and q by h and $1-h$; thus, σ becomes

$$s=[h(1-h)\,n]^{\frac{1}{2}}=\left(\frac{kl}{n}\right)^{\frac{1}{2}}. \tag{5}$$

This is reasonable for large n, since then h is near p according to the Law of Large Numbers. However, when k or l is small, we must be careful. If h or $1-h$ is near zero, $h(1-h)$ may be considerably smaller than $p(1-p)$. If this happens, we underestimate σ if we substitute s for σ.

This is especially striking in the extreme case where $k=0$. For example, suppose that out of 90 patients, not one dies after surgery. It is legitimate to conclude from these figures that the true mortality rate p is very low. However, we cannot conclude that $p-h$ is necessarily smaller than the sample standard deviation $s=(kl/n)^{\frac{1}{2}}$, for in this example s is zero, and no one would want to maintain that the mortality rate is exactly zero.

A good way out of this difficulty is discussed in the next section[3]. For the moment we shall suggest a (small) correction to counteract the possible underestimation of the standard deviation for moderate k and l. If we compare $s^2 = k\,l/n$ with $\sigma^2 = p\,q\,n$, we see that the mean of s^2 is not σ^2, but $[(n-1)/n]\,\sigma^2$. The calculation runs as follows:

$$\mathscr{E}\,s^2 = \mathscr{E}\,\frac{k(n-k)}{n} = \frac{1}{n}(\mathscr{E}\,k\,n - \mathscr{E}\,k^2).$$

Now, $\mathscr{E}\,k = p\,n$ and

$$\mathscr{E}(k^2) = (\mathscr{E}\,k)^2 + \sigma^2 = (p\,n)^2 + p\,q\,n;$$

thus,

$$\mathscr{E}\,s^2 = \frac{p\,n^2 - (p^2\,n^2 + p\,q\,n)}{n} = p\,n - p^2\,n - p\,q$$
$$= p\,q\,n - p\,q = p\,q(n-1) = \frac{n-1}{n}\,\sigma^2.$$

Therefore, to obtain an estimate with expectation σ^2, s^2 must be replaced by

$$s'^{\,2} = \frac{n}{n-1}\,s^2 = \frac{k\,l}{n-1}. \tag{6}$$

In order to determine the frequency h from the observed number (of successes) k, we have to divide by n. If we also divide the true standard deviation and the approximation s' by n, we obtain

$$s_h^2 = \frac{k\,l}{n^2(n-1)} = \frac{h(1-h)}{n-1}. \tag{7}$$

The expectation of s_h^2 is

$$\mathscr{E}\,s_h^2 = \sigma_h^2 = \frac{\sigma^2}{n^2} = \frac{p\,q}{n}. \tag{8}$$

If we use h as an estimate for p and s_h^2 as an estimate for σ_h^2, the estimates have no *bias* and no *systematic* errors.

Example 6. From 1870 to 1900, 1,359,671 boys and 1,285,086 girls were born in Switzerland. What can we state about the probability of the birth of a boy?

---

[3] Another solution consists in transforming the h-axis so that the standard deviation of the transformed h becomes almost independent of p. See R. Rao, Advanced Statistical Methods in Biometric Research, Wiley (New York) 1952, pp. 207–214.

The frequency of boys was

$$h=\frac{k}{n}=\frac{1{,}359{,}761}{2{,}644{,}757}=51.41\,\%.$$

These numbers are very large and we can certainly use the normal approximation. The standard deviation of h is

$$\sigma=\left(\frac{pq}{n}\right)^{\frac{1}{2}}\sim\left(\frac{1}{4n}\right)^{\frac{1}{2}}=0.03\,\%.$$

Using the 3σ-rule, it follows that h very likely lies between 51.32% and 51.50%.

C. Two-Sided and One-Sided Bounds for h

By inequality (3), the frequency h is bounded on both sides. The probability of observing h such that (3) holds is approximately $2\Phi(g)-1$.

If we set $\Phi(g)=1-\alpha$, the probability that (3) holds becomes

$$2\Phi(g)-1=1-2\alpha.$$

Thus, the probability that (3) does not hold is 2α. The number 2α can be made arbitrarily small when g is chosen large enough. For $g=2.58$ we have $2\alpha=0.01$, as mentioned above.

If inequality (3) does not hold, then

$$h-p>g\left(\frac{pq}{n}\right)^{\frac{1}{2}} \quad \text{or} \quad h-p<-g\left(\frac{pq}{n}\right)^{\frac{1}{2}}.$$

The probabilities of the two cases are almost equal; thus, both are approximately α. It is true that we have to neglect the additional z, z^3 terms in (1) in order to get this approximation, so it is not as good as (4). However, if we are satisfied with this rough approximation, then we can say:

$$h\leqq p+g\left(\frac{pq}{n}\right)^{\frac{1}{2}}$$

holds with probability $1-\alpha$ *and likewise*

$$h\geqq p-g\left(\frac{pq}{n}\right)^{\frac{1}{2}}$$

holds with probability $1-\alpha$.

These are bounds for one-sided confidence intervals for h. Table 3 gives the values of g corresponding to various one-sided or two-sided error levels. The relation between α and g is given by

$$\Phi(g)=1-\alpha. \tag{9}$$

§ 7. Confidence Bounds for Unknown Probabilities

A. The Problem

Suppose we have observed a frequency $h=k/n$ under the conditions of the last section. Within what bounds can the underlying probability lie? Naturally, we can't assign bounds with absolute certainty; we must always allow a certain probability of error. Let us take 2α as the admissible error probability of a two-sided confidence interval for p.

What probability we shall admit depends on what purpose we have in mind. For example, the premiums for a life insurance company have to be calculated so that bankruptcy due to a random fluctuation in the mortality rate is extremely improbable; in this case an error probability of 0.01 would be too large because then on the average one out of every hundred companies would go bankrupt. On the other hand, in biological and medical investigations there are so many possible sources of error because of uncertain theoretical hypotheses and simplifying assumptions that an error probability of 0.01 from using statistics seems relatively harmless. Very often research workers are even satisfied with an error probability of $2\alpha=0.05$ or 5%.

Today the usual "significance levels" or error probabilities for two-sided tests are $2\alpha=0.01$ and 0.05. However, the theoretical results hold in complete generality for any arbitrary α.

B. Approximate Solution for Large *n*

From formula (3), § 6,

$$|h-p| \leqq g\left(\frac{pq}{n}\right)^{\frac{1}{2}} \tag{1}$$

holds for large n with a probability of $W=1-2\alpha$. In place of (1), we can also write

$$(h-p)^2 \leqq \frac{g^2}{n} p(1-p). \tag{2}$$

For given α, we can read g from Table 3. The last column gives g^2 also. For example, if $2\alpha=0.01$, then we find $g=2.58$ and $g^2=6.63$.

Consider a point Q with coordinates p (the true probability) and h (the empirical frequency) in the (h, p)-plane. Formula (2) represents the interior and the boundary of an ellipse which is tangent to the horizontal sides of the unit square at the end points of the diagonal (see Fig. 5). The width of the ellipse depends on g and n: the larger the number of trials the more narrow the ellipse. The location of Q depends on chance because the h coordinate depends on chance. The probability that Q lies inside or on the ellipse is approximately $1-2\alpha$ for each value of p. Or, in other words (using $2\alpha=0.01$), if we assert that Q does not fall outside the ellipse, we shall be wrong only once in a hundred times on average.

Now, in practice p is unknown, but h is known. A straight line, $h=$constant, cuts the ellipse in two points, whose p coordinates can be

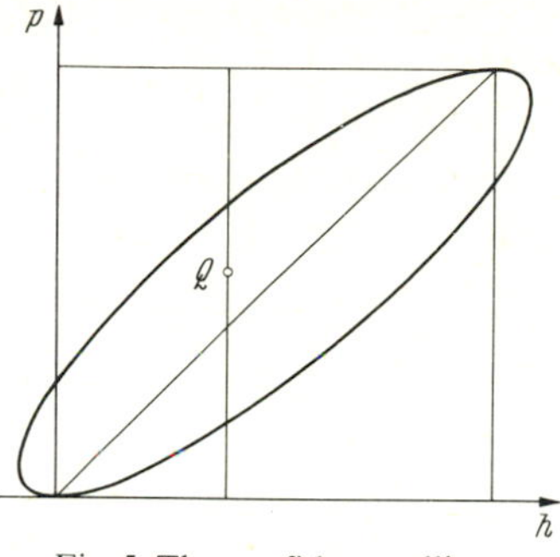

Fig. 5. The confidence ellipse

found from the solution of the quadratic equation

$$(h-p)^2=\frac{g^2}{n}p(1-p). \tag{3}$$

The interval between these two points of intersection lies in the interior of the ellipse. If we make the assertion that the true value of p lies between the two roots p_1 and p_2 of (3), then the probability of error is only 2α, for this assertion is equivalent to the one that Q does not fall outside the ellipse.

This does not mean that we can assert in each specific case, for given h and n, that p lies between p_1 and p_2 with probability 0.99. In each individual case p has a definite (though unknown) value and the statement $p_1 \leqq p \leqq p_2$ is either true or false. It does mean that if a statistician calculates p_1 and p_2 and makes the assertion that $p_1 \leqq p \leqq p_2$ for each series of experiments he encounters, then he will be in error only one time in a hundred. He is not allowed to pick experiments that have resulted in a certain frequency h, but has to accept the frequencies as they result from chance.

The solution of the quadratic Eq. (3) gives the confidence bounds p_1, p_2 between which the true value p presumably lies. The bounds are

$$\begin{aligned} p_1 &= \frac{hn+\frac{1}{2}g^2-g\left[h(1-h)n+\frac{1}{4}g^2\right]^{\frac{1}{2}}}{n+g^2} \\ p_2 &= \frac{hn+\frac{1}{2}g^2+g\left[h(1-h)n+\frac{1}{4}g^2\right]^{\frac{1}{2}}}{n+g^2}. \end{aligned} \tag{4}$$

The calculation of these formulas is quite involved. The following graphical method is more practical.

If we substitute

$$h-p=\frac{g}{\sqrt{n}}x \tag{5}$$

in (3), then the equation for the ellipse is transformed into that of a circle

$$x^2=p(1-p). \tag{6}$$

For given h, formula (5) represents a straight line in the (x, p)-plane, which will intersect circle (6). This straight line passes through the point $(x=0, p=h)$ and has slope $-(g/\sqrt{n})$. Draw circle (6) on millimeter paper with

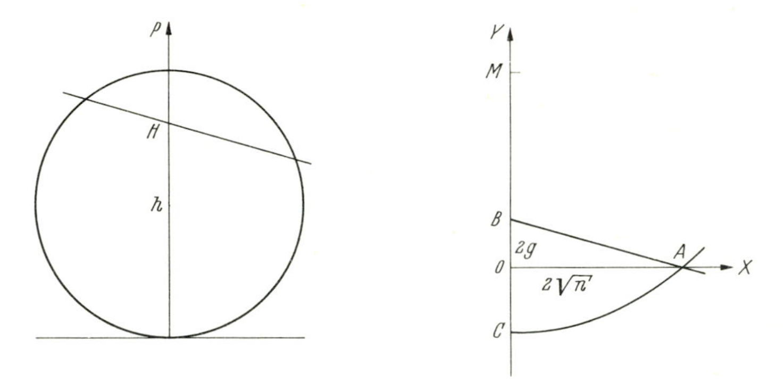

Fig. 6. Construction of confidence bounds

Fig. 7. Auxiliary figure

a diameter of (say) 10 cm. For each outcome of an experiment we can find the point H on the vertical diameter at height h. Line (5) passes through H (see Fig. 6). In order to obtain the slope of this line, draw an auxiliary figure (see Fig. 7). Construct a coordinate system OXY. Plot on the axes $OA=2\sqrt{n}$ and $OB=2g$. The line AB has the desired slope.

The square root of n can be found in a square root table or can be determined, for moderate n, from the drawing as follows. On the Y axis, plot the point C, 2 cm below the origin, and the point M, $(n-1)$ cm above the origin. Draw a circle about M passing through point C. The intersection of this circle with the X axis is the point A. For, if we denote by D the upper intersection of the circle with the Y axis, then $OC=2$ and $OD=2n$; hence $OA^2=OC\cdot OD=4n$ as required.

Now, for each case, draw a line parallel to AB through H. This line intersects the circle in two points. The heights of these points are p_1 and p_2 and they can be read directly from the millimeter paper. Hence, we have the desired confidence bounds. If we only use one of the bounds, the probability of error is α.

The formulae applied here are unreliable whenever one of the expected values pn or qn is small. Therefore, we recommend using formulae (4) or the equivalent geometrical construction only when the observed numbers $k=hn$ and $n-k=(1-h)n$ are both at least four.

Example 7. In the Zürich University Surgical Clinic, 79 patients suffering from bronchiecstasis were treated by lung resection during the years from 1948 to 1952. Out of the 79 patients, 3 died within a week of the operation[4]. Therefore, the observed mortality rate is

$$h=\frac{3}{79}=3.8\%.$$

By (4) or by means of the construction in Fig. 6, we find

$$p_1=1.3\% \quad \text{and} \quad p_2=10.6\%$$

as the 5% confidence bounds for the true mortality rate. Since the observed number of deaths is smaller than four, we shall take a somewhat larger confidence interval as a precaution and say that *the true mortality rate probably lies between* 1% *and* 11%.

We see from this example how inaccurate the determination of a probability from a small or middle-sized sample can be.

C. Exact Solution

The solution we have given to the confidence bound problem by approximation was based on replacing the exact binomial distribution ((2) § 5) by a continuous distribution ((1) § 6). Fig. 4 shows that the approximating curve lies above the exact step function sometimes and below sometimes. The result is that the error probability of the confidence bounds turns out to be somewhat larger or somewhat smaller than 2α depending on the true value p of the probability[5].

However, we can determine the exact confidence limits by Clopper and E. S. Pearson's method[6]. These limits guarantee a total error probability less than or equal 2α; the probability that p falls outside any one of the two limits is less than or equal to α.

For this purpose, we use the exact binomial distribution

$$W_k(p)=\binom{n}{k} p^k q^{n-k}. \tag{7}$$

As Fig. 3 shows, the W_k are largest in the middle and decrease rapidly to both sides, provided p doesn't lie too near zero or one. The sum of

---

[4] F. Wegmann, Die operative Behandlung der Bronchektasien, Diss. Zürich 1955, Zusammenfassung p. 39.

[5] A curve which illustrates the shape of the error probability as a function of p in the limiting case of a rare event is shown in my note "Vertrauensgrenzen für unbekannte Wahrscheinlichkeiten", Sitzungsber. sächs. Akad. Wiss. 91 (1939) p. 213.

[6] C. J. Clopper and E. S. Pearson, Biometrika 26 (1934), p. 404.

all the W_k is

$$\sum_0^h W_k(p) = 1. \tag{8}$$

The one-sided confidence bound with error probability less than or equal to α is determined in the following manner. Let K be an integer $(0 \leqq K \leqq n)$. Construct from (8) the partial sum from 0 to K

$$S_K(p) = \sum_0^K W_k(p). \tag{9}$$

$S_K(p)$ is the probability that k takes one of the values from 0 to K.

Differentiating with respect to p, all of the terms cancel out except a single negative one

$$\begin{aligned} \frac{d}{dp} S_K(p) &= -n\, q^{n-1} + \sum_1^K \left[\frac{n!}{(k-1)!\,(n-k)!} p^{k-1} q^{n-k} \right. \\ &\qquad \left. - \frac{n!}{k!\,(n-k-1)!} p^k q^{n-k-1} \right] \\ &= -\frac{n!}{K!\,(n-K-1)!} p^K q^{n-K+1}. \end{aligned} \tag{10}$$

Incidentally, notice that an interesting integral representation of $S_K(p)$ as an incomplete beta function follows from (10),

$$\begin{aligned} S_K(p) &= \int_p^0 \frac{n!}{K!\,(n-K-1)!} x^K (1-x)^{n-K-1}\, dx \\ &= \int_0^{1-p} \frac{n!}{K!\,(n-K-1)!} (1-y)^K y^{n-K-1}\, dy. \end{aligned} \tag{11}$$

Here, we need only the fact that S_K is a decreasing continuous function of p, which takes the value one for $p=0$ and zero for $p=1$. From this it follows that S_K assumes each intermediate value exactly once. Therefore, we can determine p_K (for each $K<n$) so that the function $S_K(p)$ assumes the value α for $p=p_K$,

$$S_K(p_K) = \alpha.$$

Clopper and Pearson established the following rule. *If the experiment results in the frequency k/n, then we take p_k as the upper confidence bound for p, that is, we reject all values of p which are larger than p_k. The probability that the true value of p is rejected by mistake is less than α.*

Proof. If the true value p is rejected, it means that $p > p_k$. From this it follows that

$$S_k(p) < S_k(p_k) = \alpha$$

because S_k is a decreasing function of p.

Now, suppose K is the largest index so that $S_K(p)$ is still less than α. Then $k \leqq K$, that is to say, k is one of the values $0, 1, \ldots, K$. The probability that k assumes one of the values from 0 to K is, however, exactly $S_K(p)$, and thus less than α, which was to be shown.

Thus, the upper confidence bound p_k is the solution of

$$W_0(p) + W_1(p) + \cdots + W_k(p) = \alpha$$

for $k < n$. Analogously, the lower confidence bound is the solution of

$$W_k(p) + W_{k+1}(p) + \cdots + W_n(p) = \alpha$$

for $k > 0$. Of course, the lower confidence bound is zero for $k = 0$ and the upper confidence bound is one for $k = n$.

The exact confidence interval calculated according to Clopper and Pearson is considerably wider than that bounded by the p_1 and p_2 defined in B. This is because the error probability of the approximated limits p_1 and p_2 fluctuates about 2α, whereas for the exact limits it is guaranteed to be at most 2α and is usually much smaller.

Should we use the exact or the more convenient approximate confidence bounds? If we are willing to risk an error probability of 2α, I think that we can also admit an error probability fluctuating about 2α. If we have several probabilities to estimate in the course of time and use approximate limits in each case, then the deviations in the error probability compensate for each other and we are in error on average only about $2\alpha \cdot 100$ times in 100 cases. For extremely small numbers k ($k < 4$, say), we can reduce the lower confidence bound for p somewhat in order to remain on the safe side.

§ 8. Sampling

From an urn with K white and L black balls ($K + L = N$), draw n balls (without replacement). What is the probability of finding k white and l black balls among them?

The number of ways of choosing n balls out of N is $\binom{N}{n}$. If the balls are well mixed, all of these ways are equally likely; each one of them has probability $\binom{N}{n}$. The number of possible ways of choosing k white balls out of K and l out of L is $\binom{K}{k}\binom{L}{l}$. Thus, the required probability is

$$W_k = \binom{K}{k}\binom{L}{l} \Big/ \binom{N}{n} = \frac{K!}{k!\,(K-k)!} \cdot \frac{L!}{l!\,(L-l)!} \cdot \frac{n!\,(N-n)!}{N!}. \quad (1)$$

Now, if we define a random variable $\mathbf{x}$ (as in § 5), with values equal to the number k of white balls drawn, then $\mathbf{x}$ is the sum of n random variables $\mathbf{x}_1 + \cdots + \mathbf{x}_n$, where each $\mathbf{x}_i$ depends on the color of the ith ball drawn. If the ith ball is white, we have $\mathbf{x}_i = 1$ and $\mathbf{x}_i = 0$ if the ith ball is black. The probability that $\mathbf{x}$ takes the value k is given by (1). The distribution defined by (1) is called the *hypergeometric distribution.*

Now, let us calculate the expectation and standard deviation of $\mathbf{x}$. According to § 1, Example 3, the probability that $\mathbf{x}_i = 1$ is K/N. In the same way, the probability that $\mathbf{x}_i \mathbf{x}_j = 1$ for $i \neq j$ is $\dfrac{K(K-1)}{N(N-1)}$ and $\mathscr{P}(\mathbf{x}_i \mathbf{x}_j = 1)$ for $i = j$ is K/N.

Therefore,

$$\mathscr{E}(\mathbf{x}_i^2) = \mathscr{E}\,\mathbf{x}_i = \frac{K}{N},$$

$$\mathscr{E}(\mathbf{x}_i \mathbf{x}_j) = \frac{K(K-1)}{N(N-1)} \qquad (i \neq j).$$

From this it follows that

$$\mathscr{E}\,\mathbf{x} = \sum \mathscr{E}\,\mathbf{x}_i = n \frac{K}{N}$$

$$\mathscr{E}\,\mathbf{x}^2 = \mathscr{E}\left(\sum \mathbf{x}_i\right)^2 = \mathscr{E}\left(\sum \mathbf{x}_i^2 + 2 \sum \mathbf{x}_i \mathbf{x}_j\right)$$

$$= n \frac{K}{N} + n(n-1) \frac{K(K-1)}{N(N-1)}$$

$$\sigma^2 = \mathscr{E}\,\mathbf{x}^2 - (\mathscr{E}\,\mathbf{x})^2 = n \frac{K}{N} + n(n-1) \frac{K(K-1)}{N(N-1)} - n^2 \frac{K^2}{N^2} \tag{2}$$

$$= n K \frac{N(N-1) + N(n-1)(K-1) - n(N-1) K}{N^2 (N-1)}$$

$$= \frac{n K (N-K)(N-n)}{N^2 (N-1)} = \frac{K L n (N-n)}{N^2 (N-1)}.$$

Therefore, the actual values of k will most probably lie in a neighborhood of the expected value $n \dfrac{K}{N}$, and the deviation

$$k - n \frac{K}{N} = z$$

will be of order

$$\sigma = \frac{1}{N} \left(\frac{K L n (N-n)}{N-1} \right)^{\frac{1}{2}}. \tag{3}$$

Using Stirling's formula for $n!$ (derived in § 12), we can expand probability (1) asymptotically for large K, L, n, and $N-n$. We shall omit the somewhat tedious calculations and present only the result.

$$W_k \sim \sigma^{-1}(2\pi)^{-\frac{1}{2}} e^{-\frac{1}{2}\sigma^{-2}z^2}\left[1+(K-L)(N-2n)N^{-2}\left(\tfrac{1}{2}\sigma^{-2}z-\tfrac{1}{6}\sigma^{-4}z^3\right)\right]. \quad (4)$$

From this, it follows (as in § 6) that the probability that k lies between $n\frac{K}{N}-g\sigma$ and $n\frac{K}{N}+g\sigma$ is given approximately by

$$\frac{2}{\sqrt{2\pi}}\int_0^g e^{-\frac{1}{2}t^2}\,dt=2\Phi(g)-1. \quad (5)$$

As before, we can choose g so that integral (5) is a given value $1-2\alpha$ (see Table 3).

The inequality

$$\left|k-n\frac{K}{N}\right| \leqq g\,\sigma, \quad (6)$$

whose probability is given by (5), can also be written

$$(kN-nK)^2 \leqq g^2 N^2 \sigma^2$$

or, if the value of σ is substituted from (3),

$$\frac{(kN-nK)^2(N-1)}{KLn(N-n)} \leqq g^2. \quad (7)$$

Therefore, the probability of inequality (7) is approximately $1-2\alpha$. The approximation holds uniformly in the following sense. As soon as the expectations of the four random variables k, l, $K-h$, and $L-l$ exceed a bound $M(\varepsilon)$, the probability of (7) will differ from $1-2\alpha$ by less than ε. We shall make use of this fact in the next section.

Example 8. Sampling in demography and economics covers only a part of the population that can be viewed as representative of the whole, in that this part is selected in roughly the same proportions as represented in the total population from the various regions, cities, towns, *etc.* The frequencies determined from the sample (for example, mortality rate according to cause of death) are taken as approximations for the frequencies in the whole population. What deviations between the population frequency H and the sample frequency h are expected?

If the sample is chosen from the population completely at random, then the problem is identical to our urn problem. The frequencies in question are

$$h=\frac{k}{n} \quad \text{and} \quad H=\frac{K}{N}. \quad (8)$$

The expectation of h is H and the standard deviation is

$$\sigma_h=\frac{\sigma}{n}=\frac{1}{N}\left(\frac{KL(N-n)}{n(N-1)}\right)^{\frac{1}{2}}=\left(\frac{H(1-H)}{n}\cdot\frac{N-n}{N-1}\right)^{\frac{1}{2}}. \quad (9)$$

The deviations $|h-H|$ which we should expect are less than 2.58 in 99% of the cases.

A fuller treatment of the problems connected with sampling is given in Chapter 2 of Einführung in die mathematische Statistik by L. Schmetterer published by Springer-Verlag in Vienna, 1956.

§ 9. Comparison of Two Probabilities

Even more important than the estimation of a single probability is the comparison of two probabilities, particularly in medicine and biology. Suppose, for instance, that a surgeon has tried a new operating method on a series of patients and finds a much lower mortality rate than for the usual procedure; has the mortality rate really become smaller? Or, suppose a new method of cure has been found for an illness. Previously 40 out of 400 patients died; after administration of the new cure to 50 patients only one died. Has the new cure worked, or is the difference due to chance?

Suppose that the empirical frequencies are

$$h_1=\frac{k_1}{n_1}, \qquad h_2=\frac{k_2}{n_2}, \tag{1}$$

and the true probabilities are p_1 and p_2. If we have found that $h_1>h_2$, how large must h_1-h_2 be in order to assert that $p_1>p_2$ with an agreed degree of probability?

From § 5, we know that the random variable h_1 has expectation p_1 and standard deviation $\sigma=(p_1 q_1/n_1)^{\frac{1}{2}}$. The distribution of h_1 can be represented quite accurately by a Gaussian error curve, using the correction factor (1) in § 6. Likewise for h_2.

In Example 5, § 4, we saw that the difference of two random variables with normal distributions also has a normal distribution. If h_1 and h_2 are only approximately normal, then the approximation for the distribution of h_1-h_2 by the normal distribution is even better than for h_1 and h_2 individually. The height of the jumps becomes smaller, and taking differences cancels out the skewness in the individual distributions.

Therefore, we can assume a normal distribution for h_1-h_2 with expectation p_1-p_2 and standard deviation

$$\sigma=(\sigma_1^2+\sigma_2)^{\frac{1}{2}}. \tag{2}$$

It then follows that the values of $(h_1-h_2)-(p_1-p_2)$ that are larger than $g\sigma$ are very rare, when g depends on the assumed error probability 2α (for example, $g=2.58$ for $2\alpha=0.01$). If h_1-h_2 is larger than $g\sigma$, then we can assume that p_1-p_2 is positive.

However, the difficulty arises again that the exact value of σ is not known. There are two ways out of this dilemma. The first, and less advisable, consists of replacing σ_1^2 and σ_2^2 by their approximate values

$$\begin{aligned} s_1^2 &= \frac{h_1(1-h_1)}{h_1-1} \\ s_2^2 &= \frac{h_2(1-h_2)}{h_2-1} \end{aligned} \tag{3}$$

and forming the approximation

$$s^2 = s_1^2 + s_2^2. \tag{4}$$

The expectation of s^2 is σ^2, but it can differ from σ^2 considerably in individual cases. Now, instead of requiring that $h_1 - h_2$ be larger than $g\sigma$ (before we decide $p_1 < p_2$), we require that $h_1 - h_2$ be larger than $g s$. If n_1 and n_2 are large and h_1 and h_2 do not lie near zero or one, then, on average, we shall be wrong only one time in a hundred using this rule with $g = 2.58$.

B. The χ^2-Test

The second method is theoretically preferable and requires somewhat simpler calculations. We proceed as follows. The hypothesis that we want to test and would like to reject is that the differences between h_1 and h_2 are purely chance and that in fact $p_1 = p_2$. We are acting like Socrates when he wanted to disprove an assertion in a discussion. We tentatively assume the hypothesis $p_1 = p_2$ to be correct and look at its consequences. If these stand in contradiction to the facts, then this tentative hypothesis has to be rejected.

Under the assumption $p_1 = p_2$ the variances are

$$\sigma_1^2 = \frac{pq}{n_1}, \qquad \sigma_2^2 = \frac{pq}{n_2}$$

and

$$\sigma^2 = \sigma_1^2 + \sigma_2^2 = \frac{pq(n_1+n_2)}{n_1 n_2} = \frac{pqN}{n_1 n_2} = \frac{pq}{N} \cdot \frac{N^2}{n_1 n_2},$$

where $N = n_1 + n_2$. Now we replace the unknown p by the observed frequency again. For its determination we have the total number of trials $N = n_1 + n_2$ at our disposal. Out of the N, suppose there are $K = k_1 + k_2$ successful trials and $L = l_1 + l_2$ failures. Thus, replacing N by $N-1$ (for reasons discussed in § 6 B) and p and q by

$$H = \frac{K}{N} \quad \text{and} \quad 1 - H = \frac{L}{N}, \tag{5}$$

we obtain the approximation

$$s^2 = \frac{KL}{N-1} \cdot \frac{1}{n_1 n_2} \tag{6}$$

for σ^2. This formula is more reliable than (4). Now, if

$$|h_1 - h_2| > g\,s,$$

or equivalently

$$(h_1 - h_2)^2 > g^2 s^2,$$

or

$$\frac{(h_1 - h_2)^2 n_1 n_2 (N-1)}{KL} > g^2,$$

then the hypothesis $p_1 = p_2$ is *rejected.*

We set

$$\chi^2 = \frac{(h_1 - h_2)^2 n_1 n_2 (N-1)}{KL} \tag{7}$$

and call this test the *χ^2-test for comparison of two probabilities.* Substituting for h_1 and h_2, we can also write

$$\chi^2 = \frac{(k_1 n_2 - k_2 n_2)^2 (N-1)}{KL n_1 n_2}. \tag{8}$$

C. Verification

To justify this test we have to show that the event

$$[\chi^2 \leqq g^2] \tag{9}$$

has probability

$$1 - 2\alpha = \frac{2}{\sqrt{2\pi}} \int_0^g e^{-\frac{1}{2}t^2} dt \tag{10}$$

under the hypothesis $p_1 = p_2$.

For this, first, suppose that n_1 and n_2 are large numbers and $p_1 = p_2 = p$ does not lie near zero or one, so that the expected values $p n_1$, $p n_2$, $q n_1$, and $q n_2$ are also large. Then k_1, k_2, $l_1 = n_1 - k_1$, and $l_2 = n_2 - k_2$ are very probably large as well. The cases in which one of the four numbers is small do not play a major role in the calculation of the probability of (9).

Let $\mathscr{P}(K)$ be the probability that $k_1 + k_2$ has a certain integral value K. Let $\mathscr{P}_K(\chi^2 \leqq g^2)$ be the conditional probability that $\chi^2 \leqq g^2$ under the hypothesis that $k_1 + k_2$ takes the value K. Then,

$$\mathscr{P}(\chi^2 \leqq g^2) = \sum_K \mathscr{P}(K) \cdot \mathscr{P}_K(\chi^2 \leqq g^2). \tag{11}$$

Therefore, if we can prove that the conditional probabilities $\mathscr{P}_K(\chi^2 \leqq g^2)$ are all approximately equal to $1-2\alpha$, then we are through. Namely, if each separate factor $\mathscr{P}_K(\chi^2 \leqq g^2)$ on the right side of (11) lies between $1-2\alpha-\varepsilon$ and $1-2\alpha+\varepsilon$, where ε is arbitrarily small, then the left side of (11) also lies between $1-2\alpha+\varepsilon$ and $1-2\alpha-\varepsilon$.

The probability of a particular pair of values (k_1, k_2) with $k_1+k_2=K$ is

$$\begin{aligned}\mathscr{P}(k_1, k_2) &= \binom{n_1}{k_1} p^{k_1} q^{l_1} \binom{n_2}{k_2} p^{k_2} q^{l_2}\\ &= \binom{n_1}{k_1}\binom{n_2}{k_2} p^K q^L.\end{aligned}$$

The probability $\mathscr{P}(K)$ that k_1 and k_2 have a certain sum is equal to the total probability that out of N trials K are positive

$$\mathscr{P}(K) = \binom{N}{K} p^K q^L.$$

According to the definition (5) § 1, the conditional probability of a particular pair of values (k_1, k_2) under the assumption that k_1+k_2 has a given value K is

$$\begin{aligned}\mathscr{P}_K(k_1, k_2) &= \mathscr{P}(k_1, k_2)/\mathscr{P}(K)\\ &= \binom{n_1}{k_1}\binom{n_2}{k_2}\Big/\binom{N}{K}\\ &= \frac{n_1!\, n_2!\, K!\, L!}{k_1!\, l_1!\, k_2!\, l_2!\, N!}.\end{aligned}$$

Now, if we set

$$n_1 = n, \quad k_1 = k, \qquad l_1 = l$$

in order to simplify the expression, then

$$\mathscr{P}_K(k, K-k) = \frac{n!\,(N-n)!\,K!\,L!}{k!\,l!\,(K-k)!\,(L-l)!\,N!}. \tag{12}$$

The factors p and q are cancelled in the division and the result agrees with (1), § 8. That is to say, the *conditional probability* $\mathscr{P}_K(k, K-k)$ *is exactly equal to the probability that out of an urn with* K *white and* L *black balls,* k *white and* l *black are drawn in* N *trials.*

Therefore, the inequality $\chi^2 \leqq g^2$ agrees exactly with inequality (7) § 8. However, from § 8, the probability of this inequality is approximately $1-2\alpha$. This completes the verification.

The idea of this proof was originated by M. P. Geppert. Formula (7) with $N-1$ in the numerator (instead of the usual N) was first stated by H. von Schelling.

D. One-Sided and Two-Sided Applications of the χ^2-Test

In practical applications, the χ^2-test is not only used to test the hypothesis $p_1=p_2$, but also to decide which of the two unknown probabilities is larger. The decision is made in the following manner: if χ^2 exceeds g^2 and if $h_1>h_2$, we assume that p_1 is larger than p_2; but if χ^2 exceeds g^2 and $h_1<h_2$, we assume that p_1 is less than p_2. If χ^2 does not exceed g^2, no decision is made. What is the error probability of this decision procedure?

We have to distinguish three cases. First, suppose that $p_1=p_2$. In this case the probability of rejecting the hypothesis $p_1=p_2$ and assuming $p_1>p_2$ or $p_1<p_2$ is approximately 2α, as we have seen before. The proof shows that the probabilities that $h_1>h_2$ and that $h_1<h_2$ are nearly equal. Hence, the probability of making the incorrect decision $p_1>p_2$ is approximately α, and the probability of making the incorrect decision $p_1<p_2$ is also approximately α.

Next, suppose that $p_1<p_2$. The incorrect conclusion $p_1>p_2$ is drawn only if $h_1>h_2$ and

$$\frac{(h_1-h_2)^2\, n_1\, n_2 (N-1)}{KL}>g^2,$$

or

$$h_1-h_2>\left[\frac{KLg^2}{n_1\, n_2 (N-1)}\right]^{\frac{1}{2}}. \tag{13}$$

We have proved that if $p_1=p_2$ the probability for (13) to hold is approximately α. If p_2 is kept fixed and p_1 is diminished, the larger values of h_1 become less probable, and hence the probability for (13) to hold becomes even less than α. Likewise, in the case of $p_1>p_2$, the error probability of the decision procedure is less than α. Hence, in all three cases the error probability of this decision procedure is at most 2α.

E. Reliability for Small Numbers

We can also apply the χ^2-test for small values of N. The error probability of the test for $2\alpha=0.01$ is graphed as a function of p for a few typical cases in Fig. 8. The solid curves correspond to the test with N in the numerator, the broken curves to the test with $N-1$. We see that the curves exceeds the $1\,\%$ level only rarely, and then only by a small amount.

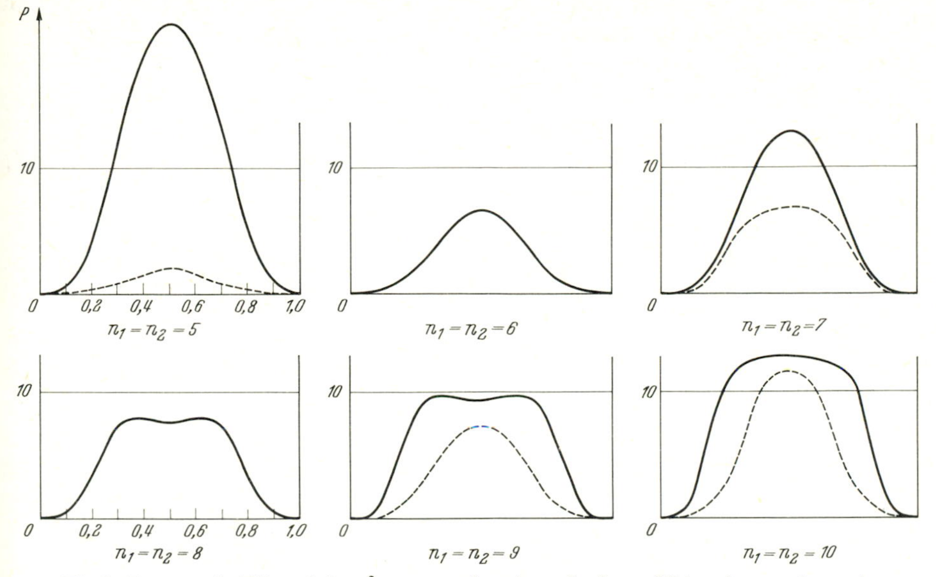

Fig. 8. Error probability of the χ^2-test as a function of p from Gildemeister and van der Waerden, Ber. sächs. Akad. Wiss. 95 (1943)

Example 9. From 1946 to 1951, 252 thrombosis cases were treated by anticoagulants at the Zürich University medical school[7]. Of these 252, seven patients died; therefore, the mortality rate was 2.8%. No anticoagulant was used from 1937 to 1942. Excluding those cases for which symptoms indicated that anticoagulant treatment would not have been suitable, there were 205 cases during that time. Of these 37 or 18.0% died. Is the favorable effect of the anticoagulant certain?

From (7) and (8), $\chi^2 = 30.2$. The 1% bound is $\chi^2 = 6.6$, and the 0.1% bound is 10.8. Both are far exceeded.

We might object that the two series of trials were drawn from two different time periods. A fanatic statistician would perhaps suggest treating patients alternately with the standard methods and the anticoagulant. However, the doctor who wants to save the greatest number of his patients naturally won't use a treatment that may be dangerous or less likely to cure the patient. In medical statistics, if two series of experiments are not carried out simultaneously, we must always question whether the results could be due to factors other than the change of treatment (fluctuations in epidemiological behavior *etc.*). Anyhow, in the thrombosis case the evidence is so overwhleming that we are bound to conclude that the new treatment was the decisive factor.

F. R. A. Fisher's Exact Test

The same reasoning that we used to verify the χ^2-test for large expectations was used by R. A. Fisher to construct an exact test with the one-sided error probability less than or equal to α. The procedure is

[7] I. Pugatsch, Zur Antikoagulantienbehandlung der Venenthrombosen in der inneren Medizin, Diss. Zürich 1954.

best explained in an example. We shall use one by K.D.Tocher (Biometrika, Vol. 37, p. 130).

Suppose we observe

$$\begin{array}{ccc} k_1=2 & l_1=5 & (n_1=\ 7) \\ k_2=3 & l_2=2 & (n_2=\ 5) \\ \hline (K=5) & (L=7) & (N=12) \end{array}$$

From these numbers, we construct a fourfold table and write down beside it all fourfold tables having the same sums for the rows and columns, but smaller k_1, as follows:

Observed			Extreme cases					
2	5	7	1	6	7	0	7	7
3	2	5	4	1	5	5	0	5
5	7	12	5	7	12	5	7	12

For these tables, calculate the conditional probabilities, given the first column sum $K=k_1+k_2$:

$$\mathscr{P}_K(k_1,k_2)=\frac{n_1!\,n_2!\,K!\,L!}{k_1!\,l_1!\,k_2!\,l_2!\,N!}. \tag{14}$$

Add these. In our case we obtain

$$\mathscr{P}=0.265+0.044+0.001=0.310.$$

The test proceeds as follows. *If* $\mathscr{P}\leqq\alpha$, *reject the hypothesis* $p_1=p_2$ *in favor of the alternative* $p_1<p_2$. For example, if $\alpha=0.05$, then the hypothesis $p_1=p_2$ is not rejected if we observe (2,5,3,2). However, if one of the two more extreme cases occurs we do reject. The conditional probability that one of these two cases occurs is

$$0.044+0.0001<0.05.$$

Therefore, the conditional probability that the hypothesis is incorrectly rejected is less than 0.05.

In general, let A be the event that the hypothesis $p_1=p_2$ is rejected in favor of $p_1<p_2$ by the above test. Suppose $\mathscr{P}_K(A)$ is the conditional probability of this event given the column sums $K=k_1+k_2$ and $L=l_1+l_2$ under the assumption that the hypothesis $p_1=p_2$ is true. Then

$$\mathscr{P}_K(A)\leqq\alpha \tag{15}$$

holds in general, since the test was set up this way.

The total probability that the hypothesis $p_1 = p_2$ is rejected on the grounds of the test when it is true is

$$\mathscr{P}(A) = \sum_K \mathscr{P}(K)\, \mathscr{P}_K(A) \tag{16}$$

summed over all K. Now, all $\mathscr{P}_K(A)$ are less than or equal to α according to (15); therefore, the entire sum (16) is also less than or equal to α

$$\mathscr{P}(A) \leqq \alpha \sum \mathscr{P}(K) = \alpha.$$

Thus, the error probability of Fisher's test is always less than or equal to α for one-sided applications, and therefore it is less than or equal to 2α for two-sided applications.

The error probability actually lies far below 2α for small or moderately large n_1 and n_2. This test is excessively cautious. Moreover, the χ^2-test requires considerably less calculation.

§ 10. Frequency of Rare Events

A. Poisson's Formula

Suppose we have a large n, but a very small probability p in the Bernoulli problem, so that np is not a large number. Then formula (2) § 5, still holds as well as (3) and (4) and their consequences; but the asymptotic approximation (1) from § 6 does not hold. Poisson has given another approximation for this limiting case, namely,

$$W_k \sim \frac{\lambda^k e^{-\lambda}}{k!}. \tag{1}$$

Here, W_k is the probability that in a long sequence of n independent trials the rare event A occurs k times. The expectation of k is $\lambda = np$. In (1), the number of trials n is not explicit. The formula is applicable for all rare events such as accidents, nuclear fission, and so on.

The proof of (1) is very simple. The exact formula is

$$\begin{aligned} W_k &= \binom{n}{k} p^k (1-p)^{n-k} \\ &= \frac{n(n-1)\cdots(n-k+1)}{k!} \frac{\lambda^k}{n^k} \left(1 - \frac{\lambda}{n}\right)^{n-k} \\ &= \frac{\lambda^k}{k!} \left(1 - \frac{\lambda}{n}\right)^n \left(1 - \frac{1}{n}\right)\left(1 - \frac{2}{n}\right) \cdots \left(1 - \frac{k-1}{n}\right)\left(1 - \frac{\lambda}{n}\right)^{-k}. \end{aligned}$$

The first factor does not depend on n. The second tends to $e^{-\lambda}$ as $n \to \infty$. All of the other factors tend to 1 as $n \to \infty$. Therefore, W_k tends to the right side of (1) as $n \to \infty$.

Hitherto, n was a finite, although large, number and the formula was only an approximation. Now, we want to consider an idealized experiment for which all of the integers $k=0, 1, 2, \ldots$ are possible numbers of successes and for which formula (1) holds exactly. Therefore, we shall consider a random variable $\mathbf{x}$ whose possible values are $k=0, 1, 2, \ldots$ with probabilities

$$W_k = \frac{\lambda^k e^{-\lambda}}{k!}. \tag{2}$$

Then $\mathbf{x}$ has a Poisson distribution function (which is a step function):

$$F(t) = \mathscr{P}(\mathbf{x} < t) = \sum_{k<t} \frac{\lambda^k e^{-\lambda}}{k!}.$$

As $t \to \infty$, $F(t)$ tends to 1 as it should.

The expectation of $\mathbf{x}$ is

$$\begin{aligned} \mathscr{E}\mathbf{x} &= \sum_{k=0}^{\infty} k \frac{\lambda^k e^{-\lambda}}{k!} = \sum_{1}^{\infty} k \frac{\lambda^k e^{-\lambda}}{k!} \\ &= \lambda e^{-\lambda} \sum_{1}^{\infty} \frac{\lambda^{k-1}}{(k-1)!} = \lambda e^{-\lambda} e^{\lambda} = \lambda. \end{aligned} \tag{3}$$

In the same way,

$$\begin{aligned} \mathscr{E}(\mathbf{x}^2) &= \sum_{1}^{\infty} k^2 \frac{\lambda^k e^{-\lambda}}{k!} = \sum_{1}^{\infty} (k-1+1) \frac{\lambda^k e^{-\lambda}}{(k-1)!} \\ &= \sum_{2}^{\infty} \frac{\lambda^k e^{-\lambda}}{(k-2)!} + \sum_{1}^{\infty} \frac{\lambda^k e^{-\lambda}}{(k-1)!} \\ &= \lambda^2 e^{-\lambda} \sum_{2}^{\infty} \frac{\lambda^{k-2}}{(k-2)!} + \lambda e^{-\lambda} \sum_{1}^{\infty} \frac{\lambda^{k-1}}{(k-1)!} \\ &= \lambda^2 e^{-\lambda} e^{\lambda} + \lambda e^{-\lambda} e^{\lambda} = \lambda^2 + \lambda. \end{aligned}$$

Therefore, the variance is

$$\sigma^2 = \mathscr{E}(\mathbf{x}^2) - (\mathscr{E}\mathbf{x})^2 = \lambda^2 + \lambda - \lambda^2 = \lambda$$

and the standard deviation is $\sigma = \sqrt{\lambda}$. We can derive the same from the earlier form by taking the limit of

$$\sigma = \sqrt{n p q}$$

as $n p \to \lambda$, $q \to 1$.

For most purposes, the values of **x** lie between $\lambda - g\sqrt{\lambda}$ and $\lambda + g\sqrt{\lambda}$, where g is not larger than three or four. Let us be more specific. For large λ and correspondingly large $k = \lambda + z$, we can approximate W_k with

$$W_k \sim \frac{1}{\sqrt{2\pi\lambda}} e^{\frac{-z^2}{2\lambda}} (1 - \tfrac{1}{2}\lambda^{-1} z + \tfrac{1}{6}\lambda^{-2} z^2). \tag{4}$$

The approximation agrees with (1) § 6 for $q = 1$. Fig. 9 shows that the approximation is quite good for $\lambda = 4$. (Again, four turns out to be a "large" number.) Therefore, for moderately large λ, the Poisson distribution (2) can be quite well approximated by a normal curve with a correction term for skewness.

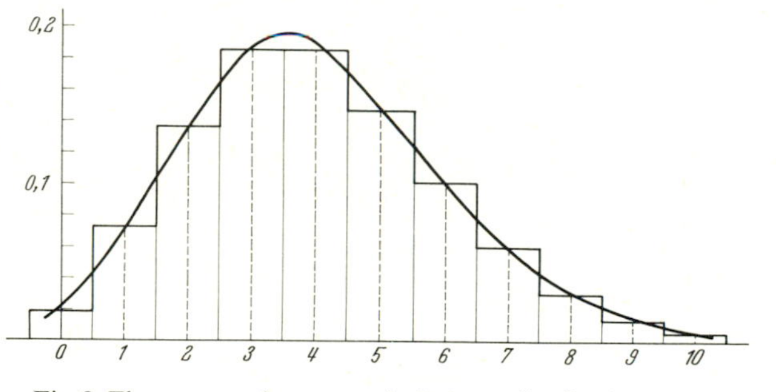

Fig. 9. The exact and asymptotic Poisson distribution ($\lambda = 4$)

From this, it follows that the probability that k lies between $\lambda - g\sqrt{\lambda}$ and $\lambda + g\sqrt{\lambda}$ is given approximately by

$$2\Phi(g) - 1 = \left(\frac{2}{\pi}\right)^{\frac{1}{2}} \int_0^g e^{-\frac{1}{2}t^2}\, dt.$$

With this established, we can construct confidence bounds for λ for given k as in § 7. We use the quadratic equation $|k - \lambda| = \pm g\sqrt{x}$ or

$$(k - \lambda)^2 = g^2 \lambda. \tag{5}$$

The value of g^2 is to be taken from Table 3. The solution of (5) gives two bounds

$$\lambda_1 = k + \tfrac{1}{2} g^2 - g(k + \tfrac{1}{4} g^2)^{\frac{1}{2}}, \qquad \lambda_2 = k + \tfrac{1}{2} g^2 + g(k + \tfrac{1}{4} g^2)^{\frac{1}{2}}. \tag{6}$$

If we use only one of these bounds, the error probability is approximately $1 - \Phi(g)$. For calculation of exact bounds, see F. Garwood, Biometrika, Vol. 28, p. 437 (1936).

Example 10. A cosmic radiation counter records 80 hits (radiation quanta) in five hours. Naturally, the number depends on chance. An identical apparatus placed beside the first one may record 70 or 90 hits for the same radiation intensity. What we are concerned with is not the number of hits k which happened to be observed, but the expectation λ of the number of hits during this time, as a measure for the radiation intensity. Can we assign confidence bounds for λ?

Since the particles of cosmic radiation registered by the counter are only a small fraction of those hitting a larger area, we may assume a Poisson distribution. Hence, $\sigma=\sqrt{\lambda}$ is the standard deviation of k. Since k lies near λ, $S=\sqrt{k}=\sqrt{80}\cong 9$ is a good estimate for σ. Let us record our observations in the conventional manner:

$$\text{number of hits in 5 hours} \quad 80\pm 9$$

or

$$\text{number of hits per hour} \quad 16\pm 1.8.$$

From the $3s$-rule, we would conclude that the average number of hits λ in five hours lies between

$$k-3s=80-27=\ 53$$

and

$$k+3s=80+27=107.$$

Using (6), we get somewhat higher limits

$$\lambda_1=84.5-3\sqrt{82}=\ 57$$

and

$$\lambda_2=84.5+3\sqrt{82}=112$$

with error probability 0.27 for $g=3$.

B. Comparison of Two Frequencies of Rare Events

For the comparison of the frequencies of two rare events, we can use a method similar to that in § 9. For example, suppose we have values k_1 and k_2 of the numbers of successes $\mathbf{x}_1$ and $\mathbf{x}_2$ in time periods t_1 and t_2, respectively. Let

$$m_1=\frac{k_1}{t_1}, \qquad m_2=\frac{k_2}{t_2} \tag{7}$$

be the average number of hits per unit time. If m_1 differs from m_2, we wonder if this difference is purely chance. Assuming it is, let μ, the true mean, be the same in both cases. Then the expected number of successes in t_1 and t_2 are

$$\lambda_1=\mu t_1, \qquad \lambda_2=\mu t_2.$$

The standard deviations are $(\mu t_1)^{\frac{1}{2}}$ and $(\mu t_2)^{\frac{1}{2}}$. The distributions of $\mathbf{x}_1$ and $\mathbf{x}_2$ are approximately normal. Therefore, the difference

$$\frac{\mathbf{x}_1}{t_1}-\frac{\mathbf{x}_2}{t_2} \tag{8}$$

is approximately normal with zero expectation and variance

$$\sigma^2=\sigma_1^2+\sigma_2^2=\frac{\mu}{t_1}+\frac{\mu}{t_2}=\frac{\mu}{t_1 t_2}(t_1+t_2). \tag{9}$$

Now, if the value of (8) exceeds g times the standard deviation in absolute value,

$$\left|\frac{k_1}{t_1}-\frac{k_2}{t_2}\right|>g\,\sigma$$

or

$$\left(\frac{k_1}{t_1}-\frac{k_2}{t_2}\right)^2>\frac{g^2\mu}{t_1 t_2}(t_1+t_2), \tag{10}$$

then the hypothesis that the difference is purely chance should be rejected.

However, we don't know μ. We may proceed as in § 9, substituting the best possible estimate of μ using all of the data. There are k_1+k_2 successes recorded in time t_1+t_2. From these observations we obtain

$$m=\frac{k_1+k_2}{t_1+t_2} \tag{11}$$

as an estimate for μ. Substituting this estimate into (10), we obtain the following practical test. *The observed difference in the number of successes per unit time is considered significant, as soon as*

$$\left(\frac{k_1}{t_1}-\frac{k_2}{t_2}\right)^2>\frac{g^2}{t_1 t_2}\cdot(k_1+k_2). \tag{12}$$

The g^2 is determined from Table 3.

If a one-sided test is used (that is, a decision is made only for either a positive difference in the parentheses or a negative one), then the error probability is half as large.

We can also write this test as a χ^2 test, setting

$$\chi^2=\left(\frac{k_1}{t_1}-\frac{k_2}{t_2}\right)^2\frac{t_1 t_2}{k_1+k_2}=\frac{(k_1 t_2-k_2 t_1)^2}{t_1 t_2(k_1+k_2)}. \tag{13}$$

Then, if $\chi^2>g^2$, the "chance" hypothesis is rejected. The verification of this χ^2 test is analogous to that in § 9 C. We shall dispense with it for now as we shall return to this test in a general context later (§ 56 F).

Chapter Three

Mathematical Tools

This supplementary chapter can be omitted for the time being and consulted later when required.

§ 11. Multiple Integrals. Transformations to Polar Coordinates

Any open set in the space of real variables $x, y, \ldots$ will be called a region. As is well known a double integral,

$$I = \iint_G f(x, y)\, dx\, dy,$$

over a bounded region G in the plane can be evaluated by two successive simple integrations,

$$I = \int_c^d dy \int f(x, y)\, dx.$$

The limits for the inner integration are found in the following way. The integration with respect to x extends over the straight line segments that the line $y = \text{constant}$ has in common with G. The limits for y are the largest and smallest value of y in the region G.

Likewise, we can evaluate an $(m+n)$-tuple integral

$$I = \int \underset{G}{\cdots} \int f(x_1, \ldots, x_m, y_1, \ldots, y_n)\, dx_1 \cdots dx_m\, dy_1 \cdots dy_n$$

by two successive integrations, first with respect to $x_1, \ldots, x_m$, then with respect to $y_1, \ldots, y_n$,

$$I = \iint dy_1 \cdots dy_n \iint f(x_1, \ldots, x_m, y_1, \ldots, y_n)\, dx_1 \cdots dx_n. \tag{1}$$

The integration limits for the $x_1, \ldots, x_m$ are given by the x values which belong to G for fixed $y_1, \ldots, y_n$; that is, they are defined by the same inequalities as G except that only the x_i are considered variable. On the other hand, the region of integration for the y's is the set of all y values for which $(x_1, \ldots, x_m, y_1, \ldots, y_n)$ belongs to G.

The *transformation of a multiple integral* is made by formula (2):

$$\int_G \cdots \int f(u_1, \ldots, u_n)\, du_1 \cdots du_n = \int_{G'} \cdots \int f(u_1, \ldots, u_n) \left| \frac{\partial(u_1, \ldots, u_n)}{\partial(x_1, \ldots, x_n)} \right| dx_1 \cdots dx_n. \tag{2}$$

The Jacobian which appears as a factor is the determinant of the partial derivatives of the u_i with respect to the x_k, and G' is the transformed domain.

The transformation to *polar coordinates* is especially important to us. It is defined for n variables as follows:

$$\begin{aligned}
x_1 &= r \cos \varphi_1 && (0 \leqq \varphi_1 \leqq \pi) \\
x_2 &= r \sin \varphi_1 \cos \varphi_2 && (0 \leqq \varphi_2 \leqq \pi) \\
&\vdots && \vdots \\
x_{n-1} &= r \sin \varphi_1 \sin \varphi_2 \cdots \cos \varphi_{n-1} && (0 \leqq \varphi_{n-1} < 2\pi) \\
x_n &= r \sin \varphi_1 \sin \varphi_2 \cdots \sin \varphi_{n-1} && (0 \leqq r),
\end{aligned} \tag{3}$$

from which

$$r^2 = x_1^2 + x_2^2 + \cdots + x_n^2$$

follows.

We prove the one-to-one uniqueness of the transformation in the region

$$r \sin \varphi_1 \sin \varphi_2 \cdots \sin \varphi_{n-1} \neq 0$$

easily by mathematical induction on n. It is clear for $n=2$. If we assume one-to-one uniqueness of the transformation for $(n-1)$ variables $x_2, \ldots, x_n$, that is,

$$\begin{aligned}
x_2 &= r_1 \cos \varphi_2 && (0 \leqq \varphi_2 \leqq \pi) \\
x_3 &= r_1 \sin \varphi_2 \cos \varphi_3 && (0 \leqq \varphi_3 \leqq \pi) \\
&\vdots && \vdots \\
x_{n-1} &= r_1 \sin \varphi_2 \sin \varphi_3 \cdots \cos \varphi_{n-1} && (0 \leqq \varphi_{n-1} < 2\pi) \\
x_n &= r_1 \sin \varphi_2 \sin \varphi_3 \cdots \sin \varphi_{n-1} && (0 \leqq r_1),
\end{aligned} \tag{4}$$

then we need only combine (4) with the two-dimensional transformation

$$\begin{aligned}
x_1 &= r \cos \varphi_1 && (0 \leqq r) \\
r_1 &= r \sin \varphi_1 && (0 \leqq \varphi_1 \leqq \pi \text{ because } 0 \leqq r)
\end{aligned}$$

in order to obtain (3).

By the same method we shall also prove that the Jacobian of transformation (3) has the value

$$\left|\frac{\partial(x_1, \ldots, x_n)}{\partial(r, \varphi_1, \ldots, \varphi_{n-1})}\right| = r^{n-1}\,\Theta, \tag{5}$$

where

$$\Theta = \sin^{n-2}\varphi_1 \sin^{n-3}\varphi_2 \cdots \sin^2\varphi_{n-3} \sin\varphi_{n-2}$$

is a function of the φ's alone.

The induction from $(n-1)$ to n for the conclusion of the proof of (5) proceeds as follows:

$$\begin{aligned}
\frac{\partial(x_1, \ldots, x_n)}{\partial(r, \varphi_1, \ldots, \varphi_{n-1})} &= \frac{\partial(x_1, x_2, \ldots, x_n)}{\partial(x_1, r_1, \varphi_2, \ldots, \varphi_{n-1})} \cdot \frac{\partial(x_1, r_1, \varphi_2, \ldots, \varphi_{n-1})}{\partial(r, \varphi_1, \varphi_2, \ldots, \varphi_{n-1})} \\
&= \frac{\partial(x_2, \ldots, x_n)}{\partial(r_1, \varphi_2, \ldots, \varphi_{n-1})} \cdot \frac{\partial(x_1, r_1)}{\partial(r, \varphi_1)} \\
&= r_1^{n-2} \sin^{n-3}\varphi_2 \cdots \sin^2\varphi_{n-3} \sin\varphi_{n-2} \cdot r \\
&= (r \sin\varphi_1)^{n-2} \sin^{n-3}\varphi_2 \cdots \sin^2\varphi_{n-3} \sin\varphi_{n-2} \cdot r \\
&= r^{n-1}\,\Theta .
\end{aligned}$$

Therefore, the transformation to polar coordinates is given by

$$\int \cdots \int f\, dx_1 \cdots dx_n = \int \cdots \int f\, r^{n-1}\,\Theta\, dr\, d\varphi_1 \cdots d\varphi_{n-1}, \tag{6}$$

or, in short, setting $\Theta\, d\varphi_1 \cdots d\varphi_{n-1} = d\Omega$, by

$$\int \cdots \int f\, dx_1 \cdots dx_n = \int \cdots \int f\, r^{n-1}\, dr\, d\Omega . \tag{7}$$

If a region G extends to infinity or if the integrand of a multiple integral becomes infinite at the boundary of G, we define an *improper multiple integral* over G as the limit of integrals with bounded integrands over a sequence of bounded regions $G_1, G_2, \ldots$ whose union is G. The integral converges if this limit exists and is finite for any sequence of regions $G_1, G_2, \ldots$.

In the integrals with which we are concerned in this book, the integrand decreases so rapidly as its argument goes to infinity that the convergence is obvious. For non-negative functions, such as probability densities, there are only two possibilities. We have either convergence or divergence to infinity for each choice of the sequence of domains $G_1, G_2 \ldots$. Therefore, we can restrict ourselves to the simplest sequences of regions when checking for convergence.

Now, transformation (2) as well as the formula for successive integration (1) hold for improper integrals, provided that the integrals on both sides of the equation converge.

§ 12. Beta and Gamma Functions

A. Gamma Function

Euler's gamma function $\Gamma(z+1)$ is defined for $z+1>0$ (or for $R(z+1)>0$, for complex variables) by

$$\Gamma(z+1)=\int_0^\infty x^z e^{-x} dx. \tag{1}$$

The improper integral is always defined as the limit of the proper integral

$$\int_0^t x^z e^{-x} dx \tag{2}$$

which we call the *incomplete gamma function.*

The integral can be transformed by substitution of variables. If we set $x=a t$, we obtain

$$\int_0^\infty t^z e^{-at} dt = a^{(z-1)} \Gamma(z+1). \tag{3}$$

If we substitute $x=t^2/2$ in (1), we have

$$\int_0^\infty t^{2z+1} e^{-\frac{1}{2}t^2} dt = 2^z \Gamma(z+1),$$

or if $2z+1=n$ and t is replaced by $\sigma^{-1} t$,

$$\int_0^\infty t^n e^{-\frac{1}{2}\sigma^{-2}t^2} dt = 2^{\frac{n-1}{2}} \sigma^{n+1} \Gamma\left(\frac{n+1}{2}\right). \tag{4}$$

In particular,

$$\int_{-\infty}^\infty e^{-\frac{1}{2}t^2} dt = 2\int_0^\infty e^{-\frac{1}{2}t^2} dt = \sqrt{2}\cdot\Gamma(\tfrac{1}{2}). \tag{5}$$

B. Functional Equation of the Gamma Function

By integrating the indefinite integral (2) by parts, we obtain

$$\int x^z e^{-x} dx = -x^z e^x + z\int x^{z-1} e^{-x} dx.$$

Therefore, assuming $z>0$ and substituting the limits 0 and ∞, we have

$$\Gamma(z+1)=z\,\Gamma(z). \tag{6}$$

Clearly $\Gamma(1)=1$. From (6) we obtain

$$\Gamma(2)=1\cdot\Gamma(1)=1,$$
$$\Gamma(3)=2\cdot\Gamma(2)=2!,$$

and so, in general, for integer n

$$\Gamma(n+1)=n!. \tag{7}$$

In order to compute $\Gamma(\frac{1}{2})$, we consider the double integral

$$I=\iint e^{-\frac{1}{2}(x^2+y^2)}\,dx\,dy \tag{8}$$

over the whole plane. We can integrate with respect to x and y successively and obtain

$$I=\int_{-\infty}^{\infty} e^{-\frac{1}{2}x^2}\,dx\cdot\int_{-\infty}^{\infty} e^{-\frac{1}{2}y^2}\,dy=2\{\Gamma(\tfrac{1}{2})\}^2 \tag{9}$$

using (5). We could also transform (8) to polar coordinates. Then we would have

$$\begin{aligned} I&=\iint e^{-\frac{1}{2}r^2}\,r\,dr\,d\varphi=\int_0^{2\pi} d\varphi\cdot\int_0^{\infty} e^{-\frac{1}{2}r^2}\,r\,dr\\ &=2\pi\,\Gamma(1)=2\pi. \end{aligned} \tag{10}$$

Comparing (9) with (10), we see

$$\{\Gamma(\tfrac{1}{2})\}^2=\pi; \tag{11}$$

therefore, since $\Gamma(\frac{1}{2})$ is positive,

$$\Gamma(\tfrac{1}{2})=\sqrt{\pi}.$$

From this we can evaluate the functional Eq. (5), as well as $\Gamma(1\frac{1}{2})$, $\Gamma(2\frac{1}{2})$, and so forth. For example,

$$\Gamma(1\tfrac{1}{2})=\tfrac{1}{2}\sqrt{\pi}. \tag{12}$$

C. Surface Area of the *n*-Dimensional Sphere

If we evaluate the n-dimensional integral over the whole space,

$$I=\int\cdots\int e^{-\frac{1}{2}(x_1^2+\cdots+x_n^2)}\,dx_1\cdots dx_n, \tag{13}$$

we obtain

$$I=\left\{\int_{-\infty}^{\infty} e^{-\frac{1}{2}x^2}\,dx\right\}^n=\{\sqrt{2}\,\Gamma(\tfrac{1}{2})\}^n=(\sqrt{2\pi})^n. \tag{14}$$

On the other hand, using n-dimensional polar coordinates, we have

$$I=\int_0^\infty e^{-\frac{1}{2}r^2}\, r^{n-1}\, dr\cdot\int d\Omega=2^{\frac{n-2}{2}}\,\Gamma\left(\frac{n}{2}\right)\int d\Omega, \tag{15}$$

where the region of integration for $\int d\Omega$ extends over the whole $\varphi_1,\ldots,\varphi_{n-1}$ space. Comparison of (14) with (15) gives

$$\int d\Omega=\frac{2}{\Gamma(n/2)}\,\pi^{n/2}. \tag{16}$$

For example, for $n=3$, we find the Archimedes' value for the surface area of the unit sphere

$$\int d\Omega=\frac{4}{\pi^{\frac{1}{2}}}\,\pi^{\frac{3}{2}}=4\pi.$$

Likewise, we can interpret (16) geometrically as the surface area of a unit sphere in n-space.

D. Stirling's Formula

We want to expand the gamma function

$$\Gamma(\lambda+1)=\int_0^\infty x^\lambda\, e^{-x}\, dx$$

asymptotically for large λ. The maximum of the integrand

$$f(x)=x^\lambda\, e^{-x}$$

lies at $x=\lambda$. In the neighborhood of the maximum, we can expand the integrand in an infinite series,

$$\begin{aligned}\ln f(x)&=\lambda\ln\lambda+\lambda\ln\frac{x}{\lambda}-x\\&=\lambda\ln\lambda+\lambda\left(\frac{x-\lambda}{\lambda}-\frac{(x-\lambda)^2}{2\lambda^2}+\cdots\right)-x\\&=\lambda\ln\lambda-\lambda-\frac{(x-\lambda)^2}{2\lambda}+\cdots.\end{aligned}$$

Therefore,

$$f(x)=\lambda^\lambda\, e^{-\lambda}\, e^{-\frac{1}{2\lambda}(x-\lambda)^2+\cdots}. \tag{17}$$

As long as $x-\lambda$ is small compared with λ, the remainder terms (indicated by the dots) are small compared with the main term and can be neglected. However, even if $x-\lambda$ is of the same order of magnitude as λ and λ is large compared to one, the remainder can still be neglected, since then both $f(x)$ and the right side of (17) tend to zero. If we drop the remainder terms and integrate both sides of (17) from 0 to ∞, we have

$$\Gamma(\lambda+1) \sim \lambda^{\lambda} e^{-\lambda} \int_0^{\infty} e^{-\frac{1}{2\lambda}(x-\lambda)^2} dx$$

$$= \lambda^{\lambda+\frac{1}{2}} e^{-\lambda} \int_{-\sqrt{\lambda}}^{\infty} e^{-\frac{1}{2}t^2} dt.$$

The symbol $\sim$ indicates that the ratio of the two sides of the equation tends to one as $\lambda \to \infty$. The lower limit of integration $-\sqrt{\lambda}$ can be replaced by $-\infty$ without loosing the validity of the asymptotic equation. Using (5) and (11), we obtain

$$\Gamma(\lambda+1) \sim \lambda^{\lambda+\frac{1}{2}} e^{-\lambda} \sqrt{2\pi}. \tag{18}$$

This is *Stirling's formula*. If we carry the expansion somewhat further, we obtain a more exact approximation

$$\Gamma(\lambda+1) = \lambda^{\lambda+\frac{1}{2}} e^{-\lambda} \sqrt{2\pi} \left(1 + \frac{1}{12\lambda} - R\right), \tag{19}$$

where the remainder $(-R)$ is negative and of order λ^{-2}. For the last factor, we can also write $\left(1+\frac{\vartheta}{12\lambda}\right)$ where ϑ is a little less than one.

In particular (18) holds for integers $\lambda = n$,

$$n! \sim n^n e^{-n} \sqrt{2\pi n}. \tag{20}$$

E. Beta Function

Euler's beta function is defined by

$$\mathrm{B}(p+q, q+1) = \int_0^1 x^p (1-x)^q \, dx. \tag{21}$$

The integral converges if p and q are both larger than -1. Substituting $u = a\,x$, we get

$$\int_0^a u^p (a-u)^q \, du = a^{p+q+1} \mathrm{B}(p+1, q+1). \tag{22}$$

Substituting $x = \sin^2 \varphi$, we have

$$\mathrm{B}(p+1, q+1) = 2 \int_0^{\pi/2} \sin^{2p+1} \varphi \cos^{2q+1} \varphi \, d\varphi. \tag{23}$$

In order to evaluate integral (21), we start with the double integral

$$I=\int_0^\infty\int_0^\infty e^{-\frac{1}{2}(x^2+y^2)}x^{2q+1}y^{2p+1}\,dx\,dy.$$

Once more, we can integrate with respect to x and y successively and obtain from (4)

$$\begin{aligned}I&=\int_0^\infty e^{-\frac{1}{2}x^2}x^{2q+1}\,dx\cdot\int_0^\infty e^{-\frac{1}{2}y^2}y^{2p+1}\,dy\\&=2^q\Gamma(q+1)\cdot 2^p\Gamma(p+1)=2^{p+q}\Gamma(p+1)\Gamma(q+1).\end{aligned}\tag{24}$$

On the other hand, using polar coordinates

$$\begin{aligned}I&=\int_0^\infty e^{-\frac{1}{2}r^2}r^{2p+2q+3}\,dr\cdot\int_0^{\pi/2}\cos^{2q+1}\varphi\sin^{2p+1}\varphi\,d\varphi\\&=2^{p+q+1}\Gamma(p+q+2)\cdot\tfrac{1}{2}\mathrm{B}(p+1,q+1)\\&=2^{p+q}\Gamma(p+q+2)\,\mathrm{B}(p+1,q+1).\end{aligned}\tag{25}$$

Comparing (24) with (25), we have

$$\Gamma(p+1)\Gamma(q+1)=\Gamma(p+q+2)\,\mathrm{B}(p+1,q+1).$$

Consequently,

$$\mathrm{B}(p+1,q+1)=\frac{\Gamma(p+1)\Gamma(q+1)}{\Gamma(p+q+2)}.\tag{26}$$

The following integral can be evaluated using the beta function:

$$K=\int_0^\infty(z^2+a)^{-l}z^k\,dz,\qquad(k>-1,2l-k>1,a>0).\tag{27}$$

If we set $(z^2+a)^{-1}a=y$, then $z^2=ay^{-1}(1-y)$, and the integral becomes

$$\begin{aligned}K&=\frac{1}{2}a^{\frac{k+1}{2}-l}\int_0^1 y^{l-\frac{k+3}{2}}(1-y)^{\frac{k-1}{2}}\,dy\\&=\frac{1}{2}a^{\frac{k+1}{2}-l}\mathrm{B}\left(l-\frac{k+1}{2},\frac{k+1}{2}\right),\end{aligned}$$

or from (26)

$$K=\frac{\Gamma\left(l-\frac{k+1}{2}\right)\Gamma\left(\frac{k+1}{2}\right)}{2\Gamma(l)}a^{\frac{k+1}{2}-l}.\tag{28}$$

In particular, for $k=0$,

$$\int_0^\infty(z^2+a)^{-l}\,dz=\frac{\Gamma(l-\frac{1}{2})\,\pi^{\frac{1}{2}}}{2\Gamma(l)}a^{\frac{1}{2}-l}.\tag{29}$$

§ 13. Orthogonal Transformations

A transformation of variables

$$\begin{aligned} y_1 &= a_{11} x_1 + a_{12} x_2 + \cdots + a_{1n} x_n \\ y_2 &= a_{21} x_1 + a_{22} x_2 + \cdots + a_{2n} x_n \\ &\cdots\cdots\cdots\cdots \\ y_n &= a_{n1} x_1 + a_{n2} x_2 + \cdots + a_{nn} x_n \end{aligned} \tag{1}$$

is called *orthogonal* if the sum of squares $x_1^2 + \cdots + x_n^2$ remains invariant, that is if

$$\sum y^2 = \sum x^2. \tag{2}$$

If we substitute (1) into (2) and compare the coefficients of x_i^2 and $x_i y_j$ $(i \neq j)$ on both sides, we find the following *orthogonality conditions:*

$$\begin{aligned} a_{1i}^2 + a_{2i}^2 + \cdots + a_{ni}^2 &= 1 \\ a_{1i} a_{1j} + a_{2i} a_{2j} + \cdots + a_{ni} a_{nj} &= 0. \end{aligned} \tag{3}$$

Multiplying the determinant matrix of the determinant

$$\Delta = \begin{vmatrix} a_{11} & a_{12} \cdots a_{1n} \\ \cdot & \cdot \quad \cdot \quad \cdot \\ a_{n1} & a_{n2} \cdots a_{nn} \end{vmatrix}$$

by its transpose from the left, using the orthogonality conditions, we obtain the determinant of the product

$$\Delta^2 = \begin{vmatrix} 1 & 0 & 0 \ldots 0 \\ 0 & 1 & \ldots 0 \\ \cdot & \cdot & \cdot \quad \cdot \\ 0 & 0 & \ldots 1 \end{vmatrix} = 1.$$

From this it follows that *the determinant of an orthogonal transformation is* ± 1. The determinant is equal to the Jacobian

$$\frac{\partial(y_1, \ldots, y_n)}{\partial(x_1, \ldots, x_n)} = \pm 1. \tag{4}$$

If we multiply the rows of Eq. (1) by $a_{1i}, a_{2i}, \ldots, a_{ni}$ and add, then all of the x's except x_i cancel because of the orthogonality conditions (3) and we have

$$x_i = a_{1i} y_1 + a_{2i} y_2 + \cdots + a_{ni} y_n. \tag{5}$$

That is, *the matrix of the inverse transformation is the transpose of the matrix of the transformation* (1).

The inverse transformation (5) is orthogonal because of (2). Therefore, the orthogonality conditions hold for the transpose matrix:

$$\begin{aligned} a_{i1}^2 + a_{i2}^2 + \cdots + a_{in}^2 &= 1 \\ a_{i1}\, a_{j1} + a_{i2}\, a_{j2} + \cdots + a_{in}\, a_{jn} &= 0. \end{aligned} \tag{6}$$

Just so, (6) implies (3).

The following theorem is very useful.

Theorem. *Any first row*

$$y = a_{11}\, x_1 + a_{12}\, x_2 + \cdots + a_{1n}\, x_n$$

can be completed to an orthogonal transformation provided that the condition

$$a_{11}^2 + a_{12}^2 + \cdots + a_{1n}^2 = 1$$

is fulfilled.

Proof. From (6), we have a linear condition

$$a_{11}\, a_{21} + a_{12}\, a_{22} + \cdots + a_{1n}\, a_{2n} = 0 \tag{7}$$

and a quadratic condition

$$a_{21}^2 + a_{22}^2 + \cdots + a_{2n}^2 = 1, \tag{8}$$

for the second row. Eq. (7) always has a nontrivial solution. Multiplying this solution by a suitable factor λ, we can also satisfy Eq. (8).

After having determined the second row, we have two linear and one quadratic condition for the third. The two linear equations have at least one nontrivial solution, because they are homogeneous and the number of equations is smaller than the number of unknowns. Again, by multiplying this solution by a suitable factor, λ_2 say, we also make the sum of squares equal to one.

We continue in this way until the last row. Then we have $(n-1)$ homogeneous equations in n unknowns and one quadratic equation, which can be satisfied by using a factor λ_n. This completes the proof.

§ 14. Quadratic Forms and Their Invariants

A. Vectors and Tensors

We call a sequence of numbers $(x^1, \ldots, x^n)$ a *vector* x. If the indices are written as superscripts, then we are talking about a *contravariant vector*.

A linear form $L = \sum u_i\, x^i$ in the variables $x^1, \ldots, x^n$ is determined by a *covariant vector* u with components $u_1, \ldots, u_n$. Similarly, a quadratic form

$$G = g_{ik}\, x^i\, x^k \qquad (g_{ik} = g_{ki})$$

is defined by a symmetric *tensor* g_{ik}. (Note that we are using Einstein's convention of dropping the double summation, that is, writing $g_{ik}x^i x^k$ rather than $\sum_i \sum_k g_{ik} x^i x^k$.) The quadratic form uniquely defines the *polar form*

$$G_{xy} = g_{ik} x^i y^k,$$

a bilinear form of the vectors x and y.

Suppose that the vector components x^i and y^i undergo a linear transformation

$$\begin{aligned} x^i &= e^i_{j'} x^{j'} \\ y^i &= e^i_{j'} y^{j'} \end{aligned} \tag{1}$$

that has an inverse. Then the u_i and g_{ik} will be transformed in such a way that the formulas $L = u_i x^i$ and $G_{xy} = g_{ik} x^i y^k$ remain invariant. That is,

$$u_i x^i = u_i e^i_{j'} x^{j'} = u_{j'} x^{j'},$$
$$g_{ik} x^i y^k = g_{ik} e^i_{j'} e^k_{l'} x^{j'} y^{l'} = g_{j'l'} x^{j'} y^{l'}.$$

Obviously, $G_{xx} = G$ remains invariant with G_{xy}. Thus, we obtain the transformation rule for covariant vectors and tensors:

$$u_{j'} = u_i e^i_{j'}, \tag{2}$$

$$g_{j'l'} = g_{ik} e^i_{j'} e^k_{l'}. \tag{3}$$

Transformation (2) is called *contragredient to* (1).

If a fixed quadratic form G is given, then we can define a covariant vector v by

$$v_i = g_{ik} y^k \tag{4}$$

for each contravariant vector. The polar form $G_{xy} = g_{ik} x^i y^k$ now may be written $v_i x^i$. Since G_{xy} remains invariant, $v_i x^i$ also remains invariant, that is, v is transformed like a covariant vector.

B. The Inverse Matrix (g^{ij})

If we assume that the form G is nonsingular (that is, its determinant g is different from zero), then we can solve for the y^k:

$$y^i = g^{ij} v_j. \tag{5}$$

The g^{ij} are the minor determinants of matrix (g^{ik}), divided by the whole determinant. We also call them *elements of the inverse matrix*.

Substituting (4) into (5), we obtain

$$g^{ij} g_{jk} y^k = y^i$$

identically in y^i. Therefore, we can also write

$$g^{ij} g_{jk} = \delta^i_k = \begin{cases} 1 & \text{for } i=k \\ 0 & \text{otherwise.} \end{cases} \tag{6}$$

In (5) v_j can be chosen arbitrarily. If u_i is a second covariant vector, then $u_i y^i$ is invariant, and thus,

$$(u\,v) = g^{ij} u_i v_j \tag{7}$$

is an invariant. The three invariants

$$(x\,y) = g_{ik} x^i x^k, \qquad (u\,x) = u_i x^i, \qquad (u\,v) = g^{ij} u_i v_j$$

are called *scalar products*.

By substituting new variables $x'_1, \ldots, x'_n$ (note we are using subscripts here to simplify the notation, although strictly speaking the indices should be superscripts), we can write each quadratic form as a difference of sums of squares:

$$G = x'^2_1 + x'^2_2 + \cdots + x'^2_k - x'^2_{k+1} - \cdots - x'^2_{k+l}.$$

If $k=n$ and $l=0$, then G takes only positive values (except when all of the x's are zero) and is called *positive definite*. Analogously, if all of the signs are negative, G is called *negative definite*.

A positive definite form can be transformed into the *unit form*

$$G = x'^2_1 + x'^2_2 + \cdots + x'^2_n. \tag{8}$$

Relative to this unit form, all scalar products become particularly simple:

$$(x\,y) = \sum x'_i y'_i, \qquad (u, x) = \sum u'_i x'_i, \qquad (u, v) = \sum u'_i v'_i.$$

From (3), it follows that the determinant is

$$g' = |g_{j'l'}| = |g_{ik} e^i_{j'}| \cdot |e^k_{l'}| = |g_{ik}| \cdot |e^i_{j'}| \cdot |e^k_{l'}|$$

according to the product theorem. Or if Δ is the determinant of transformation (1), then

$$g' = g\,\Delta^2. \tag{9}$$

If the transformation is into the unit form, then $g'=1$. Thus,

$$\Delta = \pm g^{-\frac{1}{2}}.$$

C. Calculation of an Integral

We shall use these algebraic tools for calculating integrals of the following form:

$$I=(2\pi)^{-\frac{n}{2}} g^{\frac{1}{2}} \int \underset{B}{\cdots} \int e^{-\frac{1}{2}G} dx^1 dx^2 \cdots dx^n, \tag{10}$$

where $G=g_{ik} x^i x^k$ is a positive definite quadratic form and the region of integration B is defined by two linear inequalities

$$(u\,x)>0, \qquad (v\,x)>0. \tag{11}$$

Transform G into the unit form (8), by introducing new variables $x'_1, \ldots, x'_n$. The transformed integral is given by

$$I=(2\pi)^{-\frac{n}{2}} \int \underset{B}{\cdots} \int e^{-\frac{1}{2}(x_1'^2+\cdots+x_n'^2)} dx'_1 \cdots dx'_n, \tag{12}$$

where the region of integration is determined by

$$(u'\,x')>0, \qquad (v'\,x')>0.$$

Now we introduce new variables $y_1, \ldots, y_n$ by an orthogonal transformation. According to § 13, if

$$u=(u_1'^2+\cdots+u_n'^2)^{\frac{1}{2}}$$

is substituted, we can assume

$$y_1=\frac{(u'\,x')}{u}=\frac{u'_1\,x'_1+\cdots+u'_n\,x'_n}{u}.$$

Thus, $(v\,x)$ can be transformed into $(w\,y)=w_1\,y_1+\cdots+w_n\,y_n$. Finally, we introduce $z_2, \ldots, z_n$ by means of a second orthogonal transformation in place of $y_2, \ldots, y_n$, where again z_2 and w can be chosen so that

$$z_2=\frac{w_2\,y_2+\cdots+w_n\,y_n}{w} \quad \text{and} \quad w=(w_2^2+\cdots+w_n^2)^{\frac{1}{2}}.$$

Then, we obtain

$$I=(2\pi)^{-\frac{n}{2}} \int \underset{B}{\cdots} \int e^{-\frac{1}{2}(y_1^2+z_2^2+\cdots+z_n^2)} dy_1\,dz_2 \cdots dz_n. \tag{13}$$

The forms $(u\,x)=(u'\,x')$ and $(v\,x)=(v'\,x')$ are expressed in terms of the new variables by

$$(u\,x)=u\,y_1,$$

$$(v\,x)=w_1\,y_1+w\,z_2.$$

Because of the invariance of scalar products, their scalar products are

$$(u\,u)=(u'\,u')=u^2,$$
$$(u\,v)=(u'\,v')=u\,w_1,$$
$$(v\,v)=(v'\,v')=w_1^2+w^2.$$

The region of integration is given by the inequalities

$$\begin{aligned} u\,y_1&>0,\\ w_1\,y_1+w\,z_2&>0. \end{aligned} \tag{14}$$

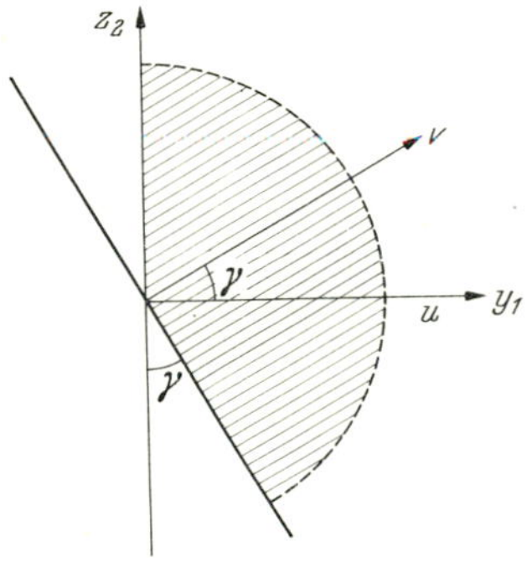

Fig. 10. The region of integration in the (y_1, z_2)-plane

Now we can perform integration (13) with respect to $dz_3\cdots dz_n$ and introduce polar coordinates for the remaining two variables y_1, z_2:

$$y_1=r\cos\varphi,$$
$$z_2=r\sin\varphi.$$

Then, we have

$$I=\frac{1}{2\pi}\int_0^\infty e^{-\frac{1}{2}r^2}r\,dr\int_\alpha^\beta d\varphi=\frac{\beta-\alpha}{2\pi}.$$

The integration limits for φ follow from the following consideration. Each of the inequalities in (14) defines a half plane in the (y_1, z_2)-plane. The normals directed into these half planes are the vectors u and v with components $(u, 0)$ and (w_1, w). The angle γ between them is given by

$$\cos\gamma=\frac{u\,w_1}{\sqrt{u^2}\sqrt{w_1^2+w^2}}=\frac{(u\,v)}{\sqrt{(u\,u)}\sqrt{(v\,v)}}.$$

The dihedral angle between the two half planes is $\pi-\gamma$. Therefore,

$$\beta-\alpha=\pi-\gamma=\arccos\frac{-(u\,v)}{\sqrt{(u\,u)}\sqrt{(v\,v)}}.$$

Thus, our integral becomes

$$I = \frac{1}{2\pi} \arccos \frac{-(u\,v)}{\sqrt{(u\,u)}\sqrt{(v\,v)}}. \tag{15}$$

The scalar products can be calculated directly using formula

$$(u\,v) = g^{ik}\,u_i\,v_k \tag{16}$$

without actually carrying out even one of the three linear coordinate transformations.

If B were defined by three linear inequalities

$$(u\,x) > 0, \qquad (v\,x) > 0, \qquad (w\,x) > 0,$$

then the same method would lead to the area of a spherical triangle. It is known that this area is proportional to the difference between the sum of the angles and π; therefore,

$$I = \frac{1}{4\pi}\left\{\arccos \frac{-(u\,v)}{\sqrt{(u\,u)}\sqrt{(v\,v)}} + \arccos \frac{-(u\,w)}{\sqrt{(u\,u)}\sqrt{(w\,w)}} + \arccos \frac{-(v\,w)}{\sqrt{(v\,v)}\sqrt{(w\,w)}} - \pi\right\}.$$

In the case of four inequalities, we would have to calculate the volume of a spherical tetrahedron, which is not as simple.

Chapter Four

Empirical Determination of Distribution Functions, Expectations, and Standard Deviations

The important sections of this chapter are § 15 and § 18.

§ 15. The Quetelet Curve

I still remember vividly how my father took me one day as a boy to the outskirts of the city, where the willows stood on the bank and had me pick a hundred willow leaves at random. After throwing out those with damaged tips, 89 were left which we took home. We arranged them very carefully according to decreasing size like soldiers in rank and file. Then my father drew a curved line through the tips and said, "This is the curve of Quetelet. From it you see how the mediocre always form the large majority and only a few stand out or are left behind."

If we place the curve upright (Fig. 11) and choose the total height of the curve as the unit of length for the ordinate axis, then the ordinate h

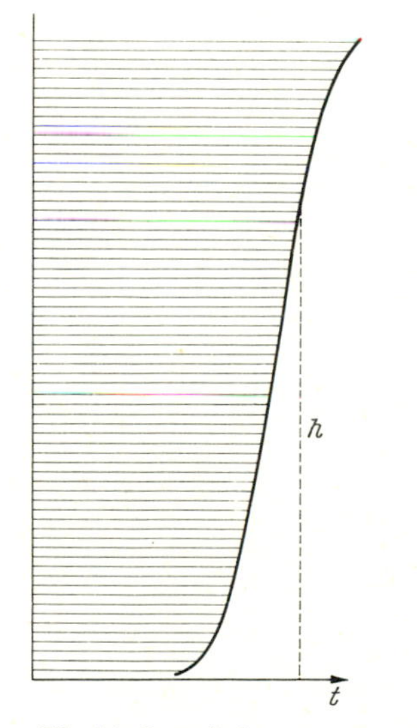

Fig. 11. Quetelet's curve

corresponding to the abscissa t obviously represents the frequency of willow leaves with lengths less than t. The frequency h of a given length is approximately equal to the probability p of finding a leaf of that length. Therefore, our curve represents an approximation for the distribution function $p=F(t)$ of leaf length.

The measured lengths $x_1, \ldots, x_n$ of the willow leaves form what we call (today) a *sample*. We can determine the distribution function $F(t)$ from a sample in the above way. From $F(t)$ we can try to determine the probability density $f(t)$ by graphical differentiation, but the results are always rather uncertain.

Another method frequently used to determine $f(t)$ and $F(t)$ is based on grouping the observed values x. The interval between t_0 and t_r, in which the observed values x lie, is subdivided into subintervals by arbitrarily chosen partition points $t_1, \ldots, t_{r-1}$. If the x values are rounded off to the nearest inch, say, then we choose the partition points appropriately as integers plus one half inch. The length of the subintervals has to be chosen small enough that the probability density $f(t)$ does not change too much within each subinterval; on the other hand, the number of observations in each subinterval must not be too small. Now the frequencies of x values in each subinterval can be determined. We mark them off as rectangles above the subintervals. Then, we plot the curve $y=f(x)$, so that the areas under the curve above the subintervals are most nearly equal to the areas of the rectangles. By numerical integration, we obtain the distribution function $F(t)$ from $f(t)$.

The first method of determining $F(t)$ is better, because all of the data is used and the arbitrary partition of the interval does not affect the result. How exact this procedure is will be examined in the next section (§ 16).

Galton and Quetelet found that very frequently the distribution of biological characteristics can be represented by the Gaussian error curve

$$f(t)=\frac{1}{\sigma\sqrt{2\pi}}\, e^{-\frac{1}{2}\left(\frac{t-a}{\sigma}\right)^2}. \tag{1}$$

Thus, such distributions are called *normal*. Nevertheless, other distributions also occur in nature. K. Pearson has made a list of types of frequently occurring distribution functions.

Example 11. In his famous experiment[1], W. Johannsen selected the 25 largest beans out of about 16,000 brown beans and bred them by self-fertilization. In the next generation there was the following distribution of weights:

Weight limits	20	25	30	35	40	45	50	55	60	65	70	75	80
Number of beans		5	18	46	144	127	70	70	63	28	15	8	4

[1] W. Johannsen, Über Erblichkeit in Populationen und reinen Linien, Jena 1903, p. 19.

Fig. 12 illustrates the results. The distribution is skew and cannot be approximated well by a normal distribution. As Johannsen's analysis shows, the deviation from the normal curve in this case is caused by the fact that the population was in fact a mixture of eleven "pure lines", each pure line consisting of the offspring from one bean. In each of the pure lines we find an approximately normal distribution, centering about a mean value that is not changed by further selection. The average weights of the eleven pure lines have the following distribution:

Weights	35	40	45	50	55	60
Number of lines	4	2	0	2	3	

The mixing of these eleven normal distributions accounts for the empirical distribution.

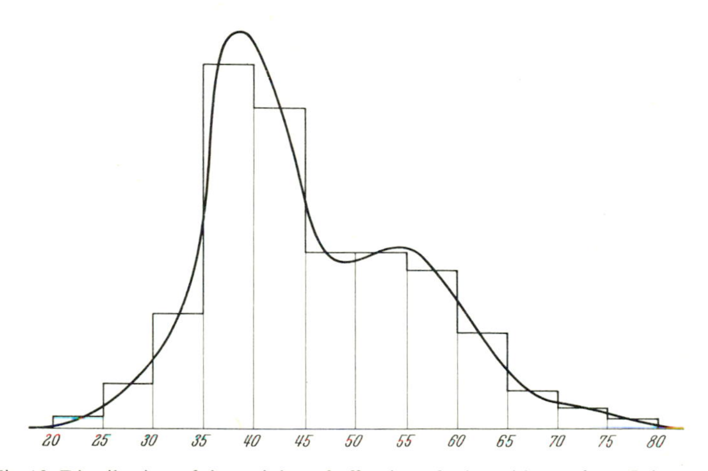

Fig. 12. Distribution of the weights of offspring of selected beans from Johannsen

§ 16. Empirical Determination of Distribution Functions

In the first reading of this Chapter §§ 16 and 17 can be omitted. The concepts explained in these two sections will not be used until much later.

Kolmogorov developed an exact theory from what was illustrated by the example of the willow leaves in the previous section. To begin with, he defined the *empirical distribution function* $F_n(t)$ of a sample $x_1, \ldots, x_n$ as the empirical frequency of the x_i that are less than t, that is, the number of x_i less than t divided by n. The graphical representation of an empirical distribution function is not a smooth curve like Quetelet and his students drew in their naive enthusiasm, but a step function with a jump in height of $\delta = 1/n$ at each x_i (Fig. 13).

The question now is how much the empirical curve $F_n(t)$ can differ from the true distribution function $y = F(t)$. First we examine the positive deviations $F - F_n$, and then the negative. In practical applications F_n is given and F is unknown; in theoretical studies, however, we have to assume that $F(t)$ is given and that $F_n(t)$ depends on chance, since the observed values do depend on chance.

Let $\Delta = \max_n \{F - F_n\}$ be the maximum of $F - F_n$; we want to know the distribution function of Δ. We assume only that the distribution function $F(t)$ is *continuous*, and nothing else. Since a continuous monotone transformation of the t-axis leaves the differences $\{F - F_n\}$ unchanged, we can replace t and x by new variables

$$t' = F(t) \quad \text{and} \quad x' = F(x)$$

without changing the maximal difference Δ at all. Let us call the new variables t and x again, so that the distribution function assumes the simple form

$$F(t) = t \qquad (0 < t < 1). \tag{1}$$

Thus, the graph of $F(t)$ is the diagonal of a unit square. Since all of the x_i lie between zero and one, we can set

$$F(t) = 0 \qquad \text{for } t \leqq 0,$$
$$F(t) = 1 \qquad \text{for } t \geqq 1.$$

(See Fig. 13.) The probability density is

$$f(t) = 1 \qquad \text{for } 0 < t < 1,$$
$$f(t) = 0 \qquad \text{otherwise}.$$

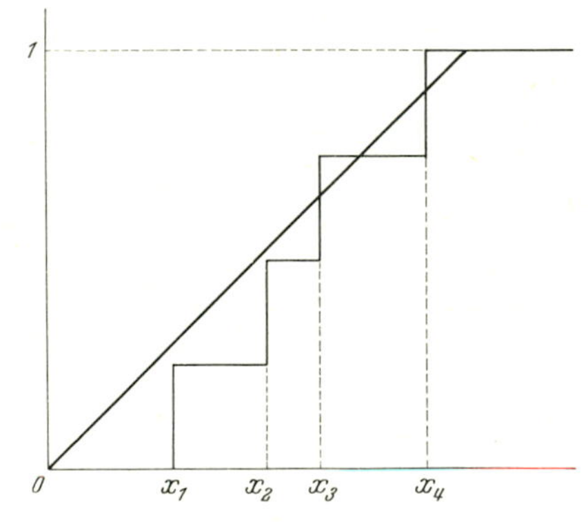

Fig. 13. Theoretical and empirical distribution curves

The graph of this function $y = f(t)$ is a rectangle. Therefore we call it a *rectangular distribution*.

We want to compute the probability Q that Δ exceeds a bound, say ε. Since the $x_1, \ldots, x_n$ have been assumed independent, all of them having probability densities $f(t)=1$, we can use Theorem II from §4 to find

$$Q=\int\cdots\int dx_1\,dx_2\cdots dx_n \tag{2}$$

integrated over the region defined by the inequalities

$$0<x_1<1,\ldots,0<x_n<1 \quad\text{and}\quad \Delta>\varepsilon.$$

It is convenient to restrict the region of integration further by the inequalities

$$x_1<x_2<\cdots<x_n.$$

By a permutation of the x_i this part of the integration region can be transformed into other similarly defined parts (*e.g.*, $x_2<x_1<\cdots<x_n$). The step function F_n and the maximum Δ are not affected by these transformations. All of these parts of the region of integration have equal volumes and therefore equal probabilities. The boundary surfaces $x_i=x_k$ have zero probability. Therefore, we have

$$Q=n!\,q, \tag{3}$$

where q is the probability of the event

$$[0<x_1<x_2<\cdots<x_n<1;\,\Delta>\varepsilon]. \tag{4}$$

At the point x_k the function $F_n(t)$ jumps from $(k-1)\,\delta$ to $k\,\delta$. It is clear that the maximum Δ will be attained at one of these points x_k. Then the value of the maximum is

$$\Delta=x_k-(k-1)\,\delta. \tag{5}$$

Therefore, the event $[\Delta>\varepsilon]$ occurs whenever one of the differences $x_k-(k-1)\,\delta$ is larger than ε. Let the probability that this event occurs for an index k, but for no $j, j<k$, be q_k. Since this can be the case for only one value of k at a time, the probability q is

$$q=q_1+q_2+\cdots+q_n. \tag{6}$$

Therefore, we only have to calculate the q_k; but q_k is the probability of the event

$$[0<x_1<x_2<\cdots<x_n<1;\,x_k-(k-1)\,\delta>\varepsilon;\,x_j-(j-1)\,\delta\leqq\varepsilon] \quad \text{for } j<k. \tag{7}$$

First suppose $k=1$. Then the event becomes

$$[x_1<x_2<\cdots<x_n<1;\,x_1>\varepsilon]. \tag{8}$$

Thus, all of the x_i lie between ε and one. The probability of this event is $(1-\varepsilon)^n$. If this is the case, all orderings of the x_i are equally likely and the probability of event (8) is

$$q_1 = \frac{1}{n!}(1-\varepsilon)^n. \tag{9}$$

Now if $k>1$, then we set $k=h+1$. Then the inequalities in (7) split into those containing only $x_1, \ldots, x_h$

$$[0<x_1<\cdots<x_h;\, x_j \leqq \varepsilon+(j-1)\,\delta] \qquad \text{for } j=1,\ldots,h, \tag{10}$$

and those containing only $x_{h+1}, \ldots, x_n$

$$[x_{h+1}<x_{h+2}<\cdots<x_n<1;\, x_{h+1}>\varepsilon+h\,\delta]. \tag{11}$$

This is so because the inequality $x_h<x_{h+1}$ which relates x_h to x_{h+1} is a consequence of (10) and (11) and can therefore be omitted.

Now, the $x_{h+1}, \ldots, x_n$ are independent of the $x_1, \ldots, x_h$. The probability of event (7) is, therefore, the product of the probabilities of events (10) and (11),

$$q_k = q_{h+1} = p_h\, r_h. \tag{12}$$

The probability r_h of event (11) is very easy to determine. The method is the same as that for (8); the result is

$$r_h = \frac{1}{(n-h)!}(1-\varepsilon-h\,\delta)^{n-h}. \tag{13}$$

Here, it is assumed that $1-\varepsilon-h\,\delta$ is positive. If this is not the case, event (11) is impossible and r_h becomes zero.

The probability p_h of event (10) is

$$p_h = \int_0^{\varepsilon} dx_1 \int_{x_1}^{\varepsilon+\delta} dx_2 \int_{x_2}^{\varepsilon+2\delta} dx_3 \ldots \int_{x_{h-1}}^{\varepsilon+(h-1)\delta} dx_h. \tag{14}$$

For $h=1$, we find easily

$$p_1 = \varepsilon.$$

If we calculate p_2 and p_3 in the same way, we are led to the conjecture

$$p_h = \frac{\varepsilon}{h!}(\varepsilon+h\,\delta)^{h-1}. \tag{15}$$

This relation is easily verified by mathematical induction on h. First it holds for $h=1$. If it is true for a value of h, then by (14), we have

$$p_{h+1} = \int_0^{\varepsilon} dx_1 \int_{x_1}^{\varepsilon+\delta} dx_2 \int_{x_2}^{\varepsilon+2\delta} dx_3 \ldots \int_{x_h}^{\varepsilon+h\delta} dx_{h+1}.$$

If we substitute new variables $y_1, y_2, \ldots, y_n$ for $x_2, x_3, \ldots, x_{h+1}$, using the transformation

$$x_j = x_1 + y_{j-1},$$

we find

$$p_{h+1} = \int_0^{\varepsilon} R\, dx_1 \tag{16}$$

with

$$R = \int_0^{\varepsilon+\delta-x_1} dy_1 \int_{y_1}^{\varepsilon+2\delta-x_1} dy_2 \ldots \int_{y_{h-1}}^{\varepsilon+h\delta-x_1} dy_h. \tag{17}$$

Now if we set

$$\varepsilon + \delta - x_1 = \varepsilon',$$

we see that integral R has the same form as the original p_h in (14) with ε' instead of ε. Therefore, according to the induction hypothesis,

$$R = \frac{1}{h!}\varepsilon'(\varepsilon' + h\,\delta)^{h-1} = \frac{1}{h!}(\varepsilon + \delta - x_1)\,[\varepsilon + (h+1)\,\delta - x_1]^{h-1}. \tag{18}$$

Substitute this in (16); then the integration is easy to perform (*e.g.*, by choosing $\varepsilon + (h+1)\,\delta - x_1$ as the new variable of integration). The result

$$p_{h+1} = \frac{\varepsilon}{(h+1)!}\,[\varepsilon + (h+1)\,\delta]^h$$

has the same form as (15) with $(h+1)$ in place of h. Thus, the induction is complete.

If we substitute (13) and (15) into (12), then we obtain

$$q_k = \frac{\varepsilon}{h!\,(n-h)!}(\varepsilon + h\,\delta)^{h-1}(1 - \varepsilon - h\,\delta)^{n-h}. \tag{19}$$

According to (19) this formula also holds for $k=1$. From (3) and (6), we obtain finally the result first found by Birnbaum and Tingey:

$$Q = \sum_{h=0}^{H} \binom{n}{h} \varepsilon(\varepsilon + h\,\delta)^{h-1}(1 - \varepsilon - h\,\delta)^{n-h} \tag{20}$$

with

$$\delta = \frac{1}{n}. \tag{21}$$

The upper summation limit H is given by the condition that $(1 - \varepsilon - h\,\delta)$ may not be negative. Therefore, H is the largest integer in $n(1-\varepsilon)$;

$$H = [n(1-\varepsilon)]. \tag{22}$$

The calculation of (20) is very tiresome. Therefore, for large n, it is better to use an asymptotic expansion

$$Q \sim e^{-2n\varepsilon^2}, \tag{23}$$

which was derived by Smirnov.

Now, for every n, we can determine an ε such that Q takes a given value α, say 0.01 or 0.05. For small n we use (20), for large n, (23). Table 4 gives the exact and asymptotic bounds ε for a few values of n and α, according to Birnbaum and Tingey. We see from the table that the exact and asymptotic bounds for $n=50$ differ very little. Further, we see that the probability for Δ to exceed the asymptotic bound is *less* than α. Therefore, we are on the safe side if we use asymptotic bounds instead of the exact ones.

With this ε, then, we have the *one-sided Δ-test* for testing a hypothetical distribution function $F(x)$; that is, *if the maximum difference $F-F_n$ is larger than ε, then the assumption that $F(t)$ is the (true) distribution function will be rejected.* We shall call the test the *Δ-test*. The error probability of the test is α.

If we replace x by $1-x$ and t by $1-t$, we invert the sign and obtain a one-sided test for the other side. The assumed distribution is rejected if

$$\Delta' = \max(F_n - F) > \varepsilon$$

occurs. The error probability is $\alpha \sim e^{-2n\varepsilon^2}$, again.

If we apply both tests, we are using *Kolmogorov's two-sided test:* The hypothetical distribution function is rejected if the maximum of $|F-F_n|$ is greater than ε. The error probability is clearly less than or equal to 2α. For small n, it suffices just to take 2α as the error probability, thus keeping on the safe side.

For large n, Kolmogorov's asymptotic expansion for the error probability is

$$2\sum_{1}^{\infty}(-1)^{j-1}\, e^{-2j^2 n\varepsilon^2}.$$

The series converges very quickly. For practical purposes we can confine ourselves to the first term $2e^{-2n\varepsilon^2}$ which corresponds to the rule of thumb above. Table 5 gives values of ε for certain error probabilities 2α.

In practice, Kolmogorov's test is applied in the following way. Suppose we have found an empirical distribution $F_n(t)$. Now look for ε in Table 5. Mark off a strip of width 2ε bounded by the step functions $y=F_n(t)+\varepsilon$ and $y=F_n(t)-\varepsilon$. The true distribution function will probably lie in this strip.

References for § 16

A. Kolmogorov, Determinazione empirica di una legga di distribuzione, Giornale Istit. Ital. Attuari 4 (1933) p. 83.

N. Smirnov, Sur les écarts de la courbe de distribution empirique, Mat. Sbornik 48 (1939) p. 3.

W. Feller, On the Kolmogorov-Smirnov limit theorems for empirical distributions, Ann. Math. Stat. 19 (1948) p. 177.

Z.W. Birnbaum, and F.H. Tingey, One-sided confidence contours for distribution functions, Ann. Math. Stat. 22 (1951) p. 592.

Z.W. Birnbaum, On the power of a one-sided test of fit for continuous probability functions, Ann. Math. Stat. 24 (1953) p. 484.

§ 17. Order Statistics

Let $(x_1, \ldots, x_n)$ be a sample consisting of n independently observed values of a random variable $\mathbf{x}$ with a continuous distribution function $F(t)$. If we arrange the x_i in increasing order (denoting the smallest by $x^{(1)}$, the next smallest by $x^{(2)}$, *etc.*),

$$x^{(1)} < x^{(2)} < \cdots < x^{(n)},$$

then we call each of them an *order statistic.*

If n is odd, that is $n = 2m-1$, then there is a *middle value* $x^{(m)}$. It is an approximation to the *median* ζ which is defined by

$$F(\zeta) = \tfrac{1}{2}.$$

Therefore, we call $x^{(m)} = Z$, the *sample median.* For a symmetric curve $y = f(t)$, in particular for a normal distribution, the median ζ coincides with the expectation $\hat{x}$; therefore, we can use Z as a convenient estimate for $\hat{x}$.

Analogously, we can define the two *sample quartiles* for $n = 4r-1$,

$$Z_1 = x^{(r)} \quad \text{and} \quad Z_3 = x^{(3r)}.$$

The sample quartiles and the sample median divide the ordered sequence of $x^{(h)}$ into four sections, with $r-1$ values each. For large n, the sample quartiles approach the true *quartiles* ζ_1 and ζ_3 which are defined by

$$F(\zeta_1) = \tfrac{1}{4} \quad \text{and} \quad F(\zeta_3) = \tfrac{3}{4}.$$

We can also define *sample sextiles* Y_1 and Y_5 similarly. They approach the points η_1 and η_5 defined by

$$F(\eta_1) = \tfrac{1}{6} \quad \text{and} \quad F(\eta_5) = \tfrac{5}{6}.$$

For a normal distribution, these points lie approximately at $\hat{x} - \sigma$ and $\hat{x} + \sigma$, since if $\Phi(x)$ is a normal distribution function with mean zero and standard deviation one, we have

$$\Phi(-1) = 0.16\ldots \quad \text{and} \quad \Phi(+1) = 0.84\ldots.$$

Therefore, for a sample from a distribution not too different from the normal, we can use the half of the distance between sextiles as a convenient rough estimate of the standard deviation σ. We can also derive an estimate for σ from the quartiles; we shall return to this in § 20.

In order to be able to judge the accuracy of these estimates, it is necessary to determine the distribution function $G^{(h)}(t)$ of the order statistic $x^{(h)}$.

$G^{(h)}(t)$ is the probability of the event $[x^{(h)}<t]$. It is equal to the probability that at the least h of the x_i are smaller than t. If W_k is the probability that exactly k of the x_i are smaller than t, then from § 5, formula (2), we have

$$W_k=\binom{n}{k} F(t)^k [1-F(t)]^{n-k}. \tag{1}$$

With this we obtain

$$G^{(h)}(t)=W_h+W_{h+1}+\cdots+W_n. \tag{2}$$

This solves the problem at hand. However, the solution is somewhat inconvenient. We shall assume $F(t)$ to be differentiable. Let

$$F'(t)=f(t).$$

By differentiating (2) with respect to t, we obtain the probability density

$$g^{(h)}(t)=W_h'+W_{h+1}'+\cdots+W_n'.$$

In the differentiation all of the terms cancel out except one; only

$$g^{(h)}(t)=n\binom{n-1}{h-1} F(t)^{h-1} [1-F(t)]^{n-h} f(t) \tag{3}$$

remains.

The product $F^{h-1}(1-F)^{n-h}$ is largest when

$$F=\frac{h-1}{n-1} \quad \text{and} \quad 1-F=\frac{n-h}{n-1}.$$

Let this be the case for $t=t_0$. Now if we set

$$\begin{aligned} F_0&=F(t_0)=\frac{h-1}{n-1},\\ f_0&=f(t_0),\\ F&=F_0+X, \end{aligned}$$

then we obtain from (3)

$$g^{(h)}(t)=g^{(h)}(t_0)\left(1+\frac{X}{F_0}\right)^{h-1}\left(1-\frac{X}{1-F_0}\right)^{n-h}\frac{f}{f_0}. \tag{4}$$

We now assume that h and $n-h$ are both large and examine the behavior of the three factors of (4) in the neighborhood of $t=t_0$. The first factor

$$g^{(h)}(t_0)=\frac{n(n-1)!}{(h-1)!\,(n-h)!}\left(\frac{h-1}{n-1}\right)^{h-1}\left(\frac{n-h}{n-1}\right)^{n-h} f_0$$

can be approximated by Stirling's formula

$$n!\sim n^n e^{-n}\sqrt{2\pi n}.$$

We obtain

$$g^{(h)}(t_0)\sim n\left[\frac{n-1}{2\pi(h-1)(n-h)}\right]^{\frac{1}{2}} f_0=\frac{\alpha}{\sqrt{2\pi}} f_0. \tag{5}$$

For the second factor

$$Q=\left(1+\frac{X}{F_0}\right)^{h-1}\left(1-\frac{X}{1-F_0}\right)^{n-h},$$

we expand the logarithm. Then we have

$$\begin{aligned}\ln Q&=(h-1)\ln\left(1+\frac{X}{F_0}\right)+(n-h)\ln\left(1-\frac{X}{1-F_0}\right)\\ &=(h-1)\left(\frac{X}{F_0}-\frac{X^2}{2F_0^2}\right)+(n-h)\left(-\frac{X}{1-F_0}-\frac{X^2}{2(1-F_0)^2}\right)+\cdots\\ &=-\frac{1}{2}X^2\frac{(n-1)^3}{(h-1)(n-h)}+\cdots,\end{aligned}$$

where the omitted terms are only of order of magnitude of X^3 and therefore, for small X, are small compared with the main term which is of order nX^2. If we now substitute $n^2(n-1)$ into the main term for $(n-1)^3$ (which makes no difference when n is large), then we have for the second factor in (4) the asymptotic expansion

$$Q\sim e^{-\frac{1}{2}\alpha^2 X^2}. \tag{6}$$

The third factor in (4) can be replaced by one, provided t is not far from t_0. With this we have

$$g^{(h)}(t)\sim\frac{\alpha}{\sqrt{2\pi}} f_0\, e^{-\frac{1}{2}\alpha^2 X^2} \tag{7}$$

where

$$\alpha=n\left[\frac{n-1}{(h-1)(n-h)}\right]^{\frac{1}{2}}. \tag{8}$$

In deriving (7), we have assumed X to be small as compared with one. If this is not the case, αX is large compared with one, and the right side of (7) becomes very small. One can prove that in this case the left side is small too. Hence, formula (7) can be applied without large error in all cases, whether X is small or not.

One can also argue as follows. What matters in the applications is not the individual value $g^{(h)}(t)$ for any particular t, but the integral of $g^{(h)}(t)$ over an interval from a to b. This interval may consist of a part in which X is less than δ (say), in which our formula (7) is valid with a sufficient degree of accuracy, and a remainder, in which αX is large as compared with one. By the Law of Large Numbers (§ 5 C) the total probability of all these values of X is arbitrarily small. Hence the integral of $g^{(h)}(t)$ over this remainder is less than any given ε, and so is the integral of the right side of (7). Thus, the application of (7) in its integrated form is justified in any case.

Now we have to express X in terms of t. Since $t-t_0$ is small and F is differentiable, $F-F_0$ is approximately equal to $(t-t_0)f_0$. Therefore, instead of (7) we can write

$$g^{(h)}(t) \sim \frac{\alpha}{\sqrt{2\pi}} f_0 \, e^{-\frac{1}{2}\alpha^2 f_0^2 (t-t_0)^2}. \tag{9}$$

That is to say, *the order statistics* $x^{(h)}$ *are asymptotically normally distributed with expectation* t_0 *and standard deviation*

$$\sigma = (\alpha f_0)^{-1} = \frac{1}{n}\left[\frac{(h-1)(n-h)}{n-1}\right]^{\frac{1}{2}} f_0^{-1} \tag{10}$$

provided that h *and* $n-h$ *are both large.*

In the case of the sample median Z, we find

$$\sigma = \frac{1}{n}\left(\frac{n-1}{4}\right)^{\frac{1}{2}} f_0^{-1} \sim \frac{1}{n} n^{-\frac{1}{2}} f_0^{-1}.$$

For a normal distribution with expectation zero and standard deviation one, we have

$$t_0 = 0 \quad \text{and} \quad f_0^{-1} = \sqrt{2\pi};$$

and hence,

$$\sigma_Z \sim \left(\frac{\pi}{2n}\right)^{\frac{1}{2}}. \tag{11}$$

For the extreme order statistics, that is, for those with small h or $n-h$, the evaluation of the distribution is much more difficult. Fisher and Tippett, Fréchet, von Mises, and Gumbel have furnished important contributions to it. (See the article on order statistics by S. S. Wilks in Bull. Amer. Math. Soc. 54 (1948).)

The probability density $f(u, v)$ of a pair of order statistics $x^{(h)}, x^{(j)}$ can also be calculated like that of a single order statistic. For example, consider the smallest and largest x_i, $x^{(1)}$ and $x^{(n)}$; then the probability that $[x^{(1)} > u$ and $x^{(n)} < v]$ is

$$\{F(v) - F(u)\}^n.$$

By differentiating with respect to u and v, we obtain the probability density

$$f(u, v) = n(n-1)\{F(v) - F(u)\}^{n-2} f(u) f(v). \tag{12}$$

An important function of the order statistics for many applications is the *range*

$$W = x^{(n)} - x^{(1)}. \tag{13}$$

Its distribution function can be obtained by integrating (12):

$$H(t) = P(W < t) = \iint_{0 < v - u < t} f(u, v)\, du\, dv = \int_{-\infty}^{\infty} du \int_u^{u+t} f(u, v)\, dv.$$

The indefinite integral of $f(u, v)$ with respect to v is clearly

$$n\{F(v) - F(u)\}^{n-1} f(u).$$

If we substitute the limits u and $u+t$, we obtain

$$\int_u^{u+t} f(u, v)\, dv = n\{F(u+t) - F(u)\}^{n-1} f(u);$$

therefore,

$$H(t) = \int_{-\infty}^{\infty} n\{F(u+t) - F(u)\}^{n-1} f(u)\, du. \tag{14}$$

§ 18. Sample Mean and Sample Variance

Among the constants (parameters) which characterize a distribution function, the most important are the *expectation* or *population mean*

$$\hat{x} = \mathscr{E} x = \int_{-\infty}^{\infty} t\, dF(t) \tag{1}$$

and the *standard deviation* σ, which is defined by its square

$$\sigma^2 = \mathscr{E}(x - \hat{x})^2 = \int_{-\infty}^{\infty} (t - \hat{x})^2\, dF(t). \tag{2}$$

The squared standard deviation is called the *variance*.

We shall assume that both integrals (1) and (2) converge. Now, how do we determine $\hat{x}$ and σ empirically?

Suppose we have a *random sample* $(x_1, \ldots, x_n)$, consisting of n observed values $x_1, \ldots, x_n$ of x. In the language of probability theory, $x_1, \ldots, x_n$ are independent random variables which all have the same distribution function $F(t)$. Now we construct the *arithmetic mean* of the sample

$$M = \frac{1}{n}(x_1 + \cdots + x_n) = \frac{1}{n}x_1 + \cdots + \frac{1}{n}x_n. \tag{3}$$

We call M the *sample mean* and $\hat{x}$ *the mean of the population*. Frequently, the notation $\bar{x}$ is used instead of M.

M itself is a random variable. Its expectation is

$$\mathscr{E}M = \hat{M} = \frac{\hat{x}}{n} + \frac{\hat{x}}{n} + \cdots + \frac{\hat{x}}{n} = \hat{x} \tag{4}$$

and from (15) § 3, its variance is

$$\sigma_M^2 = \frac{\sigma^2}{n^2} + \frac{\sigma^2}{n^2} + \cdots + \frac{\sigma^2}{n^2} = n\frac{\sigma^2}{n^2} = \frac{\sigma^2}{n}. \tag{5}$$

Thus, for large n, the standard deviation of M is considerably smaller than for the individual x_i. Since with high probability (according to the Tchebychev inequality) $M - \hat{x}$ can have only the order of magnitude

$$\sigma_M = \frac{\sigma}{\sqrt{n}},$$

M is a reasonable estimate for $\hat{x}$. The larger n, the better the approximation.

Similarly, we have

$$s_0^2 = \frac{1}{n}\sum (x_i - \hat{x})^2 \tag{6}$$

for an approximation of σ^2. The expectation of s_0^2 is clearly exactly σ^2.

Eq. (6) may be used only when $\hat{x}$ is known exactly. But this is most often not the case. We manage by substituting M for $\hat{x}$ on the right side of (6). However, by this substitution, the expectation of expression (6) is somewhat decreased, as we shall see presently.

For arbitrary a, we have the identity

$$\sum (x_i - M)^2 = \sum (x_i - a)^2 - n(M - a)^2, \tag{7}$$

which may be proved as follows:

$$\begin{aligned}\sum (x_i - M)^2 &= \sum \{(x_i - a) - (M - a)\}^2\\ &= \sum (x_i - a)^2 - 2\sum (x_i - a)(M - a) + n(M - a)^2\\ &= \sum (x_i - a)^2 - 2n(M - a)(M - a) + n(M - a)^2\\ &= \sum (x_i - a)^2 - n(M - a)^2.\end{aligned}$$

If we substitute $a = \hat{x}$ in (7), we have

$$\sum (x_i - M)^2 = \sum (x_i - \hat{x})^2 - n(M - \hat{x})^2. \tag{8}$$

Eq. (8) shows that (6) takes on its minimum value when $M = \hat{x}$. In order to compensate, it is better to divide by $(n-1)$ rather than n. Thus, we form

$$s^2 = \frac{1}{n-1} \sum (x_i - M)^2. \tag{9}$$

If we take expectations on both sides of (8), we obtain

$$(n-1)\,\mathscr{E}(s^2) = n\sigma^2 - n\frac{\sigma^2}{n} = (n-1)\,\sigma^2$$

or

$$\mathscr{E}(s^2) = \sigma^2. \tag{10}$$

The denominator $(n-1)$ in (9) causes the expectation of s^2 to be exactly equal to σ^2. We call s the *sample standard deviation* and s^2 the *sample variance*.

We can simplify the calculation of s^2 by using identity (7). We proceed in the following way.

1. To begin with, round the observed x_i to as many decimal places as necessary for the difference between the largest and the smallest x_i to become a two-place number. This rounding has practically no effect on s^2.

2. Then remove the decimal point, that is, multiply the rounded x by an appropriate power of ten so that they all become integers.

3. Calculate the mean M of the rounded x according to (3) and round to one decimal place.

4. Then choose a convenient number within the range of the x and form the differences $x - a$. Check: the sum of these differences has to be $n(M - a)$.

5. The differences $x - a$ are at most two-place integers which can be squared in one's head or with a small table of squares.

6. Now calculate $\sum (x-a)^2$ and hence $\sum (x-M)^2$ by (7) and s^2 by (9), for the rounded integers.

7. A useful check is given by formula (7) with $a=0$:

$$\sum (x-M)^2 = \sum x^2 - n M^2. \tag{11}$$

Example 12. Ninety years' rain in Rothamsted (R. A. Fisher, Statistical Methods for Research Workers, § 14). In the table below, the first column gives rainfall x in inches; the second gives the number of years k in which these amounts were observed. The third column contains the difference $x-a$, where $a=28$ was chosen as an approximate mean. The fourth column gives $k(x-a)$, the fifth $k(x-a)^2$.

x	k	$x-a$	$k(x-a)$	$k(x-a)^2$
16	1	−12	−12	144
19	3	− 9	−27	243
20	2	− 8	−16	128
21	3	− 7	−21	147
23	3	− 5	−15	75
24	2	− 4	− 8	32
25	12	− 3	−36	108
26	4	− 2	− 8	16
27	7	− 1	− 7	7
28	4	0	0	0
29	8	1	8	8
30	9	2	18	36
31	6	3	18	54
32	7	4	28	112
33	4	5	20	100
34	4	6	24	144
35	4	7	28	196
36	3	8	24	192
37	3	9	27	243
39	1	11	11	121
Total	90		56	2,106

Correction for the mean $56/90=0.62$
Mean $M \quad =28.62$
$n(M-a)^2 \quad =35$
$\sum (x-M)^2 \;=2{,}106-35=2{,}071$
Variance $s^2=2{,}071/89=23$
Standard deviation $s=4.9$

§ 19. Sheppard's Correction

The following two sections (§§ 19 and 20) can be skipped. Their main purpose is to introduce the reader to such terms as the Sheppard correction of "probable error" which frequently appear in the literature.

Frequently, the observed values $x_1, \ldots, x_n$ are rounded or grouped, that is, collected into *class intervals*. If the class numbered k includes the values between $\tau_k - \frac{1}{2}h$ and $\tau_k + \frac{1}{2}h$, then τ_k is called the *class midpoint*. Now, let n_k be the number of observed x values in the kth interval and let all of the x values be replaced by the class midpoint τ_k; then, we obtain the approximation

$$M' = \frac{1}{n} \sum n_k \tau_k \tag{1}$$

for M and the approximation

$$s_0'^2 = \frac{1}{n} \sum n_k (\tau_k - \hat{x})^2 \tag{2}$$

for s_0^2.

M' may be somewhat larger or somewhat smaller than M, but the difference $M' - M$ is zero on the average; on the other hand, $s_0'^2$ is somewhat larger than s_0^2 on the average. In order to show this and to determine the correction term to subtract from $s_0'^2$ to get s_0^2, we make the assumption that all of the class intervals are the same length. Let k go from $-\infty$ to $+\infty$. Give the class interval from $t - \frac{1}{2}h$ to $t + \frac{1}{2}h$ the number 0. Then the kth interval goes from $t + kh - \frac{1}{2}h$ to $t + kh + \frac{1}{2}h$. If we shift the origin so that $\hat{x} = 0$, then

$$M' = \frac{1}{n} \sum n_k (t + kh)$$

and

$$s_0'^2 = \frac{1}{n} \sum n_k (t + kh)^2.$$

We obtain the expectation of M' and $s_0'^2$ by replacing the n_k by $n p_k$, where

$$p_k = F(t + kh + \tfrac{1}{2}h) - F(t + kh - \tfrac{1}{2}h)$$

is the probability that an x lies between $t + kh - \frac{1}{2}h$ and $t + kh + \frac{1}{2}h$. Thus, we have

$$A(t) = \mathscr{E} M' = \sum_{-\infty}^{\infty} [F(t + kh + \tfrac{1}{2}h) - F(t + kh - \tfrac{1}{2}h)] \cdot (t - kh) \tag{3}$$

and

$$B(t) = \mathscr{E} s_0'^2 = \sum_{-\infty}^{\infty} [F(t + kh + \tfrac{1}{2}h) - F(t + kh - \tfrac{1}{2}h)] \cdot (t + kh)^2. \tag{4}$$

Both expressions (3) and (4) are periodic functions of t and are invariant under translation by h. In most cases, these periodic functions are almost constant, that is, in their Fourier expansion only the constant term, which is equal to the mean, need be considered. However, even if this is not the case, it is useful to calculate the mean values of the functions $A(t)$ and $B(t)$. The choice of the class boundaries $t + kh \pm \frac{1}{2}h$ is arbitrary and in a certain sense random; we can consider t as a random variable which is uniformly distributed between (say) 0 and h. The mean of $A(t)$ over all t values is then just the mean of the integral $\hat{A}$,

$$\hat{A}=\frac{1}{h}\int_0^h A(t)\,dt \tag{5}$$

and analogously for $B(t)$,

$$\hat{B}=\frac{1}{h}\int_0^h B(t)\,dt. \tag{6}$$

If we substitute (3) in (5), we obtain

$$\begin{aligned} h\hat{A} &= \sum_{-\infty}^{\infty}\int_0^h [F(t+kh+\tfrac{1}{2}h)-F(t+kh-\tfrac{1}{2}h)]\cdot(t+kh)\,dt \\ &= \int_{-\infty}^{\infty} [F(t+\tfrac{1}{2}h)-F(t-\tfrac{1}{2}h)]\cdot t\,dt. \end{aligned} \tag{7}$$

Integration by parts gives

$$\begin{aligned} h\hat{A} &= \int_{-\infty}^{\infty} \tfrac{1}{2}t^2\,d[F(t-\tfrac{1}{2}h)-F(t+\tfrac{1}{2}h)] \\ &= \int_{-\infty}^{\infty} \tfrac{1}{2}t^2\,dF(t-\tfrac{1}{2}h)-\int_{-\infty}^{\infty} \tfrac{1}{2}t^2\,dF(t+\tfrac{1}{2}h) \\ &= \int_{-\infty}^{\infty} \tfrac{1}{2}(t+\tfrac{1}{2}h)^2\,dF(t)-\int_{-\infty}^{\infty} \tfrac{1}{2}(t-\tfrac{1}{2}h)^2\,dF(t) \\ &= \int_{-\infty}^{\infty} \tfrac{1}{2}[(t+\tfrac{1}{2}h)^2-(t-\tfrac{1}{2}h)^2]\,dF(t) \\ &= \int_{-\infty}^{\infty} t\,h\,dF(t); \end{aligned}$$

thus,

$$\hat{A}=\int_{-\infty}^{\infty} t\,dF(t)=\hat{x}=0.$$

Likewise,

$$\begin{aligned} h\hat{B} &= \int_{-\infty}^{\infty} \tfrac{1}{3}[(t+\tfrac{1}{2}h)^3-(t-\tfrac{1}{2}h)^3]\,dF(t) \\ &= \int_{-\infty}^{\infty} (t^2h+\tfrac{1}{12}h^3)\,dF(t) \\ &= h\int_{-\infty}^{\infty} t^2\,dF(t)+\tfrac{1}{12}h^3 \end{aligned}$$

or

$$\hat{B}=\int_{-\infty}^{\infty} t^2\,dF(t)+\tfrac{1}{12}h^2=\sigma^2+\tfrac{1}{12}h^2.$$

If t is regarded as a random variable, the expectation of $s_0'^2$ is just $(\frac{1}{12})h^2$ larger than σ^2, the expectation of s^2. Thus, in order to obtain an unbiased estimate for σ^2, we have to subtract *Sheppard's correction* $(\frac{1}{12})h^2$ from $s_0'^2$.

It is usual to apply the same correction to calculation of s^2. First we calculate

$$s'^2 = \frac{1}{n-1} \sum n_k (\tau_k - M')^2$$

and subtract $(\frac{1}{12})h^2$, so that on the average we get approximately the same value as when we constructed s^2 directly from the original x_i.

Example 13. In § 18 we said that we could safely round the observed x values for the calculation of s^2 in such a way that the range becomes a two-place number. This assertion shall now be confirmed by calculation of Sheppard's correction. We assume that n is not unusually large, and that the x are approximately normally distributed.

Suppose that the x_i are rounded to the nearest integer and that the range $W = x^{(n)} - x^{(1)}$, that is, the difference between the largest and smallest x_i, is a two place number

$$10 \leqq W < 100.$$

We may assume that the standard deviation σ is larger than two. If σ were less than or equal to two, then the x_i would probably lie between $\hat{x}-5$ and $\hat{x}+5$ and the range W would probably be less than ten, contrary to assumption. Using the distribution function derived in § 17, we could sharpen the last conclusion, but an approximate estimate will suffice.

The rounding results in a grouping in which an interval covers from $g-\frac{1}{2}$ to $g+\frac{1}{2}$, where g is an integer in each case. The expectation of s^2 is greater than four; however, Sheppard's correction is only $\frac{1}{12}$. Thus, Sheppard's correction comprises less than $\frac{1}{48}$ of the total value of s^2, that is, only 2%. Thus, the difference between s' and s is only 1%, which is practically negligible.

For unusually large n or for very irregular distributions, these ratios may be less favorable. As a precaution, for very large n, we shouldn't carry the rounding so far. If W turns out to be less than 20 because of the rounding, then we should carry one more decimal place. Then, W will be a number smaller than 200 and we can always choose a so that the difference $x-a$ is a two-place number which we can square easily.

§ 20. Other Mean and Dispersion Measures

Another measure which may be used instead of the expectation is the *median* ζ. For a continuous distribution function, it is defined by

$$F(\zeta) = \tfrac{1}{2}. \tag{1}$$

For a symmetric probability density (in particular a normal density), the median ζ is equal to the population mean $\hat{x}$.

If we substitute empirical frequencies for the probabilities, then we get the *sample median* Z. For an odd number $n=2m-1$, Z is the middle term x_m in the sequence of the x_i arranged in increasing order. For an even number $n=2m$, Z lies between the two middle terms x_m and x_{m+1}.

The sample median is easier to calculate than the sample mean M. However, in the case of an approximately normal distribution, the sample mean M is more reliable. For a normal distribution, the sample median Z has, according to § 17, an approximately normal distribution with standard deviation

$$\sigma_Z = \sigma \sqrt{\frac{\pi}{2n}},$$

while the standard deviation of the sample mean M is only

$$\sigma_M = \sigma \sqrt{\frac{1}{n}}.$$

Therefore, the standard deviation of Z is larger than that of M by a factor of $\sqrt{\pi/2}$.

However, there are also distribution functions for which the sample median Z is a more accurate measure than the sample mean M. As an example, consider the Cauchy distribution

$$F(t) = \frac{1}{2} + \frac{1}{\pi} \operatorname{arc\,tan} t, \tag{2}$$

with the probability density

$$f(t) = \frac{1}{\pi(1+t^2)}. \tag{3}$$

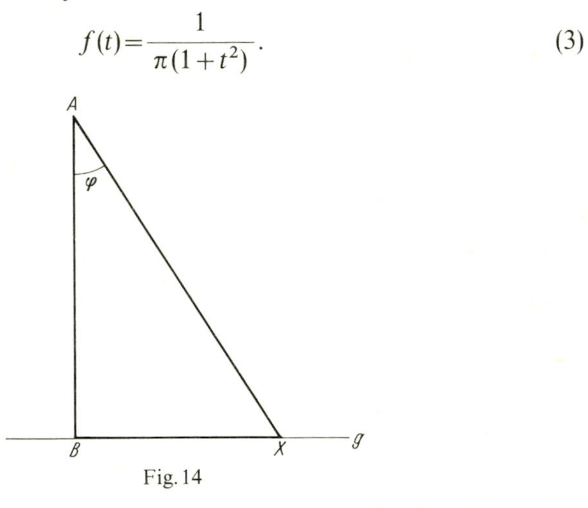

Fig. 14

To get this distribution, draw a straight line AX in any direction from a point A. Let it intersect a fixed straight line g at X. If φ is the angle AX makes with the line AB perpendicular to g and if AB is of unit length, then the distance from X to the foot of the perpendicular is

$$x = \tan \varphi.$$

Now, if the angle φ is uniformly distributed between $-\pi/2$ and $+\pi/2$ (that is, if the probability density of the random variable φ is equal to the constant π^{-1}), then $x=\tan\varphi$ has distribution function (2) and density (3).

Suppose two independent random variables x_1 and x_2 both have distribution function (2). Calculate the distribution of their sum x_1+x_2 by Theorem III, § 4. If we calculate the distribution function and then the probability density of the sample mean

$$M_2=\tfrac{1}{2}(x_1+x_2)$$

from this, we get probability density (3) exactly. The sample mean of two such sample means

$$M_4=\tfrac{1}{4}(x_1+x_2+x_3+x_4)$$

also has the same probability density, *etc.* Therefore, averaging does not cause accuracy to increase!

On the other hand, if we form the sample median from an odd number of x_i, then according to § 17 it has an approximately normal distribution with expectation zero and standard deviation

$$\sigma_Z=\frac{\pi}{2}n^{-\frac{1}{2}}.$$

Therefore, for increasing n the sample median results in a more and more exact estimate of the true median zero.

There is no expectation and no standard deviation for the Cauchy distribution (3); the integrals diverge.

Analogous to the median ζ, we define the two *quartiles* ζ_1 and ζ_3 by

$$F(\zeta_1)=\tfrac{1}{4} \quad \text{and} \quad F(\zeta_3)=\tfrac{3}{4}.$$

As approximate values to these, we have the *sample quartiles* Z_1 and Z_3 which can be defined in the following way. Let

$$q=\left[\frac{m}{4}\right].$$

Arrange the x_i in increasing order. Then Z_1 is the x_i with rank number $q+1$ and Z_3 is the x_i with rank number $n-q$, so that q of the x_i are smaller than Z_1 and q are larger than Z_3. For $n=4m-1$, this definition agrees with the one given in § 17.

The *probable deviation* w is another (old-fashioned) measure of deviation. For a continuous distribution, it is defined so that the probability that the random variable x lies between $\zeta-w$ and $\zeta+w$ is $\frac{1}{2}$, that is,

$$F(\zeta+w)-F(\zeta-w)=\tfrac{1}{2}.$$

For a symmetric distribution, $\zeta - w$ and $\zeta + w$ coincide with the quartiles

$$\zeta_1 = \zeta - w \quad \text{and} \quad \zeta_3 = \zeta + w.$$

For a normal distribution we have

$$w = 0.6745\,\sigma. \tag{4}$$

In olden times, w was frequently used instead of s or σ. However, today it has been abandoned. The additional calculation of an estimate w' for w, from

$$w' = 0.6745\,s,$$

is an unnecessary operation.

The third traditional measure of deviation is the *mean deviation* δ, which is defined as the expectation of the absolute difference of $x - \hat{x}$,

$$\delta = \mathscr{E}|x - \hat{x}| = \int_{-\infty}^{\infty} |t - \hat{x}|\, dF(t). \tag{5}$$

We obtain an empirical approximation d to δ by substituting M for $\hat{x}$;

$$d = \frac{1}{n} \sum |x_k - M|. \tag{6}$$

For a normal distribution, the ratio between δ and σ is constant. If we normalize (by taking a linear and scalar transformation of the x_i), so that $\hat{x} = 0$, $\sigma = 1$, then

$$\begin{aligned}\delta &= \frac{1}{\sqrt{2\pi}} \int_{-\infty}^{\infty} |t|\, e^{-\frac{1}{2}t^2}\, dt \\ &= \frac{2}{\sqrt{2\pi}} \int_0^{\infty} t\, e^{-\frac{1}{2}t^2}\, dt = \sqrt{\frac{2}{\pi}}.\end{aligned}$$

Consequently, for each normal distribution relation

$$\delta = \sigma \sqrt{\frac{2}{\pi}} \tag{7}$$

holds.

The approximate value d is somewhat simpler to calculate than the sample standard deviation s, but s is a more important and more accurate estimate of σ in most cases.

Later, we shall see that in a certain sense s^2 is the best possible estimate for σ^2 for a normal distribution.

Chapter Five

Fourier Integrals and Limit Theorems

§ 21. Characteristic Functions

A. Expectations of Complex Random Variables

Up to now the concept of expectation has been defined only for real random variables. However, we can construct a random variable

$$\mathbf{z} = \mathbf{x} + i\,\mathbf{y}$$

from two real-valued random variables $\mathbf{x}$ and $\mathbf{y}$ and define its expected value by

$$\mathscr{E}\mathbf{z} = \mathscr{E}\mathbf{x} + i\mathscr{E}\mathbf{y}.$$

More generally, we can define the expectation of a vector $\mathbf{v} = (\mathbf{x}_1, \ldots, \mathbf{x}_n)$ in a similar way by

$$\mathscr{E}(\mathbf{x}_1, \ldots, \mathbf{x}_n) = (\mathscr{E}\mathbf{x}_1, \ldots, \mathscr{E}\mathbf{x}_n).$$

For two arbitrary vector- or complex-valued random variables, $\mathbf{v}$ and $\mathbf{w}$, it is true that

$$\mathscr{E}(\mathbf{v} + \mathbf{w}) = \mathscr{E}\mathbf{v} + \mathscr{E}\mathbf{w}. \tag{1}$$

Two complex random variables $\mathbf{z} = \mathbf{x} + i\,\mathbf{y}$ and $\mathbf{w} = \mathbf{u} + i\,\mathbf{v}$ are said to be *independent* if $\mathbf{u}$ and $\mathbf{v}$ are independent of $\mathbf{x}$ and $\mathbf{y}$, that is, if

$$\mathscr{P}[\mathbf{x} < a, \mathbf{y} < b, \mathbf{u} < c, \mathbf{v} < d] = \mathscr{P}[\mathbf{x} < a, \mathbf{y} < b] \cdot \mathscr{P}[\mathbf{u} < c, \mathbf{v} < d]$$

holds for arbitrary a, b, c, d. If that is the case, then

$$\mathscr{E}(\mathbf{z}\,\mathbf{w}) = (\mathscr{E}\mathbf{z})(\mathscr{E}\mathbf{w}), \tag{2}$$

which can be seen immediately by separating the random variables into their real and imaginary parts.

B. The Characteristic Function

Let $\mathbf{x}$ be a real-valued random variable with distribution function $F(u) = \mathscr{P}[\mathbf{x} < u]$. We define the *characteristic function* of $\mathbf{x}$ to be the expected value of $\exp\{it\mathbf{x}\}$:

$$\varphi(t)=\mathscr{E}e^{it\mathbf{x}}=\int_{-\infty}^{\infty} e^{itu}\,dF(u). \tag{3}$$

The integral (3) always converges for real t, since the absolute value of e^{itu} is equal to one and $\int dF(u)$ converges.

If $F(u)$ is the integral of a probability density $f(u)$, we have an ordinary Fourier integral

$$\varphi(t)=\int_{-\infty}^{\infty} e^{itu} f(u)\,du. \tag{4}$$

On the other hand, if $\mathbf{x}$ is a discrete random variable taking only finitely or countably many values $x_1, x_2, \ldots$ with probabilities $p_1, p_2, \ldots$, then

$$\varphi(t)=\sum p_k \exp\{itx_k\}. \tag{5}$$

Obviously,

$$\varphi(0)=1,$$

and

$$|\varphi(t)|\leqq 1,$$

for all real t.

If $\mathbf{x}$ has characteristic function $\varphi(t)$, then $a\mathbf{x}+b$ has the characteristic function

$$\mathscr{E}e^{it(a\mathbf{x}+b)}=e^{itb}\mathscr{E}e^{iat\mathbf{x}}=e^{ibt}\varphi(at).$$

C. Continuity of the Characteristic Function

An important continuity property of the Lebesgue integral was proved by Lebesgue himself[1]. It can be stated as follows: *If $\{g_\nu(u)\}$ is a sequence of integrable functions, where the index ν may also be replaced by a continuously variable parameter t and if*

$$|g_\nu(u)|\leqq G(u), \tag{6}$$

where the integral $\int G(u)\,dF(u)$ is finite, and

$$\lim g_\nu(u)=g(u), \tag{7}$$

for all real u, then

$$\lim \int_{-\infty}^{\infty} g_\nu(u)\,dF(u)=\int_{-\infty}^{\infty} g(u)\,dF(u). \tag{8}$$

From this theorem, it follows immediately that the characteristic function $\varphi(t)$ is a continuous function of t.

[1] H. Lebesgue, Bull. Soc. Math. France 36 (1908) p. 12.

D. Moments

We define the nth *moment* of a random variable $\mathbf{x}$ to be the expected value of $\mathbf{x}^n$

$$\alpha_n = \mathscr{E}\mathbf{x}^n = \int_{-\infty}^{\infty} u^n \, dF(u). \tag{9}$$

More generally, we define moments about a point c by

$$\mathscr{E}(\mathbf{x}-c)^n = \int_{-\infty}^{\infty} (u-c)^n \, dF(u). \tag{10}$$

Moments about the mean $\hat{\mathbf{x}}$ are especially important:

$$\mu_n = \mathscr{E}(\mathbf{x}-\hat{\mathbf{x}})^n = \int_{-\infty}^{\infty} (u-\hat{\mathbf{x}})^n \, dF(u). \tag{11}$$

Obviously, $\mu_0 = 1$ and $\mu_1 = 0$, while μ_2 is the variance or square of the standard deviation

$$\mu_2 = \sigma^2 = \mathscr{E}(\mathbf{x}-\hat{\mathbf{x}})^2.$$

The integrals of (9) only converge if $F(u)$ tends to zero sufficiently rapidly as $u \to -\infty$ and to one sufficiently rapidly as $u \to +\infty$. If (9) converges for some n, we say that the moment α_n *exists*. Then all lower moments $\alpha_1, \ldots, \alpha_{n-1}$ will also exist as well as integrals (10) and (11).

If α_n exists, integral (3) *can be differentiated n times under the integral sign:*

$$\varphi^{(n)}(t) = i^n \int_{-\infty}^{\infty} u^n e^{itu} \, dF(u). \tag{12}$$

We shall carry out the proof for the first differentiation only; proofs for further differentiations are similar. Consider

$$\frac{\varphi(t+h) - \varphi(t)}{h} = \int_{-\infty}^{\infty} e^{itu} \frac{e^{ihu} - 1}{h} \, dF(u), \tag{13}$$

and let h go to zero. The limit of the fraction

$$\frac{e^{ihu} - 1}{h} = e^{\frac{1}{2}ihu} \frac{\sin \frac{1}{2} h u}{\frac{1}{2} h u} \, i u$$

is iu and the absolute value of the fraction is smaller than $|u|$ for all h. Therefore, the continuity theorem stated in C gives

$$\varphi'(t) = i \int_{-\infty}^{\infty} u^{itu} \, dF(u) \tag{14}$$

immediately.

It follows from (12) that $\varphi^{(n)}(t)$ is continuous. If $t=0$, then

$$\varphi^{(n)}(0)=i^n \alpha_n. \tag{15}$$

In so far as they exist, the moments α_n can therefore be obtained by differentiation of the characteristic function.

If the function is analytic in a neighborhood of $t=0$, it may be expanded inside some circle $|t|<r$ in a Taylor series with coefficients determined by the moments

$$\varphi(t)=\sum_0^\infty \frac{\alpha_n}{n!}(i\,t)^n. \tag{16}$$

The assumption that the series $\sum \alpha_n r^n/n!$ converges absolutely is sufficient for Eq. (16) to hold for $|t|<r$ (see Cramér, Mathematical Methods of Statistics, p. 177).

E. Inversion Formulae

It is well known that the Fourier integral (4) can be inverted by means of the formula

$$f(u)=\lim_{T\to\infty}\frac{1}{2\pi}\int_{-T}^{T} e^{-itu}\,\varphi(t)\,dt.$$

If we integrate from a to b, we obtain

$$F(b)-F(a)=\lim_{T\to\infty}\frac{i}{2\pi}\int_{-T}^{T}\frac{e^{-itb}-e^{-ita}}{t}\,\varphi(t)\,dt. \tag{17}$$

This formula can also be written

$$F(u+h)-F(u-h)=\lim_{T\to\infty}\frac{1}{\pi}\int_{-T}^{T}\frac{\sin h\,t}{t}\,e^{itu}\,\varphi(t)\,dt. \tag{18}$$

Paul Lévy showed that formulas (17) and (18) are true even if $F(u)$ is not differentiable provided that it is continuous at the points a and b (or $u-h$ and $u+h$). For the proof, see Cramér, Mathematical Methods of Statistics, p. 93. Yet another inverse formula is given there; namely,

$$\int_0^h [F(u+v)-F(u-v)]\,dv=\frac{1}{\pi}\int_{-\infty}^{\infty}\frac{1-\cos h\,t}{t^2}\,e^{-iut}\,\varphi(t)\,dt. \tag{19}$$

It follows from the inversion formulae that *a distribution function $F(u)$ is uniquely determined by its characteristic function $\varphi(t)$.*

Formula (17) only shows, at first sight, that $F(u)$ is uniquely determined at its points of continuity. Every point of discontinuity b is, however, the limit of an increasing sequence of points of continuity[2] $\{b_\nu\}$ and since $F(n)$ is continuous from the left, it is true that

$$\lim F(b_\nu) = F(b). \tag{20}$$

Furthermore, if $\{a_\nu\}$ is a sequence of continuity points tending to $-\infty$, then

$$\lim F(a_\nu) = 0. \tag{21}$$

From (20) and (21), it follows by subtraction that

$$\lim [F(b_\nu) - F(a_\nu)] = F(b). \tag{22}$$

Formula (22) shows that $F(b)$ is uniquely determined for arbitrary b.

F. Characteristic Function of a Sum

Let $\mathbf{x}$ and $\mathbf{y}$ be independent random variables. Then by (2),

$$\mathscr{E}\, e^{it(\mathbf{x}+\mathbf{y})} = \mathscr{E}\, e^{it\mathbf{x}}\, \mathscr{E}\, e^{it\mathbf{y}}. \tag{23}$$

In words, *the characteristic function of the sum of two independent random variables is the product of the characteristic functions of the summands.* The same also holds, of course, for a sum $\mathbf{x}_1 + \cdots + \mathbf{x}_n$ of arbitrary summands.

In many cases, this theorem used in conjunction with the uniqueness theorem provides a very useful method for determining the distribution function of a sum. This will become clear in the following examples.

§ 22. Examples

A. Binomial Distribution

Let the random variables $\mathbf{x}_1, \ldots, \mathbf{x}_n$ be independent, and let each take the values 1 and 0 with probabilities p and q, respectively. The sum $\mathbf{x}_1 + \mathbf{x}_2 + \cdots + \mathbf{x}_n$ has a binomial distribution, taking each integer value k between 0 and n with probability

$$W_k = \binom{n}{k} p^k q^{n-k}.$$

[2] The set of discontinuity points of a bounded monotone nondecreasing function is at most countable. This can be seen for a distribution function as follows. There can be only finitely many jumps $\geqq 1$, only finitely many $\geqq \frac{1}{2}$, only finitely many $\geqq \frac{1}{3}$, *etc.* Thus, we can enumerate jumps in order of decreasing size.

From (5) § 21, the characteristic function of each summand $\mathbf{x}_j$ is,

$$\varphi(t) = p\, e^{it} + q. \tag{1}$$

Therefore, the characteristic function of the sum is

$$\varphi(t)^n = (p\, e^{it} + q)^n. \tag{2}$$

If we expand it, we obtain the correct sum

$$(p\, e^{it} + q)^n = \sum \binom{n}{k} p^k\, e^{ikt}\, q^{n-k} = \sum W_k\, e^{ikt}.$$

B. Normal Distribution

If the distribution of $\mathbf{x}$ is normal with variance one and expectation zero, then

$$f(u) = (2\pi)^{-\frac{1}{2}}\, e^{-\frac{1}{2}u^2}, \tag{3}$$

and the characteristic function is

$$\begin{aligned} \varphi(t) &= (2\pi)^{-\frac{1}{2}} \int_{-\infty}^{\infty} e^{-\frac{1}{2}u^2 + itu}\, du \\ &= (2\pi)^{-\frac{1}{2}} \int_{-\infty}^{\infty} e^{-\frac{1}{2}(u-it)^2 - \frac{1}{2}t^2}\, du. \end{aligned} \tag{4}$$

We introduce $w = u - it$ as a new variable and obtain

$$\varphi(t) = (2\pi)^{-\frac{1}{2}}\, e^{-\frac{1}{2}t^2} \int e^{-\frac{1}{2}w^2}\, dw, \tag{5}$$

where the integral is taken along a line parallel to the real axis in the w-plane. By integration over suitable closed contours, it can be shown that (5) is equal to the integral of the same function along the real axis. Thus, we obtain

$$\varphi(t) = e^{-\frac{1}{2}t^2}. \tag{6}$$

The random variable $\sigma\,\mathbf{x}$ is also normal with variance σ^2 and expectation zero. Its characteristic function is

$$\mathscr{E}\, e^{it\sigma\mathbf{x}} = \varphi(t\,\sigma) = e^{-\frac{1}{2}\sigma^2 t^2}. \tag{7}$$

Therefore, the characteristic function of a normally distributed random variable with expectation zero is, up to a constant factor, a Gaussian error function.

If we add a constant a to a random variable $\mathbf{x}$, the characteristic function is multiplied by e^{ita}. Therefore, the characteristic function of a normally distributed random variable with mean a and variance σ^2 is

$$\varphi(t) = e^{ita}\, e^{-\frac{1}{2}\sigma^2 t^2}. \tag{8}$$

The product of two functions of this kind has the same form, again. Thus, our earlier result is obtained anew: *The sum of two independent normally distributed random variables also has a normal distribution.*

C. Poisson Distribution

If $\mathbf{x}$ takes the values $k=0, 1, 2, \ldots$ with probabilities (§ 10)

$$p_k = \frac{\lambda^k}{k!} e^{-\lambda}, \tag{9}$$

then, by (5), the characteristic function is

$$\begin{aligned} \varphi(t) = \sum_0^\infty p_k e^{itk} = e^{-\lambda} \sum_0^\infty \frac{[\lambda \exp(it)]^k}{k!} e^{-\lambda} e^{\lambda \exp(it)} \\ = e^{\lambda[\exp(it)-1]}. \end{aligned} \tag{10}$$

The product of two such functions with parameters λ_1 and λ_2 is a similar function with parameter $\lambda_1+\lambda_2$. Hence, it follows that: *The sum of two independent random variables, $\mathbf{x}_1$ and $\mathbf{x}_2$ having Poisson distributions with expectations λ_1 and λ_2 also has a Poisson distribution with expectation* $\lambda_1+\lambda_2$.

§ 23. The χ^2 Distribution

While extending the Gaussian theory of errors, the geodesist F. R. Helmert considered sums of squares of normally distributed random variables and thus discovered a distribution function $G(u)$ which K. Pearson later called the χ^2 distribution. For negative u, $G(u)=0$, and for non-negative u,

$$G(u) = \alpha \int_0^u y^{\lambda-1} e^{-\frac{1}{2}y} dy, \tag{1}$$

where $\lambda = f/2$ and f is an integer which we call the *number of degrees of freedom* in accordance with R. A. Fisher. The factor α is determined by the condition $G(\infty)=1$:

$$\alpha = \Gamma(\lambda)^{-1} 2^{-\lambda}. \tag{2}$$

The probability density is given by

$$g(u) = \alpha u^{\lambda-1} e^{-\frac{1}{2}u}, \qquad (u>0). \tag{3}$$

The simplest case is $\lambda=1$ (two degrees of freedom). The density is then simply an exponential function

$$g(u) = \tfrac{1}{2} e^{-\frac{1}{2}u}, \qquad (u>0). \tag{4}$$

The case $\lambda=\frac{1}{2}$ (one degree of freedom) can be deduced directly from the normal distribution using the following theorem.

If a random variable $\mathbf{x}$ *is normally distributed with expectation zero and variance one, then* $\mathbf{x}^2$ *has a* χ^2 *distribution with one degree of freedom.*

For the proof, we only need to work out the probability that $[\mathbf{x}^2<u]$. It is

$$G(u)=(2\pi)^{-\frac{1}{2}}\int_{-\sqrt{u}}^{\sqrt{u}} e^{-\frac{1}{2}z^2}\,dz=2(2\pi)^{-\frac{1}{2}}\int_0^{\sqrt{u}} e^{-\frac{1}{2}z^2}\,dz.$$

If we change the variable of integration to $y=z^2$, we obtain the required result

$$G(u)=(2\pi)^{-\frac{1}{2}}\int_0^u y^{-\frac{1}{2}}\,e^{-\frac{1}{2}y}\,dy \tag{5}$$

directly.

We now pass on to the general case. The characteristic function of the χ^2 distribution is

$$\varphi(t)=\int_0^\infty \alpha\, u^{\lambda-1}\,e^{-\frac{1}{2}u+itu}\,du. \tag{6}$$

Changing variables to $v=(\frac{1}{2}-it)\,u$, we obtain

$$\varphi(t)=2^{\lambda}(1-2it)^{-\lambda}\,\alpha\int v^{\lambda-1}\,e^{-v}\,dv. \tag{7}$$

The integral is taken along a straight path from zero to infinity which lies in the right half of the w-plane. In

$$v=(\tfrac{1}{2}-it)\,u,$$

t is kept fixed while u ranges from 0 to ∞. The required integral can be shown to be equal to the integral of the same function along the positive real axis by considering integrals around suitable closed contours in the w-plane. Therefore, the integral is equal to $\Gamma(\lambda)$ and we obtain

$$\varphi(t)=(1-2it)^{-\lambda}. \tag{8}$$

The first and second moments of χ^2 are easy to calculate either by the definition of the moments

$$\alpha_n=\int_0^\infty u^n\,dG(u)=\int_0^\infty u^n\,g(u)\,du$$

or by formula (15) of the previous section. We obtain

$$\alpha_1=2\lambda=f,$$

$$\alpha_2=4\lambda(\lambda+1)=f^2+2f.$$

Hence, the expectation and variance of a random variable $\mathbf{y}$ with a χ^2 distribution are

$$\mathscr{E}\mathbf{y} = \alpha_1 = f, \tag{9}$$

$$\sigma^2 = \mathscr{E}\mathbf{y}^2 - (\mathscr{E}\mathbf{y})^2 = \alpha_2 - \alpha_1^2 = 2f. \tag{10}$$

Now, let $\mathbf{y}$ and $\mathbf{z}$ be independent random variables having χ^2 distributions with f and f' degrees of freedom. By (8), the characteristic functions are

$$(1-2it)^{-\frac{1}{2}f} \quad \text{and} \quad (1-2it)^{-\frac{1}{2}f'}.$$

The product has the same form. Hence it follows that: *If two independent random variables have χ^2 distributions with f and f' degrees of freedom, then their sum has a χ^2 distribution with $f+f'$ degrees of freedom.*

This result can also be verified by direct evaluation of the integral

$$h(v) = \int_0^v g_1(u)\, g_2(v-u)\, du \tag{11}$$

derived from Eq. (7) § 4B. The calculation leads to a beta function. In this way use of the characteristic function can be avoided; however, there will be more calculations to do then.

Clearly, the result is also true for sums $\mathbf{y}_1 + \cdots + \mathbf{y}_n$ of more than two independent random variables. If we apply it to the sum of squares of several normally distributed random variables with expectation zero and variance one, we obtain the following theorem.

If $\mathbf{x}_1, \mathbf{x}_2, \ldots, \mathbf{x}_n$ are independent normally distributed random variables with expectation zero and variance one, the sum of squares

$$\chi^2 = \mathbf{x}_1^2 + \mathbf{x}_2^2 + \cdots + \mathbf{x}_n^2 \tag{12}$$

has a χ^2 distribution with n degrees of freedom.

Thus, we have reached Helmert's point of departure. If the variance is not one, but σ^2, it is clear that we have to set

$$\chi^2 = \frac{\mathbf{x}_1^2 + \cdots + \mathbf{x}_n^2}{\sigma^2} \tag{13}$$

in order to obtain a χ^2 distribution.

§ 24. Limit Theorems

A. The Lévy-Cramér Limit Theorem

The Lévy-Cramér Limit Theorem follows from the inversion formulae for the characteristic function (§ 21 E).

Theorem. *If a sequence of characteristic functions $\varphi_1(t), \varphi_2(t), \ldots$ has for each value of t a limit $\varphi(t)$ which is continuous at $t=0$, then $\varphi(t)$ is the characteristic function of a distribution function $F(u)$ and the sequence of distribution functions $F_1(u), F_2(u), \ldots$ (corresponding to $\varphi_1(t), \varphi_2(t), \ldots$) converges to $F(u)$ for all u at which $F(u)$ is continuous.*

For the proof, we may refer to H. Cramér, Mathematical Methods of Statistics, p. 96.

If $F(u)$ is a distribution function and if $\lim F_n(u)=F(u)$ for all u at which $F(u)$ is continuous, then in the future we shall say more briefly: *the F_n tend to F.*

B. Illustration of the Limit Theorem: Binomial Distribution

The characteristic function of the binomial distribution is

$$\varphi(t)=(p\,e^{it}+q)^n. \tag{1}$$

The expectation of the random variable $\mathbf{x}=\mathbf{x}_1+\cdots+\mathbf{x}_n$ is np, the variance σ^2 is npq. If we introduce

$$\frac{\mathbf{x}-np}{\sigma} \tag{2}$$

as a new random variable, the characteristic function becomes

$$\varphi_n(t)=\exp\left\{-\frac{itnp}{\sigma}\right\}\cdot\left[p\left(\exp\left\{\frac{it}{\sigma}\right\}-1\right)+1\right]^n. \tag{3}$$

Its logarithm is

$$\ln\varphi_n(t)=-\frac{itnp}{\sigma}+n\ln\left[1+p\left(\exp\left\{\frac{it}{\sigma}\right\}-1\right)\right]. \tag{4}$$

For fixed t, (it/σ) tends to zero as $n\to\infty$. The exponential function in the last bracket can be expanded in a power series

$$\exp\left\{\frac{it}{\sigma}\right\}-1=\frac{it}{\sigma}-\frac{1}{2}\left(\frac{t}{\sigma}\right)^2+\cdots.$$

If we multiply by p, the result is small compared with one, for large n. Therefore, the logarithm on the right side of (4) can be expanded in a power series:

$$\ln\left[1+p\left(\exp\frac{it}{\sigma}-1\right)\right]=\frac{itp}{\sigma}-\frac{1}{2}(p-p^2)\left(\frac{t}{\sigma}\right)^2+\cdots.$$

Consequently, (4) leads to

$$\ln\varphi_n(t)=-\frac{npq}{2}\left(\frac{t}{\sigma}\right)^2+\cdots=-\frac{1}{2}t^2+\cdots. \tag{5}$$

Now, if we let n tend to ∞, we have the limit

$$\lim \varphi_n(t) = \exp(-\tfrac{1}{2}t^2). \tag{6}$$

The right side is the characteristic function of a normal distribution. Hence, it follows by the Lévy-Cramér Limit Theorem that the distribution function of the random variable (2) converges to the normal distribution function. Or, in other words: *The random variable* $\mathbf{x}$ *is asymptotically normally distributed with expectation* $n\,p$ *and variance* $n\,p\,q$.

We already know this result, but the present derivation requires less computation. Also, the additional terms of order $n^{-\frac{1}{2}}$ are easy to obtain from the characteristic function.

C. The Law of Large Numbers

In § 5 the Law of Large Numbers was stated: With high probability, in n independent trials the relative frequency $\mathbf{h}$ of an event of probability p will be close to p.

We can express the same thing by saying: *The relative frequency* $\mathbf{h}$ *converges to* p *in probability as* $n \to \infty$. Or alternatively: $\mathbf{h}$ *is a consistent estimate of* p *as* $n \to \infty$.

All these assertions amount to the same thing, namely, the probability that $|\mathbf{h}-p| < \varepsilon$ approaches one arbitrarily closely if n is large enough.

The random variable $\mathbf{h}$ was defined to be the quotient $\mathbf{x}/n$, where

$$\mathbf{x} = \mathbf{x}_1 + \cdots + \mathbf{x}_n \tag{7}$$

is the sum of n independent random variables each of which takes the values 1 and 0 with probabilities p and q. However, we can generalize the Law of Large Numbers if we take $\mathbf{x}_1 + \cdots + \mathbf{x}_n$ to be any n independent random variables all having the same distribution $F(u)$. Khinchin showed that all that needs to be assumed about $F(u)$ is that it should have a finite expectation

$$a = \mathscr{E}\mathbf{x}_1 = \int_{-\infty}^{\infty} u\, dF(u). \tag{8}$$

Dugué showed that the somewhat weaker condition that the characteristic function of $\mathbf{x}_1$ has a finite derivative for $t=0$,

$$\varphi'(0) = i\,a, \tag{9}$$

is sufficient. The Law of Large Numbers generalized by Khinchin and Dugué can be stated in the following way.

Theorem. *If* $\mathbf{x}_1, \ldots, \mathbf{x}_n$ *are independent random variables, which all have the same distributions, and if* (9) *is satisfied, the sample mean*

$$\mathbf{m} = \frac{1}{n}(\mathbf{x}_1 + \cdots + \mathbf{x}_n) \tag{10}$$

converges in probability to a *as* $n \to \infty$.

The proof is extremely simple. In some neighborhood of $t=0$, $\varphi(t)$ is close to one; therefore, we can set

$$\varphi(t) = \exp \psi(t) \tag{11}$$

in that region. The characteristic function of the sample mean **m** is

$$\varphi\left(\frac{t}{n}\right)^n = \exp n\,\psi\left(\frac{t}{n}\right) = \exp t\,\frac{n}{t}\,\psi\left(\frac{t}{n}\right). \tag{12}$$

Since $\varphi(t)$ is differentiable at $t=0$, $\psi(t)$ is too; the derivative is

$$\psi'(0) = \frac{\varphi'(0)}{\varphi(0)} = i\,a. \tag{13}$$

As $n \to \infty$, by definition of the derivative $(n/t)\,\psi(t/n)$ tends to $\psi'(0)$ for each fixed value of t; therefore, as $n \to \infty$, equation (12) has the limit

$$\exp\{t\,\psi'(0)\} = e^{iat}. \tag{14}$$

The right side is, however, the characteristic function of a random variable **a** which takes the value a with probability one. Thus, for each t, the characteristic function of m tends to the characteristic function of the constant random variable **a**. The distribution function of **a** is a step function $E(u)$ with a single point of increase at a. For $u<a$, $E(u)=0$; for $u>a$, $E(u)=1$.

Using the limit theorem in *A*, we find that the distribution function of m tends to the function $E(u)$ at all points where $E(u)$ is continuous. Therefore, the distribution function $H(u)$ of m tends to zero if $u<a$ and to one if $u>a$. This is precisely the assertion which was to be proved.

The theorem just proved is called the Weak Law of Large Numbers. A stronger theorem is the "Strong Law of Large Numbers" which, however, hardly plays any role in mathematical statistics. See A. Khinchin, Sur la loi des grands nombres, Comptes Rendus de l'Acad. des Sciences Paris 188 (1929) p. 477.

D. The Central Limit Theorem

A random variable **x** with a distribution function that depends on a parameter n is said to be asymptotically normal if there exists two numbers a and c which may depend on n, such that the distribution of the random variable

$$\frac{\mathbf{x}-a}{c} \tag{15}$$

tends to the standard normal distribution function $\Phi(u)$ as $n \to \infty$. From A, a necessary and sufficient condition for this is that the characteristic function of the random variable (15) tends to the characteristic function of the normal distribution

$$\exp(-\tfrac{1}{2}t^2). \tag{16}$$

In many cases a is the expected value and c the standard deviation of $\mathbf{x}$. However, it may happen that the standard deviation diverges or that both the expectation and the standard deviation do not exist; nevertheless, there may be numbers a and c with the property mentioned.

In B we saw that the number of successes $\mathbf{x}=\mathbf{x}_1+\cdots+\mathbf{x}_n$ in n independent trials with constant probability of success is asymptotically normally distributed. There, each $\mathbf{x}_j$ takes the values 1 and 0 with probabilities p and q, respectively.

The Central Limit Theorem states that subject to certain conditions, every sum of independent random variables

$$\mathbf{x}=\mathbf{x}_1+\cdots+\mathbf{x}_n \tag{17}$$

has an asymptotically normal distribution.

Laplace and Gauss guessed that this theorem was true and gave reasons for their conjectures. The first complete proof is due to Liapounov (1901). Paul Lévy constructed a proof based on the use of characteristic functions. Later, A. Khinchin, P. Lévy, and W. Feller proved the theorem under considerably weaker conditions. For the literature, see P. Lévy, Theorie de l'addition des variables aléatoires, Paris, 1954.

In all events, certain conditions are necessary, first to prevent a single term from providing too great a contribution to the whole sum and secondly to insure that the distribution functions of the $\mathbf{x}_j$ tend to 0 and 1 sufficiently rapidly at $\pm\infty$. For example, if the individual $\mathbf{x}_i$ all have Cauchy distributions (§ 20), the sum $\mathbf{x}$ also has the same distribution and the Central Limit Theorem does not hold.

A very weak sufficient condition was given by Lindeberg (Math. Z. 15, 1922), but Feller's conditions (Math. Z. 40 and 42) are weaker still, since he does not require the variances to be finite.

We do not want to go into these delicate points here, but only wish to deal with the case where all the $\mathbf{x}_i$ have the same distribution function with finite expectation and variance. We shall, therefore, show the following.

Theorem. *If $\mathbf{x}_1, \ldots, \mathbf{x}_n$ are independent and have the same distribution function with expectation μ and variance σ^2, then the sum* (17) *is asymptotically normally distributed with expectation μn and variance $\sigma^2 n$.*

Proof. We can assume $\mu=0$. Let the characteristic function of $\mathbf{x}_1$ be $\varphi(t)$. Then $[\varphi(t)]^n$ is the characteristic function of $\mathbf{x}$. We have to prove that

$$\left[\varphi\left(\frac{t}{\sigma\sqrt{n}}\right)\right]^n \tag{18}$$

tends to $\exp\{-\frac{1}{2}t^2\}$ as $n\to\infty$.

The first and second derivatives of $\varphi(z)$ at $z=0$ are $i\mu=0$ and $i^2\sigma^2=-\sigma^2$. Therefore, we can expand $\varphi(z)$ in a Taylor series

$$\varphi(z)=1-\tfrac{1}{2}\sigma^2 z^2+R$$

where the remainder term R is small compared with z^2. Thus,

$$\varphi\left(\frac{t}{\sigma\sqrt{n}}\right)=1-\frac{t^2}{2n}+R \tag{19}$$

where R is small compared with n^{-1}. The logarithm becomes

$$\ln\varphi\left(\frac{t}{\sigma\sqrt{n}}\right)=\frac{-t^2}{2n}+R' \tag{20}$$

where R' is small compared to n^{-1}. If we multiply by n, we obtain the logarithm of (18). Now, if we let n tend to ∞, we obtain the limit $-\frac{1}{2}t^2$; therefore,

$$\lim\left(\frac{t}{\sigma\sqrt{n}}\right)=\exp\{-\tfrac{1}{2}t^2\} \tag{21}$$

which was to be proved.

E. Example: χ^2 Distribution

All the assumptions of the theorem just proved are satisfied for the case of a sum of squares of normally distributed random variables

$$\chi^2=\mathbf{x}_1^2+\mathbf{x}_2^2+\cdots+\mathbf{x}_n^2, \tag{22}$$

where the $\mathbf{x}_j$ all have expectation zero and variance one. The expectation of $\mathbf{x}_1^2$ is one and the variance two. Therefore, the sum (22) is asymptotically normally distributed with expectation n and standard deviation $\sqrt{2n}$ (see § 23).

It is easy to prove that $\sqrt{2\chi^2}$ is also asymptotically normally distributed. The approximation is even better for the square root than for χ^2 itself (see R. A. Fisher, Statistical Methods for Research Workers § 20). The expected value of $\sqrt{2\chi^2}$ is approximately $\sqrt{2n-1}$ and its variance very nearly one.

F. The Second Limit Theorem

The "Second Limit Theorem" of Fréchet and Shohat[3] is very useful and can be stated as follows.

Theorem. *If a sequence of distribution functions* $\{F_n(t)\}$ *has finite moments* $\{\alpha_k(n)\}$ *and if* $\alpha_k(n)$ *tends to* β_k *for every* k *as* $n\to\infty$, *then the* β_k *are the moments of a distribution function* F. *Moreover, if* F *is uniquely determined by its moments, the* $F_n(t)$ *converge to* $F(t)$ *at every point of continuity of* $F(t)$.

The proof can be found in Fréchet and Shohat's article cited in footnote 1 or in M. G. Kendall, Advanced Theory of Statistics I (1945) 4.24.

The most important case is that where the β_k are the moments of the standard normal distribution function $\Phi(t)$:

$$\begin{aligned} \beta_{2r+1} &= 0 \\ \beta_{2r} &= \frac{(2r)!}{2^r r!}. \end{aligned} \tag{23}$$

The normal distribution function is continuous everywhere and is uniquely determined by its moments. Therefore, *if the* $\alpha_k(n)$ *tend to the moments of the standard normal distribution* (23), *then the* $F_n(t)$ *converge to* $\Phi(t)$.

G. An Elementary Limit Theorem

The limit theorems dealt with up to now all depend on the Fourier integral transform. However, the following theorem (the statement of which comes from Cramér, Mathematical Methods of Statistics 20.6) is quite elementary.

Theorem. *Let* $\mathbf{x}_1, \ldots, \mathbf{x}_n$ *be random variables with distribution functions* $F_1, F_2, \ldots$ *which tend to* $F(u)$. *Further, let* $\mathbf{y}_1, \mathbf{y}_2$ *be random variables which tend to a constant* c *in probability. Then the sums*

$$\mathbf{z}_n = \mathbf{x}_n + \mathbf{y}_n \tag{24}$$

have distribution functions which tend to $F(u-c)$. *If it is assumed that* $c>0$, *corresponding results hold for products* $\mathbf{x}_n\,\mathbf{y}_n$ *and for the quotients* $\mathbf{x}_n/\mathbf{y}_n$.

[3] M. Fréchet and J. Shohat, A Proof of the Generalized Second Limit Theorem, Trans. Amer. Math. Soc. 33 (1931) p. 533.

It is worth noting that nothing need be assumed about the dependence or otherwise of the $\mathbf{x}_n$ and $\mathbf{y}_n$.

We carry out the proof for the sums. The proofs for products and quotients are similar.

Let u be a point of continuity of $F(u-c)$. For each $\varepsilon>0$, there exists a $\delta>0$ such that $F(u-c-\delta)$ and $F(u-c+\delta)$ differ from $F(u-c)$ by at most ε. An earlier remark (§ 21 E, footnote) showed that the points of continuity of F are countable. Therefore, we can choose δ such that, in addition, F is continuous at the points $u-c+\delta$ and $u-c-\delta$.

Let the probability of the event $[z_n<u]$ be $G_n(u)$. We have to show that $G_n(u)$ tends to $F(u-c)$ as $n\to\infty$. If $\mathbf{x}_n<u-c-\delta$ and $\mathbf{y}_n\leqq c+\delta$, then $\mathbf{z}_n<u$. This implies: if $\mathbf{x}_n<u-c-\delta$, then $z_n<u$ or $y_n>c+\delta$. Therefore,

$$\mathscr{P}[\mathbf{x}_n<u-c-\delta]\leqq\mathscr{P}[\mathbf{z}_n<u]+\mathscr{P}[\mathbf{y}_n>c+\delta]$$

or

$$F_n(u-c-\delta)\leqq G_n(u)+\mathscr{P}[\mathbf{y}_n>c+\delta]. \tag{25}$$

Since $\mathbf{y}_n$ tends to c in probability, the probability of the event $[\mathbf{y}_n>c+\delta]$ is smaller than ε for sufficiently large n. Therefore, (25) leads to

$$F_n(u-c-\delta)<G_n(u)+\varepsilon.$$

Since F_n tends to F, provided n is sufficiently large, it follows that

$$F(u-c-\delta)<G_n(u)+2\varepsilon$$

and further that

$$F(u-c)<G_n(u)+3\varepsilon. \tag{26}$$

Just so, one proves, interchanging the roles of $\mathbf{x}_n$ and $\mathbf{z}_n$,

$$G_n(u)<F(u-c)+3\varepsilon. \tag{27}$$

It follows from (26) and (27) that $\lim G_n(u)=F(u-c)$. With this the theorem is proved.

§ 25. Rectangular Distribution. Rounding Errors

A random variable $\mathbf{x}$ is said to be *uniformly distributed* on the interval (a, b), if the probability density is constant within the interval and zero outside:

$$f(x)=\begin{cases}\dfrac{1}{b-a} & \text{for } a<x<b,\\[2ex] 0 & \text{for } x<a \text{ or } x>b.\end{cases} \tag{1}$$

Such a distribution is often called *rectangular* because of the form of the function $f(x)$ (see Fig. 15). It is immaterial how the values $f(a)$ and $f(b)$ are defined or indeed whether they are defined at all. The distribution function is

$$F(x)=\begin{cases} 0 & \text{for } x \leqq a, \\ \dfrac{x-a}{b-a} & \text{for } a<x<b, \\ 1 & \text{for } x \geqq b. \end{cases} \tag{2}$$

The expectation of $\mathbf{x}$ is $(a+b)/2$, the variance $(b-a)^2/12$. In order to fix our ideas, let us take $a=-\frac{1}{2}$ and $b=\frac{1}{2}$. Let the random variable $\mathbf{x}_1$ be uniformly distributed on the interval $(-\frac{1}{2}, +\frac{1}{2})$ with expectation zero and variance $\frac{1}{12}$.

Fig. 15. Uniform distribution

Such a distribution arises if the result of a numerical calculation is rounded to the nearest whole number. The precise result of the calculation may depend on chance and vary between wide limits. If its density function does not vary greatly in an interval of length one, the rounding error will be approximately uniformly distributed on the interval $(-\frac{1}{2}, \frac{1}{2})$.

The characteristic function of the rectangular distribution is

$$\varphi(t)=\int_{-\frac{1}{2}}^{\frac{1}{2}} e^{itx}\,dx=\frac{2}{t}\sin\frac{1}{2}t. \tag{3}$$

We now consider the distribution of the sum of n independent random variables

$$\mathbf{x}=\mathbf{x}_1+\cdots+\mathbf{x}_n \tag{4}$$

which are all uniformly distributed on $(-\frac{1}{2}, +\frac{1}{2})$. The expected value of the sum is zero and the standard deviation $(n/12)^{\frac{1}{2}}$. If we standardize $\mathbf{x}$ so that it has unit standard deviation and let n increase, the characteristic function

$$\varphi_n(t)=\left[\varphi\left(t\sqrt{\frac{12}{n}}\right)\right]^n \tag{5}$$

approaches the Gauss function $\exp\{-\frac{1}{2}t^2\}$ very rapidly. Therefore, it is to be expected that the distribution function of $\mathbf{x}$ also approaches a normal distribution function very quickly.

Computation confirms this expectation. If $f_n(x)$ is the corresponding density function, the following recursion formulae hold

$$f_1(x)=\begin{cases}1 & \text{for } -\frac{1}{2}<x<\frac{1}{2}\\ 0 & \text{for } x<-\frac{1}{2} \text{ or } x>\frac{1}{2},\end{cases} \tag{6}$$

$$f_n(x)=\int_{x-\frac{1}{2}}^{x+\frac{1}{2}} f_{n-1}(u)\,du, \tag{7}$$

One finds

$$f_2(x)=\begin{cases}x+1 & \text{for } -1\leqq x\leqq 0\\ (x+1)-2x & \text{for } \quad 0\leqq x\leqq 1,\end{cases} \tag{8}$$

and

$$f_3(x)=\begin{cases}\frac{1}{2}(x+\frac{3}{2})^2 & \text{for } -\frac{3}{2}\leqq x\leqq -\frac{1}{2}\\ \frac{1}{2}[x+\frac{3}{2})^2-3(x+\frac{1}{2})^2] & \text{for } -\frac{1}{2}\leqq x\leqq \frac{1}{2}\\ \frac{1}{2}[x+\frac{3}{2})^2-3(x+\frac{1}{2})^2+3(x-\frac{1}{2})^2] & \text{for } \quad \frac{1}{2}\leqq x\leqq \frac{3}{2}.\end{cases} \tag{9}$$

In general,

$$\begin{aligned} f_n(x)=\frac{1}{(n-1)!}\bigg[\Big(x+\frac{n}{2}\Big)^{n-1} &-\binom{n}{1}\Big(x+\frac{n}{2}-1\Big)^{n-1} \\ &+\binom{n}{2}\Big(x+\frac{n}{2}-2\Big)^{n-1}+\cdots\bigg], \end{aligned} \tag{10}$$

where the sum continues as long as the arguments $x+(n/2)$, $x+(n/2)-1$, ... are still positive.

Shapes of the curves are shown in Figs. 16 to 18. The density function $f_2(x)$ is known as a "triangular distribution". The nonzero part of $f_2(x)$ consists of a straight line from $(-1, 0)$ to the maximum at $(0, 1)$ and another line from the maximum to $(1, 0)$. The $f_3(x)$ curve is made up of parts of three parabolas and already looks very much like a Gaussian curve. The $f_4(x)$ curve can hardly be distinguished from a Gaussian curve.

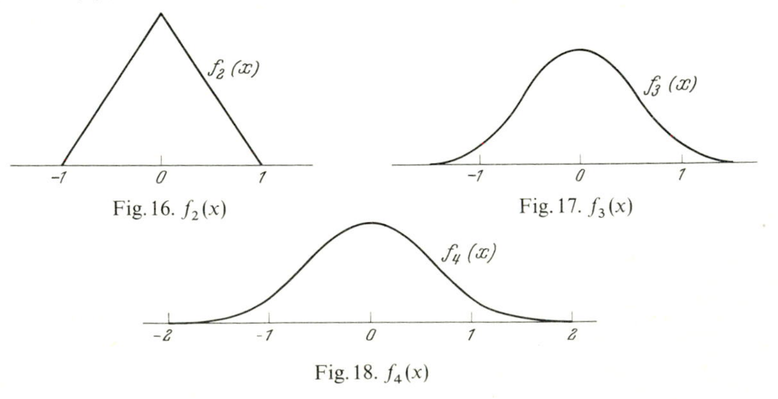

Fig. 16. $f_2(x)$

Fig. 17. $f_3(x)$

Fig. 18. $f_4(x)$

Example 14. The "fundamental argument" z in Hill's[4] Saturn tables was found as the sum of 24 terms, each of which was obtained by interpolation from four place tables. We wish to examine the behavior of the accumulated error in z.

First of all, let us make the tabulated values into integers by multiplying by 10^4. Now, if we interpolate linearly between two values y_n and y_{n+1} in the table by means of the formula

$$y = y_n x(y_{n+1} - y_n) = y_n(1-x) + y_{n+1} x, \qquad 0 \leqq x < 1 \tag{11}$$

and if y_n and y_{n+1} have rounding errors u and v, the error in y becomes

$$w = u(1-x) + v x. \tag{12}$$

If we assume that u, v, and x are independent random variables, that $\mathbf{u}$ and $\mathbf{v}$ are uniformly distributed on $(-\frac{1}{2}, +\frac{1}{2})$, and that $\mathbf{x}$ is uniformly distributed on $(0, 1)$, then (12) has a distribution with expectation zero and variance

$$\sigma^2 = \mathscr{E}\mathbf{w}^2 = \int_0^1 dx \int_{-\frac{1}{2}}^{\frac{1}{2}} du \int_{-\frac{1}{2}}^{\frac{1}{2}} dv [u(1-x) + v x]^2. \tag{13}$$

If we integrate first with respect to u and v and then with respect to x, we obtain

$$\sigma^2 = \int_0^1 [\tfrac{1}{12}(1-x)^2 + \tfrac{1}{12} x^2]\, dx = \tfrac{1}{18}. \tag{14}$$

For double entry tables and linear interpolation we find the variance to be still smaller. However, since in Hill's tables only 3 of the 24 summands are taken from double entry tables, it is not worthwhile working it out accurately.

In some tables, the tabulated values y_n are constant over long stretches. In that case u and v are no longer independent and result (12) loses its validity. The rounding error in the interpolated result becomes approximately equal to the rounding error in a table value. Hence, its variance becomes

$$\sigma^2 \sim \tfrac{1}{12}. \tag{15}$$

In most tables, linear interpolation is not sufficient and we must apply quadratic interpolation using, for example, the formula

$$y = y_n + (y_{n+1} - y_n)\, x + \frac{y_{n+1} - 2 y_n + y_{n-1}}{2} x(x-1)$$

for $-\frac{1}{2} < x \leqq \frac{1}{2}$. Similar calculations to those above, then give a somewhat greater value for the variance, namely,

$$\sigma^2 = \tfrac{47}{384} = 0.12\ldots. \tag{16}$$

If we use (16) in fifteen of the cases, and (15) in the other nine cases, we shall keep on the safe side. By the Central Limit Theorem, the sum of the 24 values has an approximately normal distribution with variance 2.58. We still have to take account of the rounding error in the sum whose variance has the value $\frac{1}{12}$. Altogether we obtain a practically normal distribution with variance

$$\sigma^2 = 2.67.$$

Therefore, the standard deviation of the final result is about 1.6×10^{-4}. Thus, a total error of more than 4 units in the fourth decimal place will occur only very rarely.

If we were to add the theoretical maximum errors of the individual summands, they would amount to 14 units in the fourth decimal place. The theoretical maximal error calculated in this way is proportional to m, the number of terms, whereas the more realistic maximal error calculated above is proportional to $\sqrt{m}$.

---

[4] Astron. Papers Amer. Ephemeris VIII (1898) p. 145–282.

Chapter Six

Gauss Theory of Errors and Student's Test

Chapters 1 to 4, the Central Limit Theorem, and the definition of the χ^2 distribution from Chapter 5 are assumed known for this chapter.

§ 26. Gauss Theory of Errors

A. Equally Accurate Observations

Repeated measurements of a physical quantity do not always yield the same value, even if the quantity itself remains unchanged. Instead, the observed values spread around an expected value $\hat{x} = \mathscr{E}(x)$. Gauss called the difference $(x - \hat{x})$ the *random error* (zufälliger Fehler) in an observation.

The mean value $\hat{x}$ need not be equal to the true value of the quantity measured, for it can have a *systematic error* resulting from the method of measurement. In some cases systematic error may be reduced by improvements in the measuring apparatus or by use of correction terms. The statistical theory of errors is not concerned with these matters, but only with random errors $(x - \hat{x})$.

In the theory of errors, we always assume that the random variable x has a finite expectation $\hat{x}$ and a finite standard deviation σ. Frequently, further assumptions are made about the distribution of the error; first of all, however, we should like to see how far we can go without assuming a particular law for the error.

Gauss calls the standard deviation σ the *mean error* (mittlerer Fehler) of an observation. We shall call it *standard error*. It can be reduced by taking the mean of several observations. For the sample mean M of n independent observations

$$M = \bar{x} = \frac{1}{n} \sum x_i \tag{1}$$

has (§ 18) the same value as that of a single observation, but a standard error which is smaller by a factor of $\sqrt{n}$;

$$\sigma_M = \frac{\sigma}{\sqrt{n}}. \tag{2}$$

In order to make some inference about the precision of M, we need an approximate value for σ^2. As was already explained in § 18, we can take the quantity

$$s^2 = \frac{1}{n-1} (x_i - M)^2 \tag{3}$$

as an estimate.

For the variance of the sample mean M,

$$\sigma_M^2 = \frac{\sigma^2}{n}, \tag{4}$$

we have the corresponding approximation

$$s_M^2 = \frac{s^2}{n}. \tag{5}$$

If the number of observations (sample size) is large, s_M is a good approximation for the standard error σ_M. The values of the sample mean M and its sample standard error s_M are often combined in one expression $M \pm s_M$.

Example 15. Repeated determination of the latitude of Cape Town between 1892 and 1894 gave the fifteen values in the table below (taken from Czuber, Wahrscheinlichkeitsrechnung, Example LVI).

First of all, discarding the degrees and minutes, we find

$$M = \frac{48.92''}{15} = 3.261''$$

as mean, rounding off to $a = 3.26''$.

No.	x	$x-a$	$(x-a)^2$
1	33° 56′ 3.48″	+22	484
2	3.50″	+24	576
3	3.50″	+24	576
4	3.32″	+ 6	36
5	3.09″	−17	289
6	2.98″	−28	784
7	3.07″	−19	361
8	3.28″	+ 2	4
9	3.27″	+ 1	1
10	3.20″	− 6	36
11	3.30″	+ 4	16
12	3.25″	− 1	1
13	3.11″	−15	225
14	3.30″	+ 4	16
15	3.27″	+ 1	1
	48.92″	+ 2	3,406

In order to work out s^2, we take 0.01″ as a new unit (§18,2). The correction term $-n(M-a)^2$ is so small in the present instance that it may be neglected. Hence, we find

$$s^2 = \frac{3{,}406}{14} = 243, \qquad s = 16$$

and

$$s_M^2 = \frac{243}{15} = 16, \qquad s_M = 4.$$

The result can be reported as

$$\varphi = -33° \, 56' \, 3.26'' \pm 0.04''.$$

B. Observations with Different Accuracy

If the individual observations $x_1, \ldots, x_n$ have differing precision, it is better to use a weighted average

$$M = g_1 x_1 + \cdots + g_n x_n \tag{6}$$

where the sum of the weights is equal to one,

$$g_1 + \cdots + g_n = 1, \tag{7}$$

in place of (1). The expectation and variance of M are given by

$$\mathscr{E} M = g_1 \hat{x}_1 + \cdots + g_n \hat{x}_n, \tag{8}$$

and

$$\sigma_M^2 = g_1^2 \sigma_1^2 + \cdots + g_n^2 \sigma_n^2. \tag{9}$$

Henceforth, we shall always assume that all the x_i have the same expectation $\hat{x}$ so that (8) simplifies to $\mathscr{E} M = \hat{x}$.

We now ask: how should we choose the weights g_i in order to minimize the standard error σ_M?

It follows from (7) and (9) that

$$\sigma_M^2 = g_1^2 \sigma_1^2 + \cdots + g_{n-1}^2 \sigma_{n-1}^2 + (1 - g_1 - \cdots - g_{n-1})^2 \sigma_n^2.$$

It is well known that such a positive definite quadratic polynomial in $g_1, \ldots, g_{n-1}$ attains its minimum when all the partial derivatives with respect to $g_1, \ldots, g_{n-1}$ are equal to zero. Differentiation with respect to g_1 gives

$$2 g_1 \sigma_1^2 - 2(1 - g_1 - \cdots - g_{n-1}) \sigma_n^2 = 0$$

or

$$g_1 \sigma_1^2 = g_n \sigma_n^2.$$

Similarly, by differentiation with respect to g_2, we find that

$$g_2 \sigma_2^2 = g_n \sigma_n^2,$$

and so forth. *That is to say, the weights $g_1, \ldots, g_n$ should be inversely proportional to the variances of the corresponding observations.*

For calculations, it is convenient to drop the side condition (7) and replace the g by other proportional weights. Then, we have to take

$$M = \frac{g_1 x_1 + \cdots + g_n x_n}{g_1 + \cdots + g_n} = \frac{\sum g x}{\sum g}. \tag{10}$$

instead of (6). The g_i must be chosen inversely proportional to the σ_i^2,

$$g_1 : g_2 : \cdots : g_n = \sigma_1^{-2} : \sigma_2^{-2} : \cdots : \sigma_n^{-2}. \tag{11}$$

We now set

$$g_i \sigma_i^2 = \sigma^2 \tag{12}$$

and call σ the "standard error of an observation of weight one". In place of (9), we now obtain

$$\sigma_M^2 = \left(\sum g_i\right)^{-2} \left(\sum g_i^2 \sigma_i^2\right),$$

or, using (12),

$$\sigma_M^2 = \frac{\sigma^2}{\sum g}. \tag{13}$$

We can work out an approximate value for σ^2 from the observations, namely,

$$s^2 = \frac{1}{n-1} \sum g_i (x_i - M)^2. \tag{14}$$

Once again the justification of formula (14) lies in the fact that the expected value of s^2 is σ^2. The proof is quite similar to the proof of formula (10) § 18 given earlier. First of all, to simplify calculations, we shift the origin of the x-axis so that $\hat{x}$ becomes zero (in the new coordinates) and then we have

$$\begin{aligned} \mathscr{E} \sum g_i (x_i - M)^2 &= \mathscr{E}\left(\sum g_i x_i^2 - 2 \sum g_i x_i M + \sum g_i M^2\right) \\ &= \sum g_i \mathscr{E} x_i^2 - \mathscr{E}(M^2) \sum g_i \\ &= \sum g_i \sigma_i^2 - \sigma_M^2 \sum g_i \\ &= n \sigma^2 - \sigma^2 = (n-1) \sigma^2; \end{aligned} \tag{15}$$

therefore, by (14)

$$\mathscr{E} s^2 = \sigma^2. \tag{16}$$

Using (13), we now have the approximate value

$$s_M^2 = \frac{s^2}{\sum g} \tag{17}$$

for σ_M^2, the variance of the weighted average M.

Formula (17) can only be applied if we know that the weights are correct, that is, inversely proportional to the σ_i^2. If the weights are only estimated, care must be taken.

For the purpose of calculating s^2, once again we may replace M by some neighboring value a, but then we must subtract $(\sum g)(M-a)^2$:

$$s^2 = \frac{1}{n-1}\left[\sum g(x-a)^2 - \left(\sum g\right)(M-a)^2\right]. \tag{18}$$

Example 16. In order to determine the period of oscillation of a pendulum, twenty times of passage through the equilibrium, in the same direction, were recorded. Let the times measured be $t_1, \ldots, t_{20}$. These values yield ten independent estimates of the period T:

$$T_1 = \tfrac{1}{19}(t_{20} - t_1),$$
$$T_2 = \tfrac{1}{17}(t_{19} - t_2),$$
$$\cdots\cdots$$
$$T_{10} = \quad (t_{11} - t_{10}).$$

If the differences $(t_i - t_k)$ all have the same variance σ^2, then $T_1, \ldots, T_{10}$ have variances

$$\sigma/19, \sigma/17, \ldots, \sigma/1.$$

Accordingly, the weights can be chosen to be

$$g_1 = 19^2,\ g_2 = 17^2, \ldots, g_{10} = 1^2.$$

The weighted average of $T_1, \ldots, T_{10}$ is thus

$$M = \frac{19^2\,T_1 + 17^2\,T_2 + \cdots + 1^2\,T_{10}}{19^2 + 17^2 + \cdots + 1^2}.$$

The standard error σ of M is given by (13);

$$\sigma_M^2 = \frac{\sigma^2}{19^2 + 17^2 + \cdots + 1^2}.$$

To estimate σ^2, we can use formula (14),

$$s^2 = \tfrac{1}{9}\left[19^2(T_1 - M)^2 + 17^2(T_2 - M)^2 + \cdots + 1^2(T_{10} - M)^2\right].$$

However, this estimate is not very accurate since it only uses the ten differences $t_{20} - t_1$, $t_{19} - t_2, \ldots, t_{11} - t_{10}$. If we fit a regression line as well as possible to the set of all twenty observations, that is to say, minimize the sum of squared deviations from the line, then we can obtain an improved estimate for the variance of a single observation. Such an estimate will lead immediately to one for the variance of the differences $(t_i - t_k)$. We shall return to this in Sections 32 and 33.

C. The Distribution Function of the Sample Mean M

Up until now, the theory has been independent of all special assumptions about the distribution of the random variable x. At this point Gauss makes the assumption that the random variables in question are *normally distributed*. As an argument for this assumption. Gauss put forward the *hypothesis of elementary errors* which states that the total error in the observations is the sum of many small independent errors, each of which has only a small variance and which arise from different sources. Thus, the Central Limit Theorem (§ 24 D) can be applied, which says: *The distribution of the sum of a large number of independent random variables, each of which has a small variance compared with the variance of the sum, is approximately normal.*

If we assume, as Gauss did, that the individual x_i have normal distributions, then so have their sum and their mean M as well. Hence, the following rule is true.

With 95% probability, the absolute value of the difference $M-\hat{x}$ is smaller than $1.96\,\sigma_M$, and with 99% probability, it is smaller than $2.58\,\sigma_M$, where σ_M is the standard error of M.

This rule holds for large n, even if the variables $x_1, \ldots, x_n$ are not normally distributed. The sample mean M is the sum of a large number of terms, each of which has a variance small in proportion to that of M. Hence, by the Central Limit Theorem, the distribution of M is well approximated by a normal distribution. If the x_i have approximately normal distributions to begin with, so much the better.

Application of the above rule requires knowledge of σ_M. Can we safely replace $\sigma_M = \sigma/\sqrt{n}$ by the empirical approximation $s_M = s/\sqrt{n}$? To answer this question, we must first investigate how much s^2 and σ^2 can differ, or put another way, we must determine the distribution of s^2.

§ 27. The Distribution of s^2

A. Introduction of the χ^2 Variable

Let $x_1, \ldots, x_n$ be independent normally distributed random variables with the same expectation $\hat{x}=a$ and the same[1] standard deviation σ. We consider

$$M = \frac{1}{n} \sum x_i \tag{1}$$

and

$$s^2 = \frac{1}{n-1} \sum (x_i - M)^2 \tag{2}$$

and wish to discover the distribution of s^2. Consider

$$\chi^2 = \frac{(n-1)\,s^2}{\sigma^2} = \frac{\sum (x_i - M)^2}{\sigma^2}$$

instead of s^2. This variable is invariant under scale changes of the x-axis. Taking the variable $(x-a)/\sigma$ instead of x, we can assume that $\hat{x}$ is zero

[1] The case where the $x_1, \ldots, x_n$ have different standard deviations $\sigma_1, \ldots, \sigma_n$ can be reduced to the one dealt with here by introducing the new variable

$$x_i' = \frac{x_i - a}{\sigma_i}.$$

and σ is one. Then,

$$\chi^2 = \sum (x_i - M)^2 = \sum x_i^2 - nM^2 = \sum x_i^2 - \frac{1}{n} \left(\sum x_i\right)^2 \tag{3}$$

and the probability density of each x_i becomes

$$f(t) = \frac{1}{\sqrt{2\pi}} e^{-\frac{1}{2}t^2}.$$

Since the $x_1, \dots, x_n$ are independent, their joint density is given by the product

$$f(t_1, \dots, t_n) = f(t_1) f(t_2) \cdots f(t_n) = (2\pi)^{-\frac{n}{2}} e^{-\frac{1}{2}(t_1^2 + \cdots + t_n^2)}.$$

Now, according to Theorem II, § 4, the required distribution function $\mathscr{P}[\chi^2 < u]$ is equal to the n-fold integral[2]

$$\begin{aligned} G(u) &= \int_{\chi^2 < u} \cdots \int f(x_1, \dots, x_n)\, dx_1 \cdots dx_n \\ &= (2\pi)^{-\frac{n}{2}} \int_{\chi^2 < u} \cdots \int e^{-\frac{1}{2}(x_1^2 + \cdots + x_n^2)}\, dx_1 \cdots dx_n. \end{aligned} \tag{4}$$

B. Evaluation of the Integral

The region of integration is the inside of the quadric $\chi^2 = u$, that is,

$$\sum x_i^2 - \frac{1}{n} \left(\sum x_i\right)^2 = u.$$

For the case $n = 3$, the surface is a cylinder as will easily be seen. By using an orthonormal transformation of coordinates, we wish to make the axis of the cylinder into the first axis of the coordinates. To this end, we generally take the first row of the orthogonal transformation to be

$$y_1 = \frac{1}{\sqrt{n}} x_1 + \frac{1}{\sqrt{n}} x_2 + \cdots + \frac{1}{\sqrt{n}} x_n = M\sqrt{n}.$$

By § 13, since the sum of squares of the coefficients of this first row is equal to one, it is possible to find $n-1$ further rows

$$\begin{aligned} y_2 &= a_{21} x_1 + a_{22} x_2 + \cdots + a_{2n} x_n \\ &\cdots\cdots\cdots\cdots\cdots \\ y_n &= a_{n1} x_1 + a_{n2} x_2 + \cdots + a_{nn} x_n, \end{aligned}$$

---

[2] It is not correct logically to denote the variables of integration in the same way as the random variables $x_1, \dots, x_n$, but it is convenient. Functions of the x_i, *e.g.*, M, χ^2, and $y_1, \dots, y_n$ introduced later, will also be found with a double meaning as random variables and as functions of the variables of integration.

in such a way that the complete transformation is orthonormal. Since $\sum x_i^2 = \sum y_i^2$, if we now write χ^2 in terms of the new variables, we obtain,

$$\chi^2 = \sum x_i^2 - \frac{1}{n}(\sum x_i)^2 = \sum y_i^2 - y_1^2 = y_2^2 + \cdots + y_n^2. \tag{5}$$

Thus, the surface $\chi^2 = u$ is indeed a cylinder. The absolute value of the Jacobian of the transformation is one; the transformed integral is therefore

$$G(u) = (2\pi)^{-\frac{n}{2}} \int_{\chi^2 < u} \cdots \int e^{-\frac{1}{2}(y_1^2 + \cdots + y_n^2)} \, dy_1 \cdots dy_n.$$

Since y_1 does not appear in the determination of the limits of integration, we can carry out the integration with respect to y_1,

$$\begin{aligned} G(u) &= (2\pi)^{-\frac{n}{2}} \int_{-\infty}^{\infty} e^{-\frac{1}{2}y_1^2} \, dy_1 \int_{\chi^2 < u} \cdots \int e^{-\frac{1}{2}(y_2^2 + \cdots + y_n^2)} \, dy_2 \cdots dy_n \\ &= (2\pi)^{-\lambda} \int_{\chi^2 < u} \cdots \int e^{-\frac{1}{2}(y_2^2 + \cdots + y_n^2)} \, dy_2 \cdots dy_n \end{aligned} \tag{6}$$

where, for shortness, we set $\lambda = (n-1)/2$.

If we think of $y_1, \ldots, y_n$ as random variables, we can obtain (6) even more simply. Since $\sum x^2 = \sum y^2$, the joint probability density of the y is the same as that of the x

$$(2\pi)^{-\frac{n}{2}} e^{-\frac{1}{2}\Sigma x^2} = (2\pi)^{-\frac{n}{2}} e^{-\frac{1}{2}\Sigma y^2}.$$

Therefore, the variables $y_1, \ldots, y_n$ are normally distributed with expectation zero and variance one. The random variable $\chi^2 = y_2^2 + \cdots + y_n^2$ does not depend on y_1 at all. Therefore, its distribution can be found by integration with respect to $y_2, \ldots, y_n$ alone. The result is (6). We have already shown in § 23 that a sum of squares such as $y_2^2 + \cdots + y_n^2$ has a χ^2 distribution with $n-1$ degrees of freedom;

$$\begin{aligned} G(u) &= \alpha \int_0^u v^{\lambda - 1} e^{-\frac{1}{2}v} \, dv, \\ \alpha &= \Gamma(\lambda)^{-1} 2^{-\lambda}, \\ \lambda &= \frac{f}{2} = \frac{n-1}{2}. \end{aligned} \tag{7}$$

This result was obtained very simply with the aid of the characteristic function. However, we can also evaluate (6) independently, by transforming to polar coordinates in the manner of § 11. The first polar coordinate usually denoted by r is χ in our case, for $\chi^2 = y_2^2 + \cdots + y_n^2$. Thus, we obtain

$$G(u) = (2\pi)^{-\lambda} \int_{\chi^2 < u} \cdots \int e^{-\frac{1}{2}\chi^2} \chi^{n-2} \, d\chi \, d\Omega.$$

Since the angular coordinates do not figure in the condition $\chi^2 < u$, integration can be carried out with respect to $d\Omega$;

$$G(u) = (2\pi)^{-\lambda} \int d\Omega \int_0^{\sqrt{u}} e^{-\frac{1}{2}\chi^2} \chi^{n-2} \, d\chi.$$

Now because $\lambda = (n-1)/2$,

$$\int d\Omega = \frac{2}{\Gamma(\lambda)} \pi^{\lambda}$$

by § 12, therefore,

$$2^{\lambda} \Gamma(\lambda) \, G(u) = 2 \int_0^{\sqrt{u}} e^{-\frac{1}{2}\chi^2} \chi^{n-2} \, d\chi.$$

If we introduce $\chi^2 = v$ as a new variable of integration, we just obtain formula (7). The integral on the right side of (7) is an *incomplete gamma function.* The corresponding density function is

$$g(u) = \alpha \, u^{\lambda - 1} e^{-\frac{1}{2}u} \qquad \text{for } u > 0. \tag{8}$$

C. Independence of M and χ^2

With the same methods we can also determine probability that $[\chi^2 < u]$ and that M lies between two given limits b and c, at the same time:

$$\mathscr{P} = \int_{\substack{\chi^2 < u \\ b \leqq M < c}} \cdots \int f(x_1, \ldots, x_n) \, dx_1 \cdots dx_n.$$

That is to say, if we use the same orthogonal transformation as the one used above, we obtain the limits $b\sqrt{n}$ and $c\sqrt{n}$ in the integral with respect to y_1 and the region of integration $\chi^2 < u$ which does not depend on y_1, in the integral with respect to $y_2, \ldots, y_n$;

$$\begin{aligned} \mathscr{P} &= (2\pi)^{-\frac{1}{2}} \int_{b\sqrt{n}}^{c\sqrt{n}} e^{-\frac{1}{2}y_1^2} \, dy_1 \cdot (2\pi)^{-\lambda} \int_{\chi^2 < u} \cdots \int e^{-\frac{1}{2}\chi^2} \, dy_2 \cdots dy_n \\ &= \mathscr{P}(b \leqq M < c) \cdot \mathscr{P}(\chi^2 < u). \end{aligned}$$

Therefore, the probability that $[b \leqq M < c$ and $0 \leqq \chi^2 < u]$ is equal to the product of the probabilities that $[b \leqq M < c]$ and that $[0 \leqq \chi^2 < u]$. That is to say, *the random variables M and χ^2 are independent.*

The joint density of the pair (M_1, χ^2) is therefore the product of the densities of M and χ^2. That of M is normal with standard deviation σ_M and that of χ^2 is given by (8).

D. Expectation and Variance of χ^2

The expectation and variance of $Q=\chi^2$ were given earlier in § 23, but they can easily be obtained directly from (8);

$$\begin{aligned}
\mathscr{E}Q &= \alpha \int_0^\infty u \cdot u^{\lambda-1} e^{-\frac{1}{2}u}\, du \\
&= 2^{-\lambda}\Gamma(\lambda)^{-1} 2^{\lambda+1}\Gamma(\lambda+1) \\
&= 2\lambda = f = n-1, \\
\mathscr{E}Q^2 &= \alpha \int_0^\infty u^2 \cdot u^{\lambda-1} e^{-\frac{1}{2}u}\, du \\
&= 2^{-\lambda}\Gamma(\lambda)^{-1} 2^{\lambda+2}\Gamma(\lambda+2) \\
&= 4(\lambda+1)\lambda = f^2+2f \\
\sigma_Q^2 &= \mathscr{E}Q^2 - (\mathscr{E}Q)^2 = 2f = 2(n-1).
\end{aligned} \tag{9}$$

Now

$$\chi^2 = f\,\frac{s^2}{\sigma^2}. \tag{10}$$

Therefore, the expectation of s^2 is equal to σ^2, which we also discovered earlier, and the standard deviation of s^2 equals

$$\sigma_{(s^2)} = \frac{\sigma^2}{f}\sqrt{2f} = \sigma^2\sqrt{\frac{2}{n-1}}. \tag{11}$$

E. Bounds for s^2

There are tables for the incomplete gamma function from which one can find values of $G(u)$. Using these tables, one can determine K such that the event $[\chi^2 < K]$ has any prescribed probability. We set

$$G(K) = \mathscr{P}[\chi^2 < K] = 1-\alpha$$

choosing, perhaps, $\alpha = 0.01$; thus, we obtain an upper bound for s^2 which is exceeded only rarely. On the other hand, setting

$$G(K') = \mathscr{P}[\chi^2 < K'] = \alpha$$

we obtain a lower bound K' which χ^2 falls below only rarely. Values of the upper bound K are given in Table 6 at the end of the book. Values of the lower bound K' can be found in most collections of statistical tables.

The quantity s^2/σ^2 is related to χ^2 by (10). Therefore, the upper bound K will give an upper bound for s^2/σ^2, that is to say, an upper bound for s^2 given σ^2 or a lower bound for σ^2 given s^2. The probability that s^2/σ^2 exceeds the new bound is α.

F. Additivity of the χ^2 Distribution

In § 23 it was proven that *the sum of two independent χ^2 variables with degrees of freedom f and g has a χ^2 distribution with $f+g$ degrees of freedom.* This can also be easily proved directly without using characteristic functions. The variables χ_1^2 and χ_2^2 have the same distributions as the sums

$$y_1^2+\cdots+y_f^2 \quad \text{and} \quad y_{f+1}^2+\cdots+y_{f+g}^2,$$

where $y_1, \ldots, y_{f+g}$ have independent normal distributions with expectation zero and variance one. Therefore, $\chi_1^2+\chi_2^2$ has the same distribution function as the sum

$$y_1^2+\cdots+y_{f+2}^2,$$

that is, a χ^2 distribution with $f+g$ degrees of freedom.

§ 28. Student's Test

We return now to the question posed at the end of § 26. Can we replace the number σ_M by s_M when applying such rules as

$$\frac{|M-\hat{x}|}{\sigma_M}<1.96 \qquad \text{with } 95\% \text{ probability}$$

and

$$\frac{|M-\hat{x}|}{\sigma_M}<2.58 \qquad \text{with } 99\% \text{ probability?}$$

To answer this question, we have to determine the distribution of the quotient

$$t=\frac{M-\hat{x}}{s_M}. \tag{1}$$

First of all, we simplify the problem. If we make $\hat{x}=0$ by a suitable displacement of the origin of the x-axis, then

$$t=\frac{M}{s_M}. \tag{2}$$

The quotient remains invariant under changes of scale on the x-axis. Therefore, we can assume that $\sigma=1$. If we multiply both numerator and denominator by $\sqrt{n}$, we obtain

$$t=\frac{M\sqrt{n}}{s_M\sqrt{n}}=\frac{y_1}{s}, \tag{3}$$

using the notation of § 27. We now multiply the numerator and denominator in (3) by $\sqrt{f}=\sqrt{n-1}$, and because $f s^2=\chi^2$, we obtain

$$t=\frac{y_1}{\chi}\sqrt{f}. \tag{4}$$

Our problem is to determine the distribution function $H(a)$ of the random variable t. Therefore, we must find the probability of the event

$$\left[\frac{y_1}{\chi}\sqrt{f}<a\right]. \tag{5}$$

If we put

$$c=\frac{a}{\sqrt{f}}, \tag{6}$$

then (5) simplifies to

$$[y_1<c\chi]. \tag{7}$$

In § 27 C, it was shown that y_1 and χ^2 are mutually independent. The joint density of y_1 and χ^2 is therefore the product of their densities

$$f(y)\,g(z)=\frac{1}{\sqrt{2\pi}}\,e^{-\frac{1}{2}y^2}\cdot\beta\,z^{\frac{1}{2}f-1}\,e^{-\frac{1}{2}z}, \tag{8}$$

$$\beta=\Gamma(\tfrac{1}{2}f)^{-1}\,2^{-\frac{1}{2}f}.$$

Therefore, the required probability is

$$\begin{aligned} H(a)&=\beta'\iint\limits_{y<c\sqrt{z}} e^{-\frac{1}{2}y^2}\,z^{\frac{1}{2}f-1}\,e^{-\frac{1}{2}z}\,dy\,dz \qquad \left(\beta'=\frac{\beta}{\sqrt{2\pi}}\right)\\ &=\beta'\int_0^\infty z^{\frac{1}{2}f-1}\,e^{-\frac{1}{2}z}\,dz\int_{-\infty}^{c\sqrt{z}} e^{-\frac{1}{2}y^2}\,dy. \end{aligned} \tag{9}$$

We make the substitution $y=x\sqrt{z}$ in the integral with respect to y in order to obtain constant limits of integration:

$$H(a)=\beta'\int_0^\infty z^{\frac{f-1}{2}}\,e^{-\frac{1}{2}z}\,dz\int_{-\infty}^{c} e^{-\frac{1}{2}x^2 z}\,dx \tag{10}$$

and reverse the order of integration since this is always permissible in the case of a positive integrand;

$$H(a)=\beta' \int_{-\infty}^{c} dx \int_{0}^{\infty} z^{\frac{f-1}{2}} e^{-\frac{1}{2}(1+x^2)z} dz. \tag{11}$$

By (12) §12 the integral with respect to z is a gamma function

$$\int_{0}^{\infty} z^{\frac{f-1}{2}} e^{-\frac{1}{2}(1+x^2)z} dz = \left(\frac{1+x^2}{2}\right)^{-\frac{f+1}{2}} \Gamma\left(\frac{f+1}{2}\right).$$

Hence

$$\begin{aligned} H(a) &= \gamma \int_{-\infty}^{c} (1+x^2)^{-\frac{f+1}{2}} dx \\ &= \frac{\gamma}{\sqrt{f}} \int_{-\infty}^{a} \left(1+\frac{t^2}{f}\right)^{-\frac{f+1}{2}} dt \end{aligned} \tag{12}$$

with

$$\gamma = \beta \pi^{-\frac{1}{2}} 2^{\frac{f}{2}} \Gamma\left(\frac{f+1}{2}\right) = \pi^{-\frac{1}{2}} \Gamma\left(\frac{f+1}{2}\right) \Gamma\left(\frac{f}{2}\right)^{-1}. \tag{13}$$

The problem under discussion is solved by (12). The integral $H(a)$ can be evaluated by standard methods. For the density of t, we obtain

$$h(t) = \frac{\gamma}{\sqrt{f}} \left(1+\frac{t^2}{f}\right)^{-\frac{f+1}{2}}. \tag{14}$$

The curve of (14) is bell-shaped like the Gauss error (normal) curve. It obviously converges to the normal curve

$$f(t) = \frac{1}{\sqrt{2\pi}} e^{-\frac{1}{2}t^2}$$

as $f \to \infty$.

If we have $H(a)$ as a function of a, we can determine the point a for which $H(a)$ assumes any given value $1-\alpha$, where α is equal to 0.025 or 0.005, say. Then, the random variable t is greater than $a=t_\alpha$ with probability α only. Since $-t$ has the same distribution as t, the probability that $-t$ is greater than t_α will also be α. Thus, the probability that the absolute value $|t|$ exceed t_α is 2α.

The formulation of the above problem and its solution by means of formula (12) are due to the English statistician Gosset in his pioneer work[3] written under the modest pseudonym Student. Hence, the distri-

[3] Student, The probable error of a mean, Biometrika 6 (1908) p. 1.

bution of the random variable t is known as *Student's distribution*. The rule which states that an assumed expectation $\hat{x}$ should be rejected whenever the absolute value of t exceeds t_α is called *Student's test*.

The cut off point t_α depends on the level of significance α and on the number of degrees of freedom $f=n-1$. Values of t_α are found in Table 7 at the end of the book.

In the one-sided form of the test we would reject a hypothetically adopted value $\hat{x}$ if t were positive and greater than t_α or, alternatively, if t were negative and less than $-t_\alpha$. For the two-sided test, we only pay attention to the absolute value $|t|$ and not to the sign of t. For the one-sided test the probability of false rejection of an adopted value $\hat{x}$ when it is correct is α, and for the two-sided test using the same absolute cut off it is 2α. In order for this to be true, the x_i must be assumed independent and normally distributed; if they are not, the error probabilities are only approximately α and 2α.

§ 29. Comparison of Two Means

The following problem is of great importance in all experimental sciences. A physical quantity x is measured several times, each time with a new and independent adjustment of the measuring instruments. The measurements have yielded the values $x_1, \ldots, x_g$ with sample mean $\bar{x}$. Under some other conditions measurements of the same quantity have yielded the values $y_1, \ldots, y_h$ with sample mean $\bar{y}$. Now, if it is found that $\bar{x}$ and $\bar{y}$ differ, does this imply that there is a real difference between the expectations $\hat{x}$ and $\hat{y}$ corresponding to the two sets of conditions? Or can the observed difference be purely due to chance? In other words, how large must the difference $D=\bar{x}-\bar{y}$ be in order to be "significant", that is, to indicate a real difference between $\hat{x}$ and $\hat{y}$?

The Gauss theory of errors answers the question in the following way. For large g and h, $\bar{x}$ and $\bar{y}$ have approximately normal distributions with variances

$$\sigma_{\bar{x}}^2=\frac{1}{g}\,\sigma_x^2 \quad \text{and} \quad \sigma_{\bar{y}}^2=\frac{1}{h}\,\sigma_y^2,$$

which can be estimated by

$$s_{\bar{x}}^2=\frac{1}{g}\,s_x^2=\frac{\sum(x-\bar{x})^2}{g(g-1)} \tag{1}$$

and

$$s_{\bar{y}}^2=\frac{1}{h}\,s_y^2=\frac{\sum(y-\bar{y})^2}{h(h-1)}. \tag{2}$$

Hence, the difference $D=\bar{x}-\bar{y}$ also has an approximately normal distribution with variance

$$\sigma_D^2=\sigma_{\bar{x}}^2+\sigma_{\bar{y}}^2$$

which can be estimated by

$$S_D^2=s_{\bar{x}}^2+s_{\bar{y}}^2. \tag{3}$$

Now, if the real difference $\hat{x}-\hat{y}=0$, the quotient $q=D/\sigma_D$ is normally distributed with expectation zero and variance one. Therefore, with probability $1-2\alpha$, $|q|$ is less than a cut off point q_α which can be read from the standard normal distribution; for example, $q_\alpha=2.58$ for $\alpha=0.01$. However, σ_D has to be replaced by s_D, since σ_D is not known. This may give rise to additional uncertainty; therefore the cut off point q_α must be increased somewhat. Thus, we often choose 3 (or if we want to be quite sure, 4) as the cut off point for the quotient D/s_D. The hypothesis $\hat{x}=\hat{y}$ is rejected if $|q|$ is larger than the cut off point chosen, and then we conclude either that $\hat{x}>\hat{y}$ or that $\hat{x}<\hat{y}$ depending on the sign of D.

Such a rule of thumb lacks a fully satisfactory justification. Should we choose 2.58, 3, or 4 as cut off point for D/s_D and what will be the corresponding probability of error? Everything is clear in the case of very large g and h since then σ_D may be replaced s_D without any great loss. For small and moderate g and h, we should like to know much more accurately how large a cut off point to choose for D/s_D in order to achieve, say, a 1% level of significance.

Unfortunately, this question does not have an exact answer in that the probability that D/s_D exceeds any given value depends (albeit in small measure) on the unknown ratio of the variances σ_x^2 and σ_y^2. Therefore, we phrase the question somewhat differently.

The hypothesis which is to be tested is that the observed difference is purely due to chance and that the true expectations $\hat{x}$ and $\hat{y}$ are equal. However, if the difference between the x and the y is solely due to chance, then we can also assume that the variance σ_x^2 and σ_y^2 are equal. Therefore, we start with the hypothesis that the x and y have not only the same expectation $\hat{x}=\hat{y}=\mu$, but also the same variance σ^2 and test whether the observed value of $D=\bar{x}-\bar{y}$ is in agreement with the hypothesis.

If the true variances σ_x^2 and σ_y^2 are equal, it doesn't make sense to calculate two separate estimates s_x^2 and s_y^2; therefore, we shall choose a simple estimate for the common value σ^2, namely, a weighted average of s_x^2 and s_y^2. By § 26 B, the weights to be given to s_x^2 and s_y^2 should be inversely proportional to the variance of s_x^2 and s_y^2. According to (11) § 27, these have the values

$$\frac{2\sigma^4}{g-1} \quad \text{and} \quad \frac{2\sigma^4}{h-1}.$$

The weights are therefore in the ratio $(g-1):(h-1)$. Hence, the weighted mean is

$$s^2 = \frac{(g-1)\,s_x^2 + (h-1)\,s_y^2}{(g-1)+(h-1)} = \frac{\sum (x-M_x)^2 + \sum (y-M_y)^2}{g+h-2}. \tag{4}$$

Using s^2, we now find the estimated variances of $\bar{x}$ and $\bar{y}$

$$S_1^2 = \frac{1}{g}\,s^2, \qquad S_2^2 = \frac{1}{h}\,s^2 \tag{5}$$

and the estimated variance of the difference D

$$S^2 = S_1^2 + S_2^2 = \left(\frac{1}{g} + \frac{1}{h}\right) s^2. \tag{6}$$

The value of S^2 derived in this way usually differs only very slightly from s_D^2 derived from (1), (2), and (3). For the case $g=h$, they are both equal. Therefore, in practice, it is fairly immaterial whether (1), (2), and (3) or (4), (5), and (6) are used to calculate the estimate for the variance of D.

Following Student and Fisher[4], we now take the quotient

$$t = \frac{D}{S} \tag{7}$$

and ask for its distribution under the hypothesis that $x_1, \ldots, x_g$ and $y_1, \ldots, y_h$ have independent normal distributions with expectation μ and variance σ^2.

Since t remains invariant under translations and scale changes, we can assume $\mu=0$ and $\sigma^2=1$. We set

$$\chi_1^2 = (g-1)\,s_x^2 = \sum (x-\bar{x})^2$$

and

$$\chi_2^2 = (h-1)\,s_y^2 = \sum (y-\bar{y})^2.$$

Then,

$$(g+h-2)\,s^2 = \chi_1^2 + \chi_2^2. \tag{8}$$

First we observe that the random variables χ_1^2, χ_2^2, $\bar{x}$, and $\bar{y}$ are independent. For χ_1^2 and $\bar{x}$ are functions of the x only and similarly, χ_2^2 and $\bar{y}$ are functions of the y only. Since the x and the y are independent, the pairs $(\chi_1^2, \bar{x})$ and $(\chi_2^2, \bar{y})$ must also be independent. But we have already seen in § 27 that χ_1^2 and $\bar{x}$ are independent and similarly that χ_2^2 and $\bar{y}$ are independent. Therefore, all four random variables are independent.

---

[4] R.A. Fisher, Applications of "Student's" distribution, Metron 5 (1926) p. 90.

Accordingly, their joint density $f(u_1, u_2, v_1, v_2)$ is the product of the four individual densities which are known from § 27. The density of χ_1^2 is

$$g_1(u) = \beta_1\, u^{\frac{g-3}{2}}\, e^{-\frac{1}{2}u}$$

(a χ^2 distribution with $g-1$ degrees of freedom), similarly that of χ_2^2 is

$$g_2(u) = \beta_2\, u^{\frac{h-3}{2}}\, e^{-\frac{1}{2}u},$$

while $\bar{x}$ and $\bar{y}$ are normally distributed with expectation zero and variance $1/g$ and $1/h$, respectively. It now follows that

$$\chi^2 = (g+h-2)\, s^2 = \chi_1^2 + \chi_2^2 \tag{9}$$

and

$$D = \bar{x} - \bar{y} \tag{10}$$

are independent.

By § 27 F, the density of χ^2 is of the form

$$g(u) = \beta\, u^{\frac{f-2}{2}}\, e^{-\frac{1}{2}u} \tag{11}$$

with $f = g+h-2$ degrees of freedom, while D is normally distributed with expectation zero and variance

$$\sigma_D^2 = \sigma_{\bar{x}}^2 + \sigma_{\bar{y}}^2 = \frac{1}{g} + \frac{1}{h}. \tag{12}$$

Therefore, from (6) and (9)

$$S^2 = \left(\frac{1}{g} + \frac{1}{h}\right) s^2 = \left(\frac{1}{g} + \frac{1}{h}\right) \frac{\chi^2}{g+h-2} = \left(\frac{1}{g} + \frac{1}{h}\right) \frac{\chi^2}{f}. \tag{13}$$

From (12) and (13), it follows that

$$S^2 = \sigma_D^2\, \frac{\chi^2}{f}. \tag{14}$$

The quotient (7) may now be written

$$t = \frac{D}{S} = \frac{D\sqrt{f}}{\sigma_D\, \chi} = Y\, \frac{\sqrt{f}}{\chi} \tag{15}$$

where

$$Y = \frac{D}{\sigma_D}. \tag{16}$$

Formula (15) is very similar to formula (4) § 28 which was

$$t = y_1 \frac{\sqrt{f}}{\chi}. \tag{17}$$

Now just as y_1 and χ_2^2 were independent, Y and χ^2 are also independent random variables, the first of which is normally distributed with zero expectation and unit variance, while the second has a χ^2 distribution with f degrees of freedom. Therefore, the quotient t has the same distribution as before.

Thus, we obtain the following test, known as *Student's test for the equality of two means.*

If the absolute value

$$|t| = \frac{|D|}{S} = \frac{|\bar{x} - \bar{y}|}{S}$$

is larger than the cut-off value t_α from Table 7, then the hypothesis $\hat{x} = \hat{y}$ is rejected and it is assumed that either $\hat{x} > \hat{y}$ or $\hat{x} < \hat{y}$ depending on whether D is positive or negative.

If the distributions are not too different from normal ones, the probability that the hypothesis will be rejected incorrectly is 2α. If, in fact, $\hat{x} > \hat{y}$ and the variances are equal, the probability of deducing erroneously that $\hat{x} < \hat{y}$ is less than α. For a one-sided form of the test, the error probability will be less than or equal to α.

Student's test can also be used to decide between two hypotheses in the following situation. Suppose that it is doubted whether a particular one of a group of observations has the same expectation as the rest. The first hypothesis is then that the difference between the observation and the average of the rest is purely due to chance so that the observation may be used for inference about the unknown common mean value. The alternative hypothesis is that the expectation of the single observation is different and so that observation should be discarded.

We need merely to apply the formulas of the present section with $h = 1$. The single observation we wish to test becomes $y_1 = x_{g+1}$, and $x_1, \ldots, x_g$ are the remainder. The probability that x_{g+1} is wrongly discarded using the test is 2α.

If we apply the same testing procedure afterwards to the sequence $x_1, \ldots, x_g$, the total probability that one of them is less than $2\alpha(g+1) = 2\alpha n$, where n is the total number of observations. If, for example, we choose $2\alpha = 0.01$, the probability for $n = 20$ that one of the observations is wrongly discarded is almost 0.20. However, that does not matter too much, since the mean of the remaining 19 observations is almost as accurate as that of all 20.

Example 17 (from M. G. Kendall, Advanced Theory of Statistics II, Example 21.4).

In a class of twenty children, ten chosen at random received apple juice every day and the other ten had milk. After a certain period of time, the increases in weight were (in pounds):

First Group	4,	$2\frac{1}{2}$,	$3\frac{1}{2}$,	4,	$1\frac{1}{2}$,	1,	$3\frac{1}{2}$,	3,	$2\frac{1}{2}$,	$3\frac{1}{2}$
Second Group	$1\frac{1}{2}$,	$3\frac{1}{2}$,	$2\frac{1}{2}$,	3,	$2\frac{1}{2}$,	2,	2,	$2\frac{1}{2}$,	$1\frac{1}{2}$,	3.

The average weight increase in the first group is 2.9 pounds, in the second 2.4 pounds. Is the difference significant?

We find

$$D = 2.9 - 2.4 = 0.5,$$

$$s^2 = \frac{13.3}{18} = 0.74,$$

$$S^2 = \left(\frac{1}{10} + \frac{1}{10}\right) s^2 = 0.148,$$

$$t = \frac{D}{S} = 1.30.$$

The 5% cut-off for t with $20-2=18$ degrees of freedom is 2.10. Therefore, the observed difference is not significant.

Chapter Seven

Method of Least Squares

§ 30. Smoothing Observational Errors

Gauss found the following problem in astronomy and geodesy.

Suppose the true values $\vartheta_1, \ldots, \vartheta_r$ of some physical constants (*e.g.*, orbital elements of a planet) are unknown. The quantities $\vartheta_1, \ldots, \vartheta_r$ are not observed directly, but rather other quantities $x_1, \ldots, x_n$ (*e.g.*, the coordinates of the planet as seen from the earth at different times), whose true values $\xi_1, \ldots, \xi_n$ depend upon the $\vartheta_1, \ldots, \vartheta_r$ in a certain way:

$$\xi_i = \varphi_i(\vartheta_1, \ldots, \vartheta_r). \tag{1}$$

Which values of the parameters ϑ_i agree best with the observations $x_1, \ldots, x_n$?

Earlier Lagrange had proposed that "best agreement" should be defined by the minimization of the sum of the squared errors

$$Q = (x_1 - \xi_1)^2 + \cdots + (x_n - \xi_n)^2. \tag{2}$$

Gauss gave this notion a theoretical probability basis with the following remark: according to the Gaussian Law of Errors, the probability that the observed values x_i lie between $t_i - \frac{1}{2}\delta t_i$ and $t_i + \frac{1}{2}\delta t_i$ is approximately

$$\delta W = \sigma^{-n}(2\pi)^{-\frac{n}{2}} \exp\left\{-\frac{1}{2}\,\frac{(t_1-\xi_1)^2 + \cdots + (t_n - \xi_n)^2}{\sigma^2}\right\} \delta t_1 \cdots \delta t_n \tag{3}$$

for small δt_i, provided no systematic errors are present and all observations have the same standard deviation σ. For given t_i and δt_i, this probability δW is the largest for that set of values ξ in the submanifold of the ξ-space defined by (1) which minimizes the quadratic form

$$(t_1 - \xi_1)^2 + \cdots + (t_n - \xi_n)^2.$$

If the t_i are replaced here by the observed values x_i, then form (2) is obtained directly. *Thus, according to Gauss, the "best values" of $\vartheta_1, \ldots, \vartheta_r$ are those which give the observed event the largest probability.*

Later, Gauss gave another argument for the least squares principle which does not depend on the assumption of a normal distribution for

$x_1, \ldots, x_n$. He compares the estimation of a parameter to a game of chance in which the gambler can only lose. If T is the estimated parameter value, then the larger $|T-\vartheta|$, the larger the loss. Gauss takes $(T-\vartheta)^2$ as a measure for the loss and requires first that the estimate have no systematic error (that is, $\mathscr{E}T=\vartheta$) and second that $\mathscr{E}(T-\vartheta)^2$, the variance of the estimate or the expected value of the loss, be as small as possible. He shows then that this minimization postulate leads precisely to the least squares method.

If the observed quantities $x_1, \ldots, x_n$ have different standard deviations $\sigma_1, \ldots, \sigma_n$, then the form

$$\sigma_1^{-2}(x_1-\xi_1)^2+\sigma_2^{-2}(x_2-\xi_2)^2+\cdots+\sigma_n^{-2}(x_n-\xi_n)^2 \tag{4}$$

appears naturally in place of formula (2), or, if "weights" $g_1, \ldots, g_n$ which are inversely proportional to the σ_i^2, are introduced as in § 26 B, then the form becomes

$$Q=g_1(x_1-\xi_1)^2+\cdots+g_n(x_n-\xi_n)^2. \tag{5}$$

In applications of the method to problems of biology and economics, the $(x_i-\xi_i)$ are not really observational errors, but chance deviations of the quantities x_i from their expected values ξ_i. The x_i are assumed to be independent random variables. Let their expected values depend on the unknown parameters $\vartheta_1, \ldots, \vartheta_r$ by means of (1). Those values of the parameters which minimize (5) are taken as estimates of these parameters.

For the solution of the minimization problem, the set of equations

$$\begin{aligned} \vartheta_1 &= \vartheta_1^0+u \\ \vartheta_2 &= \vartheta_2^0+v \\ &\cdots\cdots \end{aligned} \tag{6}$$

is formed, where the ϑ_i^0 are provisional approximations and the $u, v \ldots$ small correction terms. Now we assume that for small $u, v, \ldots$, the functions (1) can be approximated sufficiently accurately by linear functions

$$\xi_i=\xi_i^0+a_i u+b_i v+\cdots. \tag{7}$$

In these equations, the ξ_i^0 are approximations to the ξ_i which correspond to the provisional approximation ϑ^0:

$$\xi_i^0=\varphi_i(\vartheta^0).$$

The coefficients $a_i, b_i, \ldots$ of the linear approximations may be set equal to the derivatives of the exact functions evaluated at the point ϑ^0;

$$a_i=\left(\frac{\partial \xi_i}{\partial \vartheta_1}\right)^0, \quad b_i=\left(\frac{\partial \xi_i}{\partial \vartheta_2}\right)^0, \ldots. \tag{8}$$

In order to simplify the calculations, we imagine the origin of the coordinates shifted to the point ξ^0. We also replace the observed quantities x_i by the differences

$$l_i = x_i - \xi_i^0. \tag{9}$$

Their expected values are

$$\lambda_i = \xi_i - \xi_i^0 = a_i u + b_i v + \cdots. \tag{10}$$

Thus, the form Q may now be written

$$Q = \sum g_i (l_i - \lambda_i)^2 = \sum g_i (l_i - a_i u - b_i v - \cdots)^2. \tag{11}$$

The minimum of this expression is obtained by setting the partial derivatives equal to zero. After division by two, we get

$$\begin{aligned} &\sum g_i a_i (a_i u + b_i v + \cdots - l_i) = 0 \\ &\sum g_i b_i (a_i u + b_i v + \cdots - l_i) = 0 \\ &\cdots\cdots\cdots\cdots\cdots\cdots \end{aligned} \tag{12}$$

If, following Gauss, we introduce the abbreviations

$$\sum g_i a_i^2 = [g\,a\,a], \qquad \sum g_i a_i b_i = [g\,a\,b], \ldots,$$

the equations of (12) finally reduce to the *normal equations*

$$\begin{aligned} &[g\,a\,a]\,u + [g\,a\,b]\,v + \cdots = [g\,a\,l] \\ &[g\,b\,a]\,u + [g\,b\,b]\,v + \cdots = [g\,b\,l] \\ &\cdots\cdots\cdots\cdots\cdots\cdots \end{aligned} \tag{13}$$

The number of normal equations is equal to the number of unknown parameters $\vartheta_1, \ldots, \vartheta_r$. If all observations have the same precision, the weights g_i can be set equal to one, and we have

$$\begin{aligned} &[a\,a]\,u + [a\,b]\,v + \cdots = [a\,l] \\ &[b\,a]\,u + [b\,b]\,v + \cdots = [b\,l] \\ &\cdots\cdots\cdots\cdots\cdots\cdots \end{aligned} \tag{14}$$

In the notation of the normal equations, we have kept as close as possible to the tradition begun by Gauss. By introducing matrix notation, the equations can be somewhat compressed. However, the old-fashioned notation (14) is very convenient for applications. The $g_i, a_i, b_i, \ldots$, and l_i are each written in a column and the coefficients $[a\,a]$ or $[g\,a\,a]$ calculated from these columns.

The system of equations (13) or (14) always has a solution, since a positive quadratic polynomial always has a minimum. However, the solution need not be unique. It sometimes happens that these equations can be solved for certain linear combinations of the parameters, but not for the $u, v, \ldots$ themselves. Following Rao[1], we shall call these linear combinations of the parameters *estimable*.

In order to discover which functions of the parameters are estimable, we consider the linear forms

$$\lambda_i = a_i u + b_i v + \cdots. \tag{15}$$

Among them linearly independent forms $u_1, \ldots, u_p$ may be found. All of the λ_i can be expressed in terms of these; thus, (11) may be written as a quadratic polynomial in $u_1, \ldots, u_p$. The quadratic terms in the polynomial are

$$\sum g_i \lambda_i^2 = \sum g_i (a_i u + b_i v + \cdots)^2.$$

This form, then, can only be zero if $u_1, \ldots, u_p$ are all zero. Writing it as a sum of squares, we obtain the squares of p independent linear forms, $v_1^2 + \cdots + v_p^2$. Therefore, Q itself becomes

$$(v_1 - c_1)^2 + \cdots + (v_p - c_p)^2 + c_0.$$

Its minimum is attained at $v_1 = c_1, \ldots, v_p = c_p$. Thus, $v_1, \ldots, v_p$ are estimable and therefore, $u_1, \ldots, u_p$ are. Hence, it follows that: *The estimable linear functions of the parameters u, v, ..., are precisely those which can be written as linear combinations of the forms* (15).

If $u_1, \ldots, u_p$ are introduced as new parameters in place of the original $u, v, \ldots$, then we obtain a system of normal equations with a unique solution. We shall assume from now on that the normal equations do have a unique solution.

The simplest method of solving the equations in (13) is the elementary method explained by Gauss. We solve for u from the first equation and substitute the result into all the other equations, and so on. It is expedient to arrange the calculation in the following way. Substitute the numerical values of the coefficients on the left, leaving the right sides of the equations as undetermined quantities at first. We then obtain the solutions as linear functions of the right hand terms

$$\begin{aligned} u &= h^{11}[g\,a\,l] + h^{12}[g\,b\,l] + \cdots \\ v &= h^{21}[g\,a\,l] + h^{22}[g\,b\,l] + \cdots \\ &\cdots\cdots\cdots\cdots\cdots \end{aligned} \tag{16}$$

The h^{ik} form the inverse of the matrix of coefficients of the system of equations in (13).

---

[1] C. R. Rao, Advanced Statistical Methods in Biometric Research, New York, 1952.

The ϑ are calculated from the $u, v, \ldots$ by means of (6) and the λ by means of (10). Since we are concerned not with the true ϑ and λ, but only with estimated values, we denote them by $\tilde{\vartheta}$ and $\tilde{\lambda}$. The estimated ξ follow from $\tilde{\xi}_i = \xi_0 + \tilde{\lambda}_i$ and the *estimated corrections to the observations*[2] from

$$k_i = \tilde{\xi}_i - x_i = \tilde{\lambda}_i - l_i. \tag{17}$$

If the estimated $\tilde{\vartheta}$ differ greatly from the initial values ϑ^0 and if the functions in (1) are not linear, the calculation must be repeated with $\tilde{\vartheta}$ as new initial values in place of the ϑ^0.

For practical calculations, checks are necessary. One check is that according to (12), the k_i must satisfy the conditions

$$\begin{aligned} [g\,a\,k] &= 0 \\ [g\,b\,k] &= 0 \\ \cdots & \cdots \end{aligned} \tag{18}$$

Calculation of the minimum $\tilde{Q}$ of Q defined by (11) provides another check. The values of λ_i which minimize Q are the $\tilde{\lambda}_i$. Therefore, we obtain

$$\tilde{Q} = \sum g_i (l_i - \tilde{\lambda}_i)^2 = \sum g_i k_i^2 = [g\,k\,k]. \tag{19}$$

A simpler expression for $\tilde{Q}$ is obtained as follows:

$$\begin{aligned} \tilde{Q} &= \sum g_i (\tilde{\lambda}_i - l_i)(\tilde{\lambda}_i - l_i) \\ &= \sum g_i (-l_i + a_i u + b_i v + \cdots) k_i \\ &= -[g\,l\,k] + [g\,a\,k]\,u + [g\,b\,k]\,v + \cdots \\ &= -[g\,l\,k] \end{aligned}$$

by (18).

If $\tilde{\lambda}_i - l_i$ is substituted for k_i again, we obtain

$$\tilde{Q} = [g\,l\,l] - [g\,a\,l]\,u - [g\,b\,l]\,v - \cdots. \tag{20}$$

We calculate $\tilde{Q}$ from (20); Eq. (19) serves as a check.

By (16), $u, v, \ldots$ are linear functions of the observed differences $l_i = x_i - \xi_i^0$:

$$\begin{aligned} u &= \alpha_1 l_1 + \cdots + \alpha_n l_n \\ v &= \beta_1 l_1 + \cdots + \beta_n l_n \\ \cdots & \cdots \end{aligned} \tag{21}$$

The coefficients $\alpha_i, \beta_i, \ldots$ can be calculated easily from (16);

$$\alpha_i = g_i (h^{11} a_i + h^{12} b_i + \cdots). \tag{22}$$

[2] In Gauss' work, these corrections are denoted by λ_i.

Formulas (21) and (22) play no role in practical calculations; however, we need them for the derivation of the standard deviation in the next section.

Example 18 (from F. R. Helmert, Die Ausgleichungsrechnung, Leipzig 1872). From the base point D' of his network of triangles near Speyer, Schwerd found the averages of several measurements of the angles between objects $ABWHN$. The results were

$$\begin{array}{lll} BA & \text{(90 measurements)} & 19°\,25'\,59.42'', \\ BW & \text{(80 measurements)} & 34°\,18'\,43.61'', \\ AW & \text{(70 measurements)} & 14°\,52'\,44.33'', \\ HW & \text{(20 measurements)} & 15°\,34'\,58.80'', \\ BH & \text{(20 measurements)} & 18°\,43'\,45.60'', \\ NA & \text{(40 measurements)} & 12°\,26'\,24.65'', \\ BN & \text{(60 measurements)} & 6°\,59'\,34.51'', \\ NH & \text{(20 measurements)} & 11°\,44'\,11.60''. \end{array}$$

By applying the method of repetitions, the calibration errors are largely cancelled out. Therefore, we can assume that the observations had no systematic errors and take the weights g to be proportional to the numbers of repetitions. As unknowns ϑ_i, we take the four angles BN, BH, BA, and BW, in terms of which all the others can be expressed. As initial values, we take the measured values of these four angles; therefore, we set

$$\begin{aligned} \vartheta_1 &= BN = 6°\,59'\,34.51'' + u, \\ \vartheta_2 &= BH = 18°\,43'\,45.60'' + v, \\ \vartheta_3 &= BA = 19°\,25'\,59.42'' + w, \\ \vartheta_4 &= BW = 34°\,18'\,43.61'' + t. \end{aligned}$$

The eight angles $\xi_1 = BA, \ldots, \xi_8 = NH$ can now be expressed in terms of the unknown as follows:

$$\begin{aligned} \xi_1 &= BA = 19°\,25'\,59.42'' + w, \\ \xi_2 &= BW = 34°\,18'\,43.61'' + t, \\ \xi_3 &= AW = 14°\,52'\,44.19'' - w + t, \\ &\cdots\cdots\cdots\cdots \\ \xi_8 &= NH = 11°\,44'\,11.09'' - u + v. \end{aligned}$$

The weights g_i, the coefficients a_i, b_i, c_i, d_i of the last expressions, and the differences may be set out in the following table.

g	a	b	c	d	l
9	0	0	+1	0	0
8	0	0	0	+1	0
7	0	0	−1	+1	+0.14
2	0	−1	0	+1	+0.79
2	0	+1	0	0	0
4	−1	0	+1	0	−0.26
6	+1	0	0	0	0
2	−1	+1	0	0	+0.51

The normal equations become

$$
\begin{aligned}
12u-2v-\ 4w \qquad\qquad &= -0.02\\
-2u+6v \qquad\quad -\ 2t &= +0.56\\
-4u \qquad +20w-\ 7t &= -2.02\\
+2v-\ 7w+17t &= +2.56.
\end{aligned}
\tag{23}
$$

We solve the equations generally, that is, we first replace the right sides by undetermined quantities A, B, C, D, and solve for u, v, w, t;

$$
\begin{aligned}
u &= 0.00978\,A + 0.00375\,B + 0.00247\,C + 0.00146\,D\\
v &= 0.00375\,A + 0.01890\,B + 0.00178\,C + 0.00296\,D\\
w &= 0.00247\,A + 0.00178\,B + 0.00650\,C + 0.00289\,D\\
t &= 0.00146\,A + 0.00296\,B + 0.00289\,C + 0.00742\,D.
\end{aligned}
\tag{24}
$$

The coefficients on the right are the elements $h^{11}, \ldots, h^{44}$ of the inverse matrix. If we substitute the right sides of (23) into (24), we obtain

$$u=-0.032, \quad v=-0.065, \quad w=-0.067, \quad t=+0.115.$$

For the corrections we find

$$
\begin{aligned}
k_1 &= -0.067, & k_5 &= -0.065,\\
k_2 &= +0.115, & k_6 &= +0.225,\\
k_3 &= +0.042, & k_7 &= -0.032,\\
k_4 &= -0.609, & k_8 &= -0.543.
\end{aligned}
$$

Now, we can calculate $\tilde{Q}$ from (19) or (20). We find the two formulas in agreement;

$$\tilde{Q} = 1.71.$$

§ 31. Expectations and Standard Deviations of the Estimates $\tilde{\vartheta}$

The estimates $\tilde{\vartheta}$ obtained by the least squares method are linear functions of the observed variables x_k; therefore, they are also random variables. We wish to work out their expectations and standard deviations.

A. Expectations

Suppose that the normal equations in (13) § 30 have a unique solution. We now change the notation for the solutions from $u, v, \ldots$ to $u^1, \ldots, u^r$. We write the normal equations themselves as

$$\sum h_{jk}\, u^k = r_j. \tag{1}$$

The solution is given by

$$u^i = \sum h^{ij} r_j \tag{2}$$

where (h^{ij}) is the inverse of (h_{jk});

$$\sum_j h^{ij} h_{jk} = \delta^i_k = \begin{cases} 1 & \text{for } i=k \\ 0 & \text{otherwise.} \end{cases} \tag{3}$$

The estimates $\tilde{\vartheta}$ now satisfy

$$\tilde{\vartheta}_k = \vartheta^0_k + u^k, \qquad (k=1, \ldots, r). \tag{4}$$

In order to simplify calculation of the expectations, we assume the approximate values ϑ^0_k equal to the true values of the parameters ϑ_k. Of course, in practice we cannot do this, since the true values are unknown; but for calculating the theoretical expectations and standard deviations it doesn't matter. Therefore, in place of (4) we write

$$\tilde{\vartheta}_k = \vartheta_k + u^k. \tag{5}$$

If the ϑ^0_k are assumed equal to the true ϑ_k, then the corresponding ξ^0_i will be equal to the ξ_i, the expected values of the x_i. The expected values of the differences

$$l_i = x_i - \xi^0_i \tag{6}$$

will therefore be zero. Hence, it follows from (21) § 30 that the u^k will also have zero expectation. Thus, Eq. (5) yields: *The expected values of the* $\tilde{\vartheta}_k$ *are equal to the true parameter values* ϑ_k.

The same fact can also be expressed as: *The estimates* $\tilde{\vartheta}_k$ *have no systematic errors* or *the estimates* $\tilde{\vartheta}_k$ *are unbiased.*

B. Standard Deviations

In calculating the standard deviations, we take it as a basic assumption that the x_i are independent random variables with fixed standard deviations σ_i, independent of the ϑ. Let the weights g_i of § 30 be chosen inversely proportional to the variances σ_i^2. Thus, we can set

$$g_i \sigma_i^2 = \sigma^2. \tag{7}$$

The value σ so defined is the standard deviation of an observation with weight one.

By (5), the variance of $\tilde{\vartheta}_k$ is equal to the variance of u^k. In order to calculate it, we go back to (21) § 30. We have for $k=1$,

$$u^1 = u = \alpha_1 l_1 + \cdots + \alpha_n l_n. \tag{8}$$

Since the l_i are independent random variables with variances σ_i^2, the variance of u is given by

$$\sigma_u^2 = \alpha_1^2\, \sigma_1^2 + \cdots + \alpha_n^2\, \sigma_n^2. \tag{9}$$

Therefore, by (7), we can write

$$\sigma_u^2 = \sum \alpha_i^2\, g_i^{-1}\, \sigma^2,$$

or using (22) § 30,

$$\begin{aligned}\sigma_u^2 &= \sum g_i (h^{11}\, a_i + h^{12}\, b_i + \cdots)^2\, \sigma^2 \\ &= (h^{11}\, h^{11} [g\,a\,a] + 2 h^{11}\, h^{12} [g\,a\,b] + h^{12}\, h^{12} [g\,b\,b] + \cdots)\, \sigma^2.\end{aligned}$$

The $[g\,a\,a], \ldots$ are the coefficients of the normal equations, which are also denoted by h_{jk}. Thus, we obtain

$$\sigma_u^2 = \Big(\sum_j \sum_k h^{1j}\, h_{jk}\, h^{1k}\Big)\, \sigma^2, \tag{10}$$

or using (3),

$$\sigma_u^2 = h^{11}\, \sigma^2. \tag{11}$$

Similarly, we obtain for $k=2$,

$$\sigma_v^2 = h^{22}\, \sigma^2, \tag{12}$$

and so forth.

C. Geometrical Interpretation

In order to illustrate the least squares method geometrically, we shall assume there to be only one unknown parameter ($r=1$) and only three observations of equal accuracy. Then, we can conceive of the observed values x_1, x_2, x_3 as coordinates of a point X, the *observed point*, in space.

We take as origin the point ξ^0, which was taken in § 30 as a first approximation of ξ. Now, we drop the assumption that ξ^0 coincides with the true point ξ.

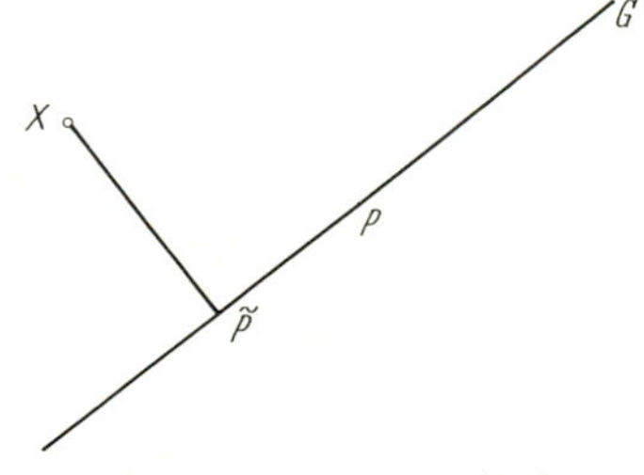

Fig. 19. Least squares method

Since we have assumed there to be only one parameter and that $\xi_i^0=0$, Eqs. (7) § 30 simplify to

$$\xi_i=a_i u \qquad (i=1,2,3). \tag{13}$$

The equations of (13) define a straight line G in parametric form. The "*true point*" P, whose coordinates are the expected values $\xi_i=\hat{x}_i$ of the observed variables, must lie on the line G. The observed point X lies somewhere in the neighborhood of P. The quantity

$$Q=(x_1-\xi_1)^2+(x_2-\xi_2)^2+(x_3-\xi_3)^2 \tag{14}$$

represents the squared distance of the observed point X from a point on G. If Q is to be minimized, it means that we look for point $\tilde{P}$ on G which is the smallest distance from X. Therefore, $\tilde{P}$ is the foot of the perpendicular from X to G.

If we carry out an orthogonal transformation of coordinates beforehand, the formulas for calculating the foot of the perpendicular can be made very much simpler. Choose G to be one coordinate axis and take the others at right angles. In the new coordinates the parametric representation of G is

$$\eta_1=a u, \quad \eta_2=0, \quad \eta_3=0 \qquad (a^2=a_1^2+a_2^2+a_3^2). \tag{15}$$

In the general case where there are r parameters and n variables x_i, let the subspace G be given by the parametric representation

$$\xi_i=a_i u+b_i v+\cdots. \tag{16}$$

The orthogonal transformation can be set up in the following way

$$x_i=\sum e_{ik}\, y_k. \tag{17}$$

The first r columns of the matrix (e_{ik}) must be linear combinations of the vectors (a_i), (b_i), ..., in order that the first r unit vectors lie in the subspace (15). We let the first column be a multiple $(\lambda\, a_k)$, the second column a linear combination $(\mu\, a_i+\nu\, b_i)$, and so forth, determining the coefficients $\lambda, \mu, \nu, \ldots$ according to the conditions for orthonormality.

If we take the expectation of both sides of (17), we obtain

$$\xi_i=\sum e_{ik}\,\eta_k,$$

where $\eta_1, \ldots, \eta_r$ are linear combinations of $u, v, \ldots$ and $\eta_{r+1}, \ldots, \eta_n$ are zero, as in (15), for $r=1$ and $n=3$.

The quantity Q is invariant under a orthonormal transformation; therefore, when $n=3$, we obtain

$$Q=(y_1-\eta_1)^2+(y_2-\eta_2)^2+(y_3-\eta_3)^2,$$

or using (15),

$$Q=(y_1-\eta_1)^2+y_2^2+y_3^2. \tag{18}$$

The minimum $\tilde{Q}$ is assumed for $\eta_1=y_1$. Thus,

$$\tilde{\eta}_1=y_1, \qquad \tilde{\eta}_2=\tilde{\eta}_3=0$$

and

$$\tilde{Q}=y_2^2+y_3^2. \tag{19}$$

Eq. (18) expresses the Pythagorian Theorem

$$PX^2=P\tilde{P}^2+X\tilde{P}^2. \tag{20}$$

The left side of (20) is Q, the first term on the right is $(y_1-\eta_1)^2$ and the second $\tilde{Q}=y_2^2+y_3^2$.

Generalization of (18) and (19) to r parameters and observed variables leads to

$$Q=(y_1-\eta_1)^2+\cdots+(y_r-\eta_r)^2+y_{r+1}^2+\cdots+y_n^2 \tag{21}$$

and

$$\tilde{Q}=y_{r+1}^2+\cdots+y_n^2. \tag{22}$$

In the general case, we can also interpret $\tilde{Q}$ as the square of the distance $X\tilde{P}$ in n-dimensional space.

The case of observations with differing accuracy can be reduced to the case illustrated here by means of the substitution

$$x_i=x_i'\,\sigma_i.$$

D. A Theorem of Gauss

The second argument given by Gauss for the method of least squares is based on the following theorem.

Theorem. *Among all unbiased estimates of the parameter ϑ_1 which are linear functions of the observations, the estimate $\tilde{\vartheta}_1$ has the smallest variance.*

There is a very short proof for this theorem by R. L. Plackett in Biometrika 36 (1949), p. 458. Here we shall give a proof based upon the orthonormal transformation (17). Again, we shall take $r=1$ and $n=3$ and leave to the reader the generalization to arbitrary r and n.

Proof. Let T be an estimate for ϑ_1, which is a linear function of the x_i, and therefore, also of the y_i;

$$T=c_0+c_1\,y_1+c_2\,y_2+c_3\,y_3. \tag{23}$$

The estimate T is called unbiased if the expectation of T, which depends on the ϑ, is equal to ϑ_1 for all ϑ. Therefore,

$$\mathscr{E}\, T = c_0 + c_1 \eta_1 = c_0 + c_1 a u. \tag{24}$$

Since for all n

$$\vartheta_1 = \vartheta_1^0 + u, \tag{25}$$

we must also have

$$c_0 = \vartheta_1^0 \quad \text{and} \quad c_1 = a^{-1}. \tag{26}$$

Moreover, it follows that

$$T - \mathscr{E}\, T = c_1 (y_1 - \eta_1) + c_2 y_2 + c_3 y_3. \tag{27}$$

In order to work out the variance of T, we have to square (27) and take the expected value;

$$\begin{aligned} \sigma_T^2 = c_1^2\, \mathscr{E}(y_1 - \eta_1)^2 + 2c_1 c_2\, \mathscr{E}(y_1 - \eta_1) y_2 + 2 c_1 c_3\, \mathscr{E}(y_1 - \eta_1) y_3 \\ + c_2^2\, \mathscr{E}\, y_2^2 + 2 c_2 c_3\, \mathscr{E}\, y_2 y_3 + c_3^2\, \mathscr{E}\, y_3^2. \end{aligned} \tag{28}$$

The individual expectations on the right are easy to calculate, for example,

$$\begin{aligned} \mathscr{E}(y_1 - \eta_1)^2 &= \mathscr{E}\left[\sum e_{i1} (x_i - \xi_i)\right]^2 \\ &= \sum_i \sum_j e_{i1} e_{j1}\, \mathscr{E}(x_i - \xi_i)(x_j - \xi_j) \\ &= \sum_i e_i^2 \sigma^2 = \sigma^2. \end{aligned}$$

Using the orthonormality of the inverse matrix (e_{ki}), we find the expected values of the squares and products in (28) to be σ^2 and zero, respectively.

Thus, we obtain

$$\sigma_T^2 = (c_1^2 + c_2^2 + c_3^2)\, \sigma^2. \tag{29}$$

Here c_1 is kept fixed by (26). Therefore, the minimum of the variance (29) is attained for

$$c_2 = c_3 = 0.$$

Hence, the estimate with the smallest variance is given by

$$T = \vartheta_1^0 + a^{-1} y.$$

This is precisely the estimate obtained by the least squares method. Therefore, the assertion is proved.

The geometrical interpretation of the theorem is the following. Any linear unbiased estimate can be obtained by constructing a plane through the observed point parallel to a given fixed plane and intersecting it with the line G. The parameter value of the point of intersection is then the estimate T. If we choose the fixed plane perpendicular to G, then we obtain the estimate $\tilde{\vartheta}$ with the smallest variance.

§ 32. Estimation of the Variance σ^2

The minimum of the quadratic form

$$Q=\sum(x_i-\xi_i)^2 \tag{1}$$

has been denoted by $\tilde{Q}$. The expression Q is transformed into

$$Q=(y_1-\eta_1)^2+\cdots+(y_r-\eta_r)^2+y_{r+1}^2+\cdots+y_n^2 \tag{2}$$

by the orthonormal transformation (17) § 31. Its minimum becomes

$$\tilde{Q}=y_{r+1}^2+\cdots+y_n^2. \tag{3}$$

The minimum is attained for $\tilde{\eta}_1=y_1, \ldots, \tilde{\eta}_r=y_r$.

The expected value of $\tilde{Q}$ is the sum of the expected values of the squares y_k^2. These are worked out as in § 31 D. We obtain

$$\mathscr{E}\tilde{Q}=(n-r)\,\sigma^2. \tag{4}$$

Thus,

$$s^2=\frac{\tilde{Q}}{n-r} \tag{5}$$

can be used as an unbiased estimate for σ^2.

If the observations have unequal standard deviations σ_i, we have to consider the expression

$$Q=\sum g_i(x_i-\xi_i)^2 \tag{6}$$

instead. However, this case can be reduced to the previous one by means of the substitution

$$x_i'=x_i\sqrt{g_i}. \tag{7}$$

Again, we obtain (5) as an unbiased estimate of the variance of an observation of unit weight.

It is clear that, for small $n-r$, the estimate (5) is not very accurate. It only becomes more accurate for large $n-r$. In order to make this assertion more precise, we must investigate the distribution function of $\tilde{Q}$. For this purpose we assume that the $x_1, \ldots, x_n$ are normally distributed with the same standard deviation σ. Their joint probability density is then

$$f(x_1, \ldots, x_n)=\sigma^{-n}(2\pi)^{-\frac{n}{2}}\exp\left\{-\frac{1}{2\sigma^2}\sum(x_i-\xi_i)^2\right\}. \tag{8}$$

Again, transforming orthonormally the x_i to the y_i leads to the joint probability density

$$f(y_1, \ldots, y_n)=\sigma^{-n}(2\pi)^{-\frac{n}{2}}\exp\left\{-\frac{1}{2\sigma^2}\sum(y_i-\eta_i)^2\right\}. \tag{9}$$

The y_i are independent normally distributed random variables with expectations

$$\eta_1, \ldots, \eta_r, \qquad 0, \ldots, 0$$

and standard deviations σ. The variables $\frac{y_{r+1}}{\sigma}, \ldots, \frac{y_n}{\sigma}$ have expectation zero and standard deviation one. *Therefore, the sum of their squares*

$$\chi^2 = \frac{\tilde{Q}}{\sigma^2} = \frac{y_{r+1}^2 + \cdots + y_n^2}{\sigma^2} \tag{10}$$

has a χ^2 distribution with $n-r$ degrees of freedom.

The expected value of χ^2 is $n-r$ in agreement with (5). For large $n-r$, χ^2 is approximately normally distributed with mean $n-r$ and standard deviation $\sqrt{2(n-r)}$. Therefore,

$$\frac{\chi^2}{n-r} = \frac{\tilde{Q}}{(n-r)\,\sigma^2} = \frac{s^2}{\sigma^2}$$

is approximately normally distributed with expectation one and variance $2/(n-r)$. Thus, for large $n-r$, s^2 is a good estimate for σ^2; for small $n-r$, the estimate is not very accurate. Confidence bounds for s^2/σ^2 can be obtained from the χ^2 table (Table 6).

In § 30 (19) and (20), we saw how to calculate $\tilde{Q}$ in practice. From $\tilde{Q}$ we can obtain s^2 by means of (5), and with the aid of s^2, we can find the approximate values for σ_u^2, σ_v^2, and so on;

$$\begin{aligned} s_u^2 &= h^{11} s^2 \\ s_v^2 &= h^{22} s^2 \\ &\cdots\cdots \end{aligned} \tag{11}$$

From (10), it follows that

$$\frac{(n-r)\,s_u^2}{\sigma_u^2} = \frac{(n-r)\,s^2}{\sigma^2} = \frac{\tilde{Q}}{\sigma^2} = \chi^2$$

has a χ^2 distribution with $n-r$ degrees of freedom. Moreover, χ^2 is independent of $y_1 = \eta_1, y_2 = \eta_2, \ldots, y_r = \eta_r$. Hence, as in § 28, it follows that

$$t = \frac{\tilde{\vartheta}_1 - \vartheta_1}{s_u} = \frac{u - \mathscr{E}u}{s_u} = \frac{u - \mathscr{E}u}{\sigma_u}\,\frac{\sigma_u}{s_u} = \frac{(u - \mathscr{E}u)\sqrt{n-r}}{\sigma_u\,\chi} \tag{12}$$

has a t distribution with $n-r$ degrees of freedom. That is to say:

Confidence limits for the individual parameters $\vartheta_1, \vartheta_2, \ldots$ can be obtained by applying Student's test with $n-r$ degrees of freedom to each least squares estimate $\tilde{\vartheta}_1, \tilde{\vartheta}_2, \ldots$.

Example 19. The times of entry of the sun into the twelve signs of the zodiac were given in a Byzantine table[3] as follows.

Libra	23 September	Day	12^h	Aries	20 March	Night	5^{20}
Scorpio	23 October	Day	3^{30}	Taurus	21 April	Night	11^h
Sagittarius	21 November	Day	10^{30}	Gemini	22 May	Night	1^{40}
Capricorn	20 December	Night	3^{20}	Cancer	23 June	Day	6^{31}
Aquarius	19 January	Day	2^{20}	Leo	24 July	Night	3^h
Pisces	18 February	Day	2^{20}	Virgo	24 August	Night	0^{30}

The hours of the day were reckoned from 6^h in the morning, the hours of the night from 6^h in the evening. If the entry into Libra is taken as zero point on the time scale, we obtain the times of entry set out in the second column of the following table. Subtracting the times of entry of the mean sun into the various signs, we obtain the corrections given the last column. For this purpose, a $365^d\,6^h$ year is assumed, which is quite permissible since with one exception the times in the text are obviously rounded to the nearest 10^m.

Longitude	Time t	Mean time	Difference l
−180	0	0	0
−150	$29^d\,15^h\,30^m$	$30^d\,10^h\,30^m$	$-\ 19^h$
−120	$58^d\,22^h\,30^m$	$60^d\,21^h$	$-\ 46^h\,30^m$
− 90	$88^d\ \ 3^h\,20^m$	$91^d\ \ 7^h\,30^m$	$-\ 76^h\,10^m$
− 60	$117^d\,14^h\,20^m$	$121^d\,18^h$	$-\ 99^h\,40^m$
− 30	$147^d\,14^h\,20^m$	$152^d\ \ 4^h\,30^m$	$-110^h\,10^m$
0	$178^d\ \ 5^h\,20^m$	$182^d\,15^h$	$-105^h\,40^m$
30	$209^d\,11^h$	$213^d\ \ 1^h\,30^m$	$-\ 86^h\,30^m$
60	$241^d\ \ 1^h\,40^m$	$243^d\,12^h$	$-\ 58^h\,20^m$
90	$272^d\,18^h\,30^m$	$273^d\,22^h\,30^m$	$-\ 28^h$
120	$304^d\ \ 3^h$	$304^d\ \ 9^h$	$-\ \ 6^h$
150	$335^d\ \ 0^h\,30^m$	$334^d\,19^h\,30^m$	$+\ \ 5^h$

Certain symmetries in the numbers (which are, in fact, only approximately fulfilled) and related texts indicated that the table might have been calculated according to the eccentric theory. According to this theory, the sun moves with constant speed on an eccentric circle around the earth (Fig. 20). We now assume that the table was calculated by means of this theory, but with certain random errors. With this hypothesis, we wish to try and determine the eccentricity and the apogee A as accurately as possible.

Let λ be the longitude of the sun on entry into one of the signs of the zodiac and let α be the longitude of the apogee. The difference $x=\lambda-\alpha$ is called the *true anomaly*. The arc length from the apogee A to the sun along the eccentric circle is called the *mean anomaly*; we denote it by $x+\omega$. The difference $-\omega$ between the true and mean anomalies is called the *equation of center*. If e is the eccentricity, the relations

$$\sin\omega=e\sin x \tag{13}$$

or

$$\omega=\arcsin(e\sin x) \tag{14}$$

follow from the Sine Rule of plane trigonometry.

---

[3] B. L. van der Waerden, Eine byzantinische Sonnentafel. Sitzungsber. Bayer. Akad. München (math.-nat.) 1954, p. 159.

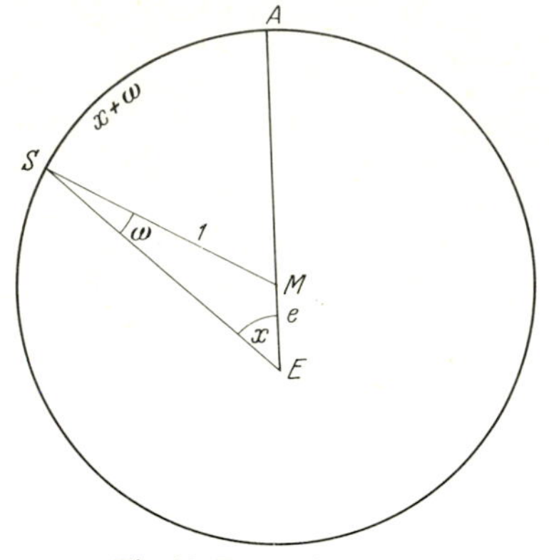

Fig. 20. Eccentric motion

Since e is small, we can expand arc sin in a power series which can be stopped at the second term:

$$\begin{aligned}\omega &= e \sin x + \tfrac{1}{6} e^3 \sin^3 x \\ &= e \sin x + \tfrac{1}{24} e^3 (3 \sin x - \sin 3x) \\ &= (e + \tfrac{1}{8} e^3) \sin x - \tfrac{1}{24} e^3 \sin 3x.\end{aligned} \tag{15}$$

The time required by the sun to cover the arc $x+\omega$ on the eccentric circle is

$$t = \frac{T}{2\pi}(x+\omega),$$

where $T = 365\frac{1}{4}^{\mathrm{d}}$ is the time of one rotation. With uniform motion, the time would be

$$t_0 = \frac{T}{2\pi} x.$$

The differences

$$t - t_0 = \frac{T}{2\pi} \omega,$$

increased by an unknown constant d, have to yield the l values of the last column of our table, increased by corrections k, also unknown, which depend on mistakes in calculation and transcription in the text and in rounding:

$$\frac{T}{2\pi} \omega + d = l + k. \tag{16}$$

If we substitute for ω the expression (15) found earlier, we obtain

$$a \sin x + b \sin 3x + d = l + k \tag{17}$$

with

$$a = \frac{T}{2\pi}(e + \tfrac{1}{8} e^3),$$

$$b = -\frac{T}{2\pi} \cdot \frac{1}{24} e^3 = -c\, e^3$$

where c is unknown. Finally, if we substitute $x = \lambda - \alpha$, we obtain the equations

$$a \sin(\lambda - \alpha) + b \sin 3(\lambda - \alpha) + d = l + k. \tag{18}$$

Since the twelve longitudes λ and the corresponding l are known from the table, we obtain twelve equations (18) with three unknowns e, α, and d. These are then determined so that the sum of the squares $k_1^2+\cdots+k_{12}^2$ is minimized.

The calculation can be carried out more conveniently if the small term with b is neglected at first. Then, after finding an approximation for e, we can calculate $b=-c\,e^3$, transpose the term with b, and make a second approximation. However, in this case, it turns out that the second approximation for a is equal to the first one because the terms with $\sin 3x$ disappear when the normal equations are formed. Therefore, the b term can be completely neglected and we can write the observational equations,

$$a \sin \lambda \cos \alpha - a \cos \lambda \sin \alpha + d = l + k. \tag{19}$$

If we introduce the new variables

$$u = a \cos \alpha,$$
$$v = -a \sin \alpha,$$
$$w = d,$$

we obtain finally twelve linear equations

$$u \sin \lambda + v \cos \lambda + w = l + k. \tag{20}$$

The normal equations lead to

$$\begin{aligned} [a\,a]\,u+[a\,b]\,v+[a\,c]\,w&=[a\,l]\\ [b\,a]\,u+[b\,b]\,v+[b\,c]\,w&=[b\,l]\\ [c\,a]\,u+[c\,b]\,v+[c\,c]\,w&=[c\,l]. \end{aligned} \tag{21}$$

The coefficients are easy to work out

$$\begin{aligned} [a\,a]&=\sum \sin^2\lambda=6,\\ [b\,b]&=\sum \cos^2\lambda=6,\\ [c\,c]&=\sum 1 \qquad =12,\\ [a\,b]&=\sum \sin\lambda\cos\lambda=0,\\ [a\,c]&=\sum \sin\lambda=0,\\ [b\,c]&=\sum \cos\lambda=0. \end{aligned}$$

Thus, the normal equations simplify to

$$\begin{aligned} 6u&=\sum l \sin\lambda\\ 6v&=\sum l\cos\lambda\\ 12w&=\sum l. \end{aligned} \tag{22}$$

From u and v, a and α are determined by

$$\begin{aligned} a\cos\alpha&=u\\ a\sin\alpha&=-v \end{aligned} \tag{23}$$

and, finally e is found from

$$e+\tfrac{1}{8}e^3=\frac{2\pi}{T}\,a. \tag{24}$$

We find

$$e=0.04157 \quad \text{and} \quad \alpha=65^\circ\,40'.$$

Hipparchus and Ptolemy assumed that

$$e=\tfrac{1}{24}=0.04167 \quad \text{and} \quad \alpha=65°\,30'.$$

The agreement is marked. Thus, indeed, the table seems to have been calculated according to Hipparchus' model of the sun's motion.

If we work out the standard error s of a single time of entry by formula (5) with $n=12$ and $r=3$, we find it to be about 20 minutes. This estimate is not, however, very precise since the denominator $n-r=9$ is not very large. Moreover, it is not certain that the individual values are independent of one another. If the individual corrections are calculated, it is seen that six of the twelve times of entry agree exactly with the Hipparchus' model, whereas large errors of between 30 and 50 minutes are involved in the remaining six. In two places successive times of entry have the same errors of 50 and 30 minutes, respectively. Therefore, it looks as if the individual values were not calculated independently.

In effect, the calculation carried out amounts to a Fourier analysis of periodic function $l(\lambda)$ of which twelve values are given. Fourier analysis proves to be a very useful tool for the investigation of astronomical tables whose law of construction is not known. In various cases this method has led to a complete clarification[4].

§ 33. Linear Regression

Let x be an independent variable and let y be a random variable whose distribution depends on x. In econometrics x is usually time and y a quantity which can be obtained statistically (for example, iron production) and which varies with time on one hand, but also depends on many other factors. However, x and y may both be random variables which exhibit some dependence, for example, the number of marriages in one year and the number of births in the following year.

Suppose that the n pairs of values $(x_1, y_1), \ldots, (x_n, y_n)$ are observed. We wish to investigate the variation of y as a function of x and for this purpose we set up a model, say, of linear regression

$$y=\vartheta_0+\vartheta_1 x+u, \tag{1}$$

where the regression line $y=\vartheta_0+\vartheta_1 x$ should fit as closely as possible to the actual graph of y values, that is, the "chance deviations" should prove as small as possible. Other models can also be used, for example, a polynomial of second degree (*quadratic regression*) where the regression function is a parabola, or a polynomial of higher degree (*r*th *order regression*)

$$y=\vartheta_0+\vartheta_1 x+\cdots+\vartheta_r x^r+u, \tag{2}$$

or, in the case of cyclical fluctuations, a trigonometric polynomial

$$y=\vartheta_0+\vartheta_1 \cos\omega x+\vartheta_2 \sin\omega x+u.$$

[4] B. L. van der Waerden, Die Bewegung der Sonne nach griechischen und indischen Tafeln. Sitzungsber. Bayer. Akad. (math.-nat.) 1952, p. 219. I. V. M. Krishna Rav, The Motion of the Moon in Tamil Astronomy, Centaurus 4 (1956).

The requirement that the residual u which is not accounted for by regression should turn out as small as possible is again achieved by means of the least squares method. Therefore, we wish to minimize the expression

$$Q = \sum u_i^2 \tag{3}$$

and determine the constants $\vartheta_0, \vartheta_1, \ldots$ from it. The calculation is exactly the same as in § 26. For example, in the case of linear regression the condition

$$\sum u_i^2 = \sum (y_i - \vartheta_0 - \vartheta_1 x_i)^2 = \min$$

leads immediately to

$$-\sum y_i + \vartheta_0 n + \vartheta_1 \sum x_i = 0,$$
$$-\sum y_i x_i + \vartheta_0 \sum x_i + \vartheta_1 \sum x_i^2 = 0,$$

or using the notation of Gauss (see § 30),

$$\begin{aligned} \vartheta_0 n + \vartheta_1 [x] &= [y] \\ \vartheta_0 [x] + \vartheta_1 [x x] &= [x y]. \end{aligned} \tag{4}$$

Similarly, for rth order regression, we obtain $(r+1)$ conditions

$$\begin{aligned} \vartheta_0 n + \vartheta_1 [n] + \cdots + \vartheta_r [x^r] &= [y] \\ \vartheta_0 [x] + \vartheta_2 [x^2] + \cdots + \vartheta_r [x^{r+1}] &= [x y] \\ \vartheta_0 [x^r] + \vartheta_1 [x^{r+1}] + \cdots + \vartheta_r [x^{2r}] &= [x^r y]. \end{aligned} \tag{5}$$

The solution of the equations in (5) is very tedious. The answer is reached with far fewer calculations if the polynomials $1, x, x^2, \ldots, x^r$ are orthogonalized beforehand. Two functions $\varphi(x)$ and $\psi(x)$, both defined for the values $x_1, x_2, \ldots, x_n$, are said to be orthogonal if

$$\sum \varphi(x_i) \psi(x_i) = 0.$$

If m functions $\varphi_1, \varphi_2, \ldots, \varphi_m$ are given, they can be replaced by an orthogonal system $\psi_1, \psi_2, \ldots, \psi_m$ which can be defined by

$$\begin{aligned} \psi_1 &= \varphi_1, \\ \psi_2 &= \varphi_2 - \alpha \psi_1, \\ \psi_3 &= \varphi_3 - \beta \psi_1 - \gamma \psi_2 \\ &\cdots\cdots\cdots \end{aligned}$$

The constant α is determined in such a way that ψ_2 is orthogonal to ψ_1; then β and γ are determined so that ψ_3 is orthogonal to ψ_1 and ψ_2, and so on.

Every linear combination $\vartheta_1 \varphi_1 + \cdots + \vartheta_m \varphi_m$ may then be written $\mu_1 \psi_1 + \cdots + \mu_m \psi_m$. Once again, if we make use of the least squares model to determine the μ, only one unknown μ_i appears in each normal equation and the solution can be written down immediately.

In the case of linear regression, the calculation takes the following form. The original functions in (1) are 1 and x. The orthogonalized functions are

$$\psi_0 = 1 \quad \text{and} \quad \psi_1 = x - \bar{x},$$

where $\bar{x}$ is the sample mean of the x_i.

The least squares rule

$$\sum (y - \mu_0 \psi_0 - \mu_1 \psi_1)^2 = \min$$

yields, by differentiation,

$$\begin{aligned} -\sum y \psi_0 + \mu_0 \sum \psi_0^2 = 0 \\ -\sum y \psi_1 + \mu_1 \sum \psi_1^2 = 0 \end{aligned} \tag{6}$$

or

$$\begin{aligned} \mu_0 n = \sum y \\ \mu_1 \sum (x - \bar{x})^2 = \sum y (x - \bar{x}). \end{aligned} \tag{7}$$

If, for the sake of simplicity, we write m_0, m_1 instead of $\tilde{\mu}_0, \tilde{\mu}_1$, the solution becomes

$$m_0 = \bar{y} = \frac{1}{n} \sum y, \tag{8}$$

$$m_1 = \frac{\sum (x - \bar{x})\, y}{\sum (x - \bar{x})^2} = \frac{\sum (x - \bar{x})(y - \bar{y})}{\sum (x - \bar{x})^2}. \tag{9}$$

The equation of the *sample regression line* is then

$$y - \bar{y} = m_1 (x - \bar{x}). \tag{10}$$

The slope of the line m_1 is called the *sample regression coefficient*. Its value, of course, depends on chance. If we assume that the x values are not random (for example, points in time), while the y values are random variables, we can determine the expectation and standard deviation as in § 30. If x is time, the regression is sometimes called the *trend*.

Example 20. Statistics of world production of pig iron from 1865 to 1910, using data given by Cassel[5], are set out in the second column of the following table. We wish to separate the fluctuations exhibited by pig iron production into trend and variation due to the business cycle, as best we can.

[5] G. Cassel, Theoret. Sozialökonomie, 3rd ed. p. 587, figure p. 532.

In the table below, t is the year, x the production of pig iron in millions of tons, and y is 1,000 log x. The number $a = 1{,}890$ is subtracted from the t values and $b = 1{,}400$ from the y values in order to obtain conveniently small numbers.

t	x	y	$t-a$	$y-b$	$(t-a)^2$	$(t-a)(y-b)$
1865	9.10	959	−25	−441	625	+11,025
1866	9.66	985	−24	−415	576	+ 9,960
1867	10.06	1,003	−23	−397	529	+ 9,131
1868	10.71	1,030	−22	−370	484	+ 8,140
1869	11.95	1,077	−21	−323	441	+ 6,783
1870	12.26	1,088	−20	−312	400	+ 6,240
1871	12.85	1,109	−19	−291	361	+ 5,529
1872	14.84	1,172	−18	−228	324	+ 4,104
1873	15.12	1,180	−17	−220	289	+ 3,740
1874	13.92	1,144	−16	−256	256	+ 4,096
1875	14.12	1,150	−15	−250	225	+ 3,750
1876	13.96	1,145	−14	−255	196	+ 3,570
1877	14.19	1,152	−13	−248	169	+ 3,224
1878	14.54	1,162	−12	−238	144	+ 2,856
1879	14.41	1,159	−11	−241	121	+ 2,651
1880	18.58	1,269	−10	−131	100	+ 1,310
1881	19.82	1,297	− 9	−103	81	+ 927
1882	21.56	1,334	− 8	− 66	64	+ 528
1883	21.76	1,338	− 7	− 62	49	+ 434
1884	20.46	1,311	− 6	− 89	36	+ 534
1885	19.84	1,298	− 5	−102	25	+ 510
1886	20.81	1,318	− 4	− 82	16	+ 328
1887	22.82	1,358	− 3	− 42	9	+ 126
1888	24.03	1,381	− 2	− 19	4	+ 38
1889	25.88	1,413	− 1	+ 13	1	− 13
1890	27.87	1,445	0	45	0	0
1891	26.17	1,418	1	18	1	18
1892	26.92	1,430	2	30	4	60
1893	25.26	1,402	3	2	9	6
1894	26.03	1,416	4	16	16	64
1895	29.37	1,468	5	68	25	340
1896	31.29	1,495	6	95	36	570
1897	33.46	1,525	7	125	49	875
1898	36.46	1,562	8	162	64	1,296
1899	40.87	1,611	9	211	81	1,899
1900	41.35	1,616	10	216	100	2,160
1901	41.14	1,614	11	214	121	2,354
1902	44.73	1,651	12	251	144	3,012
1903	46.82	1,670	13	270	169	3,510
1904	46.22	1,665	14	265	196	3,710
1905	54.79	1,739	15	339	225	5,085
1906	59.66	1,776	16	376	256	6,016
1907	61.30	1,787	17	387	289	6,579
1908	48.80	1,688	18	288	324	5,184
1909	60.60	1,782	19	382	361	7,258
1910	66.20	1,821	20	421	400	8,420
		63,413	−115	−987	8,395	147,937

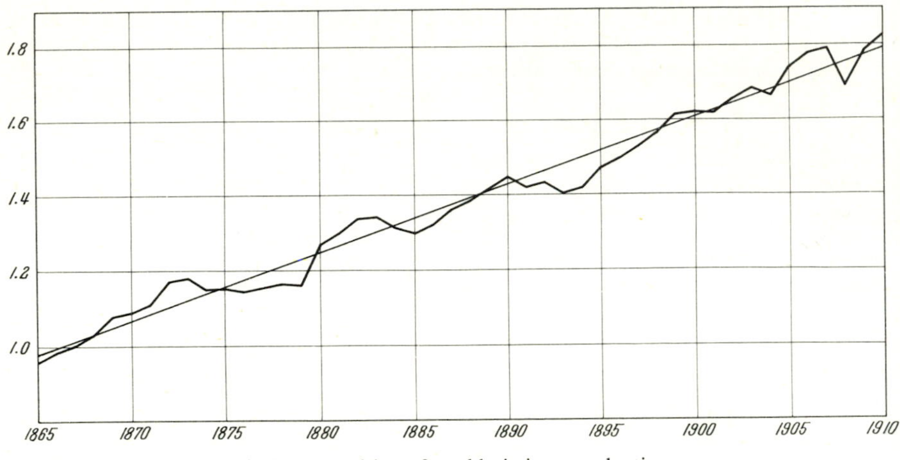

Fig. 21. Logarithm of world pig iron production

First of all, if we draw a curve through the given numbers and smooth it quite crudely, we see that its slope rapidly increases at more than a linear rate. Therefore, the curve cannot be represented very well by a straight line or a parabola.

In contrast, an exponential curve seems to be a good fit. Similarly, the fluctuations increase with time. Therefore, it seems that we should work with the logarithms y rather than the numbers x themselves and fit a straight line to them as well as possible.

We find

$$\bar{t}=1{,}890-\tfrac{115}{46}=1{,}890-2.5=1{,}887.5,$$

$$m_0=\bar{y}=1{,}400-\tfrac{987}{46}=1{,}400-21=1{,}379.$$

The regression line goes through the point $(\bar{t}, \bar{y})$. Its slope is

$$m_1=\frac{\sum(t-\bar{t})(y-\bar{y})}{\sum(t-\bar{t})^2}=\frac{147{,}937-2.5\times 987}{8{,}395-46\times 2.5^2}=\frac{145{,}470}{8{,}107.5}=17.94.$$

The equation of the regression line

$$y=m_0+m_1(t-\bar{t})=\bar{y}+m_1(t-\bar{t})$$

is therefore

$$y=1{,}379+17.94(t-1{,}887.5).$$

The regression line fits the actual graph very well. We could improve the approximation still further if we add a quadratic term $m_2\,\psi_2$, where

$$\psi_2=(t-\bar{t})^2-\gamma.$$

The constant γ is determined in such a way as to make ψ_2 orthogonal to $\psi_0=1$;

$$\sum_t \psi_0(t)\,\psi_2(t)=0$$

(ψ_2 is automatically orthogonal to ψ_1 for all γ, since the t values are symmetric about $\bar{t}$).

Therefore, we are led to the condition

$$\sum_t (t-\bar{t})^2 - 46\gamma = 0,$$

which we can solve for γ without further difficulty since $\sum(t-\bar{t})^2$ is already known to be 8,107.5.

Orthogonalization has the advantage that it is not necessary to recalculate the coefficients m_0 and m_1 already obtained. We merely work out m_2 from the third normal equation and add the new term $m_2\psi_2$ to the equation of the regression function. The reader may carry this out for himself.

§ 34. Causal Interpretation of Dependence between Economic Variables

If an economic variable w depends on the variables $x, y, \ldots$, but is also influenced by other causes, we can try to relate changes in w to changes in $x, y, \ldots$, as far as possible, and thus make w accessible to theoretical calculation.

A classic example is provided by investigation of periodic fluctuation in the price of pigs carried out by A. Hanau[6]. If the price of pigs is high, this will induce farmers to breed more pigs. Accordingly, about one and a half years later larger numbers of pigs will come onto the market and the prices will fall. Then the opposite process will take place, and so on. Therefore, insofar as the process is not disturbed by other causes, oscillations in prices with a period of three years will arise.

In general, research into economic activity does not turn out so simply. Nevertheless, we can still try to see how far we can go with causal explanations. We start with the values of $x, y, \ldots$ and w observed in a sequence of years, subtract the appropriate arithmetic mean for each variable so that the means become zero, and adjust the sequence for time trend by subtracting a fitted function of time (a linear function will do in most cases), thus leaving only the periodic and irregular fluctuations. Approximation of the causal dependence between the fluctuations of $x, y, \ldots$ and those of w by a linear relation, leads to the model

$$w = \lambda x + \mu y + \cdots + u, \tag{1}$$

where u represents an unexplained residual which, naturally, we wish to make as small as possible. Hence, we determine the coefficients $\lambda, \mu, \ldots$, so that the sum of squares of the u values

$$[u\,u] = \sum u_i^2 \tag{2}$$

is minimized.

[6] A. Hanau, Die Prognose der Schweinepreise, Sonderheft 18 der Vierteljahrshefte zur Konjunkturforschung, Berlin, 1930.

If we differentiate $[uu]$ with respect to $\lambda, \mu, \ldots$ and set the derivatives equal to zero, as in § 30, we obtain the normal equations

$$\begin{aligned} \lambda[xx]+\mu[xy]+\cdots&=[xw] \\ \lambda[yx]+\mu[yy]+\cdots&=[yw], \end{aligned} \tag{3}$$

from which we can determine the coefficients $\lambda, \mu, \ldots$.

If the conjecture that the economic variable w is influenced by the variable x after a time lag seems reasonable (as in the example above where inflated pig prices cause an increased supply of pigs after a lag of one and a half years), we can try to establish the amount of lag by means of time displacements of the x values. First of all, by means of a preliminary analysis, we settle the question of which time displacement yields the largest correlation between x and w. Therefore, we first calculate the correlation coefficient between x_i and w_i, then that between x_{i-1} and w_i, between x_{i-2} and w_i, and so forth (of course, within moderate limits corresponding to reasonable theoretical considerations). Then, we choose that lag which produces the largest correlation coefficient. Using this best possible lag, we again set up model (1). We can also improve on the delay in order to make the minimized sum of squares $[uu]$ as small as possible. That is to say, we try out different values of the lag one after the other and each time we solve the normal equations and calculate $[uu]$.

Examples of these methods are to be found in J. Tinbergen, Business Cycles in the United States, Publ. Völkerbund, Genf 1939. Since the appearance of this pioneer work, economists have become much more careful in applying the method. First of all, they would check whether there is a too strong dependence between the "independent variables" $x, y, \ldots$ by means of the bunch graphs of R. Frisch. However, we cannot go into this more refined econometric method here, but refer the reader to:

G. Tintner, Econometrics, New York and London, 1933;

G. Tintner, Econometrics, New York and London, 1952;

L. R. Klein, A Textbook on Econometrics, Evanston and New York, 1953;

W. C. Hood and T. C. Koopmans, Studies in Econometric Method, Cowles Monograph No. 14, New York (Wiley) 1953.

Chapter Eight

Estimation of Unknown Constants

Of the four parts of this chapter, the first (§§ 35 and 36) is devoted to explanation and illustration of the maximum likelihood method of estimation and is intended for the reader who is not familiar with this method. Solution of equations arising in the practical application of the method is discussed in § 36.

In the second part (§§ 37 through 39) it is shown that in the estimation of unknown parameters there exists a maximal degree of accuracy and that in certain cases the method of maximum likelihood is optimal because it attains this degree of accuracy. The mathematical proof of these results in this second part depends upon an inequality of Fréchet.

This second part serves well as an introduction to the problem, but as a practical method of finding optimal estimates it leaves much to be desired. In the third part (§§ 40 through 44) a method is developed which is more powerful. The method of the third part gives rise to an optimal unbiased estimate even in cases in which the method of maximum likelihood fails.

The short fourth part (§ 45) gives a survey of the asymptotic properties of maximum likelihood estimates.

In this chapter the observed random variables from which the unknown parameters are estimated are, for the most part, continuous variables $x_1, \ldots, x_n$. The case in which the observed quantities are frequencies is considered in the following chapter. Examples 21, 28, and 31 in this chapter are also of this kind.

§ 35. R. A. Fisher's Method of Maximum Likelihood

As we have seen in § 30, Gauss' primary justification of the method of least squares depended on the principle that the best values of the unknown parameters $\vartheta_1, \ldots, \vartheta_r$ are those which assign the observed event the greatest probability. R. A. Fisher used the same principle as the starting point for a general method of estimating unknown parameters $\vartheta_1, \ldots, \vartheta_r$, given numerical values of random quantities having a probability law depending on $\vartheta_1, \ldots, \vartheta_r$.

The observed variables $x_1, \ldots, x_n$ may be of the discrete or of the continuous type. In the discrete case let

$$g(t \mid \vartheta) = g(t_1, \ldots, t_n \mid \vartheta_1, \ldots, \vartheta_r)$$

be the probability that the variables $x_1, \ldots, x_n$ take on particular values $t_1, \ldots, t_n$. In the case of continuous random variables let $g(t \mid \vartheta) = g(t_1, \ldots, t_n \mid \vartheta_1, \ldots, \vartheta_r)$ be the probability density function of the variables $x_1, \ldots, x_n$. In the theory of least squares, $g(t \mid \vartheta)$ is a product of Gaussian probability densities

$$g(t \mid \vartheta) = \sigma^{-n} \sqrt{2\pi}^{-n} \exp\left\{-\frac{1}{2} \sum \frac{(t-\xi)^2}{\sigma^2}\right\}, \tag{1}$$

where the "true values" ξ_i are given functions of ϑ. We shall drop this particular case now and allow $g(t \mid \vartheta)$ to be any type of probability density function of ϑ and t.

Fisher replaced the t_i in $g(t \mid \vartheta)$ by the observed values x_i and called the resulting function $g(x \mid \vartheta)$ of $\vartheta_1, \ldots, \vartheta_r$ the *likelihood function.* Those parameter values $\tilde{\vartheta}$ maximizing the likelihood function – that is, those values giving the observed event the greatest probability – are called the *most likely* values of the parameters ϑ. To employ the *method of maximum likelihood* is, therefore, to use the most likely values $\tilde{\vartheta}$ as estimates of the true parameter values ϑ.

The logarithm of $g(x \mid \vartheta)$ will be denoted in the following by $L(x \mid \vartheta)$ or $L(\vartheta)$.

The likelihood function must not be confused with a probability. To be sure, it is defined in terms of a probability or probability density. But it is the probability density of the observed event and not that of the unknown parameters. The parameters have no probability density, in as much as they do not depend on chance. However, those parameter values for which the observed event is relatively more probable may seem more likely than those for which it is highly improbable.

In the case of continuous variables we may have occasion to make a transformation of variables from the t_i to new variables t'_i, in which case the probability density must be multiplied by the determinant of the transformation. Therefore the function $g(t \mid \vartheta)$ is determined only up to a factor depending on t alone. Likewise in the case of discrete quantities we may, for purposes of simplification, multiply the function $g(t \mid \vartheta)$ by a positive factor depending only on t. Obviously such operations will not affect the point at which g as a function of ϑ assumes its maximum.

The examples given in Chapter 7 to illustrate the method of least squares serve also as examples of the method of maximum likelihood. In addition we include here three more examples.

Example 21. An event with an unknown probability p has occurred x times in n independent trials. We wish to estimate the unknown probability p.

According to Bernoulli (§ 5 A), the likelihood function is

$$\binom{n}{x} p^x (1-p)^{n-x},$$

or, disregarding the binomial coefficient, which is independent of p,

$$g(x \mid p) = p^x (1-p)^{n-x}. \tag{2}$$

Instead of maximizing $g(x \mid p)$, we may just as well maximize the logarithm

$$L(p) = x \ln p + (n-x) \ln(1-p).$$

Differentiation with respect to p yields

$$L'(p) = \frac{x}{p} - \frac{n-x}{1-p} = \frac{x-np}{p(1-p)}.$$

The derivative $L'(p)$ is zero if $np = x$, positive for smaller values of p, and negative for larger values, so that $L(p)$ attains its maximum when $np = x$. The most plausible value of p is therefore

$$\tilde{p} = h = \frac{x}{n}. \tag{3}$$

The estimate (3) is both *unbiased* and *consistent*; i.e., the expected value of h is exactly equal to the true value p, and according to the *law of large numbers* (§§ 5 and 33), it converges in probability to the true value p as $n \to \infty$.

Example 22. A variable x has a normal probability density function

$$f(x) = \sigma^{-1}(2\pi)^{-\frac{1}{2}} \exp\left\{-\frac{1}{2}\left(\frac{x-\mu}{\sigma}\right)^2\right\},$$

with unknown mean μ and unknown standard deviation σ. A sample $x_1, \ldots, x_n$ consists of n independent observations of the variable x. What are the most likely values of μ and σ?

If the probability density function is multiplied by the nonessential factor $(2\pi)^n/^2$, the likelihood function is

$$g(x_1, \ldots, x_n \mid \mu, \sigma) = \sigma^{-n} \exp\{-\tfrac{1}{2}\sigma^{-2} \sum (x_i - \mu)^2\},$$

and its logarithm is

$$L(\mu, \sigma) = -n \ln \sigma - \tfrac{1}{2}\sigma^{-2} \sum (x_i - \mu)^2. \tag{4}$$

The second term is a negative definite quadratic polynomial in μ. Its maximum is found by differentiation with respect to μ to occur when

$$\sum (x_i - \tilde{\mu}) = 0,$$

or (5)

$$\tilde{\mu} = \frac{1}{n} \sum x_i = \bar{x}.$$

Hence *the most likely value of μ is the arithmetic mean of the observed x-values*, the same result as Gauss found using the method of least squares.

If the μ in (4) is replaced by $\tilde{\mu}=\bar{x}$ and the resulting function differentiated with respect to σ, the derivative is

$$\frac{d}{d\sigma}L(\tilde{\mu},\sigma)=-\frac{n}{\sigma}+\frac{1}{\sigma^3}\sum(x_i-\bar{x})^2.$$

This derivative is zero for

$$n\sigma^2=\sum(x_i-\bar{x})^2,$$

positive for smaller values of σ and negative for larger values. Therefore the most plausible value of σ is defined by

$$\tilde{\sigma}^2=\frac{1}{n}\sum(x_i-\bar{x})^2. \tag{6}$$

Previously instead of (6) we had the approximate value

$$s^2=\frac{1}{n-1}\sum(x_i-\bar{x})^2, \tag{7}$$

where the factor $1/(n-1)$ was chosen so that the mean value of s^2 would be σ^2 exactly. The expected value of (6) is obviously somewhat smaller than that of (7). Therefore the maximum likelihood estimate (6) is *biased*: its expected value is not equal to the true value σ^2.

In this example the bias of the estimate $\tilde{\sigma}^2$ is small and tends to zero as $n\to\infty$. The standard deviation of the estimate $\tilde{\sigma}^2$ likewise tends to zero as $n\to\infty$. The *consistency* of the estimate follows from these two properties and the inequality of Tchebychev (§ 3C).

Example 23. In the following example the method of maximum likelihood does not give rise to a consistent estimate.

In a laboratory each of n concentrations was measured twice. The accuracy of measurement was the same in every case, but the true values were possibly different in each of the n cases. If the $2n$ observed measurements $x_1, y_1; \ldots; x_n, y_n$ are assumed to be independent and normally distributed, the probability density function is

$$g(x_i, y_i|\sigma,\mu_i)=\sigma^{-2n}(2\pi)^{-n}\exp\left\{-\sum\frac{(x_i-\mu_i)^2+(y_i-\mu_i)^2}{2\sigma^2}\right\}. \tag{8}$$

The n mean values $\mu_1,\ldots,\mu_n$ and the standard deviation σ are unknown. The most plausible value of μ_i is again the arithmetic mean

$$\tilde{\mu}_i=\tfrac{1}{2}(x_i+y_i).$$

If the μ_i in (8) are replaced by $\tilde{\mu}_i$, the resulting function is

$$(2\pi)^n g(x_i,y_i|\sigma,\tilde{\mu}_i)=\sigma^{-2n}\exp\left\{-\sum\frac{(x_i-y_i)^2}{4\sigma^2}\right\}.$$

Use of logarithmic differentiation as above yields the solution

$$\tilde{\sigma}^2=\frac{1}{4n}\sum(x_i-y_i)^2, \tag{9}$$

which maximizes the likelihood function.

The difference (x_i-y_i) is, for each i, normally distributed with expectation zero and variance $2\sigma^2$. Therefore the expected value of $(x_i-y_i)^2$ is $2\sigma^2$ and that of $\sum(x_i-y_i)^2$ is $2n\sigma^2$. Hence the expected value of $\tilde{\sigma}^2$ is

$$\mathscr{E}(\tilde{\sigma}^2)=\tfrac{1}{2}\sigma^2,$$

so that in this case the method of maximum likelihood leads to a systematic under-estimation of the variance σ^2.

An estimate without bias would be

$$s^2 = \frac{1}{2n} \sum (x_i - y_i)^2 . \tag{10}$$

The estimate (10) is also consistent. We shall later prove that it has the smallest variance among all unbiased estimates.

We see from these examples that in some cases the maximum likelihood method leads to satisfactory unbiased estimates and in others to estimates which are, at least, consistent as $n \to \infty$, but that in still other cases it does not yield a good result.

Hence there arises the problem of determining in which cases the maximum likelihood method is good and in which cases it fails. A complete answer is difficult to give, but, nevertheless, investigating the problem yields something of a clarification of it. In general, we can say the following. If there are many independent observations $x_1, \ldots, x_n$ and at most a limited number of parameters $\vartheta_1, \ldots, \vartheta_r$, and if the distribution function satisfies certain regularity conditions, then the maximum likelihood method is good and becomes better with increasing n. If, however, n is not large or r increases simultaneously with n (as in the last example), one can not rely on the method. In such cases there exist other methods for finding the best unbiased estimates. We shall encounter one such method in § 41.

For the moment we shall continue our discussion of the maximum likelihood method and assume, for the sake of simplicity, that there is only *one* unknown parameter ϑ.

§ 36. Determination of the Maximum

Practical application of the maximum likelihood method requires above all the solution of the *likelihood equation:*

$$L'(x \mid \tilde{\vartheta}) = 0. \tag{1}$$

$L'(t \mid \vartheta)$ is the logarithmic derivative with respect to ϑ of the probability density $g(t \mid \vartheta)$. Therefore we have

$$L'(x \mid \vartheta) = \frac{g'(x \mid \vartheta)}{g(x \mid \vartheta)} .$$

We denote by $\tilde{\vartheta}$ the maximum likelihood estimate, which must fulfill in every case condition (1), and by ϑ_0 the (unknown) true value of the parameter ϑ. The expected value of a variable y is denoted by $\mathscr{E}_0\, y$ if its probability density is $g(t \mid \vartheta_0)$ and by $\mathscr{E}_\vartheta\, y$ if the density is $g(t \mid \vartheta)$. Primes always indicate differentiation with respect to ϑ.

We shall now assume that $x_1, \ldots, x_n$ are independent observations, identically distributed with probability density function $f(x \mid \vartheta)$ depending on ϑ. Hence we have

$$g(x|\vartheta)=f(x_1|\vartheta)\dots f(x_n|\vartheta),$$

and

$$L'(x|\vartheta)=\sum \varphi(x_k|\vartheta), \tag{2}$$

where $\varphi=f'/f$ is the logarithmic derivative of f.

As we have seen in § 35, there exist cases in which the Eq. (1) possesses an elementary solution. More often, however, (1) is a complicated algebraic or transcendental equation which must be solved by successive approximation.

The simplest procedure is to choose first an approximate value ϑ_1 and calculate $L'(x|\vartheta_1)$ as the sum of the contributions (scores) of the single observations x_k:

$$L'(x|\vartheta_1)=\sum \varphi(x_k|\vartheta_1). \tag{3}$$

Then we must determine a closer approximation

$$\vartheta_2=\vartheta_1+h$$

by Newton's method of successive approximation, using the first order expansion

$$L'(x|\vartheta_2)\sim L'(x|\vartheta_1)+h\,L''(x|\vartheta_1). \tag{4}$$

Setting the right-hand side equal to zero yields

$$h=\frac{L'(x|\vartheta_1)}{-L''(x|\vartheta_1)}, \tag{5}$$

the denominator of which is

$$-L''(x|\vartheta_1)=-\sum \varphi'(x_k|\vartheta_1). \tag{6}$$

The right-hand sum is n times the arithmetic mean of the φ'. A simplification of the calculation can be achieved by replacing the arithmetic mean by the expected value,

$$\mathscr{E}_0\,\varphi'(x|\vartheta_1)=\int \varphi'(t|\vartheta_1)\,f(t|\vartheta_0)\,dt,$$

integration being over the whole range of possible x-values.

In as much as we are concerned here with an approximation, the unknown parameter-value ϑ_0 may be replaced by ϑ_1. Hence we replace the denominator $-L''(x|\vartheta_1)$ in (5) by $nj(\vartheta_1)$, where $j(\vartheta)$ is defined by

$$j(\vartheta)=-\int \varphi'(t|\vartheta)\,f(t|\vartheta)\,dt. \tag{7}$$

Instead of (5) we now have

$$h_1=\frac{L'(x|\vartheta_1)}{nj(\vartheta_1)}. \tag{8}$$

The term $j(\vartheta)$ in the denominator of (8) may also be represented as

$$j(\vartheta) = -\int \left(\frac{f'}{f}\right)' f\,dt = \int \left(\frac{f'f'}{f} - f''\right) dt. \tag{9}$$

Recalling that

$$\int f(t|\vartheta)\,dt = 1, \tag{10}$$

and assuming that (10) may be differentiated twice under the integral sign, we have

$$\int f''\,dt = 0.$$

Hence (9) becomes

$$j(\vartheta) = \int \left(\frac{f'}{f}\right)^2 f\,dt = \mathscr{E}_\vartheta \left(\frac{f'}{f}\right)^2. \tag{11}$$

Multiplying this by n we get the expression which R. A. Fisher called the *information in the sample:*

$$I(\vartheta) = n\,j(\vartheta) = n\,\mathscr{E}_\vartheta \left(\frac{f'}{f}\right)^2. \tag{12}$$

As a second approximation for $\tilde{\vartheta}$ we now have

$$\vartheta_2 = \vartheta_1 + h_1 \tag{13}$$

with

$$h_1 = \frac{L'(x|\vartheta_1)}{I(\vartheta_1)}, \tag{14}$$

where $L'(x|\vartheta)$ and $I(\vartheta)$ are defined by (2) and (12) respectively.

The expression $I(\vartheta)$ may be generalized to the case in which the x_k have different distributions with densities f_k by

$$I(\vartheta) = \sum_k \mathscr{E}_\vartheta (f_k^{-1} f_k')^2.$$

The most general expression for $I(\vartheta)$ reads

$$I(\vartheta) = \mathscr{E}_\vartheta\, L'(x|\vartheta)^2 = \int L'(t|\vartheta)^2\, g(t|\vartheta)\,dt, \tag{15}$$

integrated over the entire x-space.

Information is additive in the sense that if two independent samples $x_1, \ldots, x_m$ and $y_1, \ldots, y_n$ are observed, the information I in the combined sample is the sum of the contributions of the two samples separately:

$$I(\vartheta) = I_1(\vartheta) + I_2(\vartheta). \tag{16}$$

$I(\vartheta)$ is always non-negative. Furthermore, $I(\vartheta)$ is equal to zero if and only if $g(x|\vartheta)$ does not depend on ϑ, so that the x-values contain no information about ϑ. These characteristics of $I(\vartheta)$ make the use of the word "information" intuitively appealing.

Example 24. A radioactive source which emits particles uniformly in all directions is on a foil at an unknown position. When the rays strike a screen parallel to the foil, they give rise to observable scintillations. How can the position of the source be determined from the positions of the scintillations?

Let the screen be taken as the xy-plane with the distance between the foil and the screen as unit length. Let the two parallel planes have the equations $z=0$ and $z=1$. Let the coordinates of the source be $(\vartheta, \eta, 1)$.

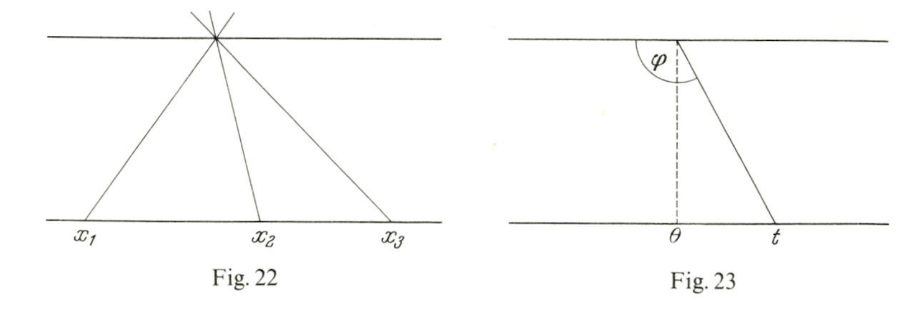

Fig. 22 Fig. 23

To simplify the problem, let us assume that we are interested only in the x-coordinate ϑ of the source and that consequently we have measured only the x-coordinates of the points of impact $x_1, \ldots, x_n$. We can then project the above model onto the xz-plane (Fig. 22).

The distribution function $F(t)$ of x is the probability that a ray, as projected, strikes to the left of the point t (Fig. 23). These rays lie in a dihedral angle bounded by the two planes which include the angle

$$\varphi = \frac{\pi}{2} + \arc\tan(t-\vartheta). \tag{17}$$

All rays which strike the screen at all form a dihedral angle with angle π. The desired probability is therefore

$$F(t) = \frac{\varphi}{\pi} = \frac{1}{2} + \frac{1}{\pi} \arc\tan(t-\vartheta). \tag{18}$$

Hence we have a Cauchy distribution, the probability density of which is

$$f(t\,|\,\vartheta) = \frac{1}{\pi} \frac{1}{(t-\vartheta)^2+1}. \tag{19}$$

The likelihood function is

$$\begin{aligned} g(x\,|\,\vartheta) &= \pi^n f(x_1\,|\,\vartheta) \ldots f(x_n\,|\,\vartheta) \\ &= \prod_1^n \{(x_k-\vartheta)^2+1\}^{-1}, \end{aligned} \tag{20}$$

and its logarithm is

$$L(x\,|\,\vartheta) = -\sum \ln\{(x_k-\vartheta)^2+1\}. \tag{21}$$

At the maximum, the derivative must be zero:

$$\sum \frac{2(x_k-\vartheta)}{(x_k-\vartheta)^2+1} = 0. \tag{22}$$

For $n=1$, the solution of (22) is obviously

$$\vartheta = x_1.$$

For $n=2$, we get a third degree equation

$$(x_1-\vartheta)\{(x_2-\vartheta)^2+1\}+(x_2-\vartheta)\{(x_1-\vartheta)^2+1\}=0,$$

or

$$(x_1+x_2-2\vartheta)\{(x_1-\vartheta)(x_2-\vartheta)+1\}=0,$$

which always has a solution

$$\vartheta_1=\bar{x}=\tfrac{1}{2}(x_1+x_2). \tag{23}$$

The remaining two solutions must satisfy the quadratic equation

$$\vartheta^2-2\vartheta\,\bar{x}+x_1\,x_2+1=0,$$

which may also be written as

$$(\vartheta-\bar{x})^2=\left(\frac{x_1-x_2}{2}\right)^2-1. \tag{24}$$

If the distance between the points of impact x_1 and x_2 is smaller than 2, Eq. (24) has no real root and solution (23) maximizes the likelihood. If the distance is exactly 2, the three roots of the likelihood equation coincide at $\vartheta_1=\bar{x}$. If, however, the distance is larger than 2, solution (23) yields a minimum and the two real solutions to Eq. (24) yield maxima. One of the two solutions is close to x_1 and the other to x_2. The maximum likelihood method makes no provision for choosing between the two solutions. In applications, one would probably choose that one which lies nearer the center of the foil.

For $n>2$, one must solve Eq. (22) by successive approximation. A first approximation ϑ_1 is, say, the empirical median Z (i.e., for n odd, the middle one of the n points $x_1,\ldots,x_n$). A better approximation is $\vartheta_2=\vartheta_1+h_1$ with

$$h_1=\frac{L'(x\,|\,\vartheta_1)}{I(\vartheta_1)}. \tag{25}$$

The numerator is the left-hand side of (22) for $\vartheta=\vartheta_1$. The denominator is the information $I(\vartheta_1)=n\,j(\vartheta_1)$, where $j(\vartheta)$, as defined by (11), is found to be

$$j(\vartheta)=\int_{-\infty}^{\infty}\left(\frac{f'}{f}\right)^2 f\,dt=\frac{4}{\pi}\int_{-\infty}^{\infty}\left\{\frac{t-\vartheta}{(t-\vartheta)^2+1}\right\}^2\frac{1}{(t-\vartheta)^2+1}\,dt=\frac{4}{\pi}\int_{-\infty}^{\infty}\frac{u^2\,du}{(u^2+1)^3}=\frac{1}{2}.$$

Hence the information

$$I(\vartheta)=n\,j(\vartheta)=\frac{n}{2} \tag{26}$$

is independent of ϑ. From (25) we now have

$$h_1=\frac{2}{n}\,L'(x\,|\,\vartheta_1)=\frac{4}{n}\sum\frac{(x_k-\vartheta_1)}{(x_k-\vartheta_1)^2+1}. \tag{27}$$

The successive approximations converge very quickly. The variance of the estimate obtained in this way is, asymptotically for large n,

$$I(\vartheta)^{-1}=\frac{2}{n}. \tag{28}$$

The empirical median has (see § 20) asymptotically the variance

$$\frac{1}{4n[f(\vartheta)]^2}=\frac{\pi^2}{4n}. \tag{29}$$

Comparison of (28) with (29) shows that asymptotically the maximum likelihood estimate is much better than the median. Much worse than the median, however, is the arithmetic mean $\bar{x}$, since the variance of $\bar{x}$ is infinite and the distribution function of $\bar{x}$ is the same as that of the single observations x_k.

§ 37. An Inequality Due to Fréchet

We shall consider an estimate T of an unknown parameter ϑ to be a good estimate if the T-values are concentrated as closely as possible about the true value ϑ. The two quantities which are used primarily as criteria for judging the optimality of an estimate are the expected value $\hat{T} = \mathscr{E}\, T$ and the variance

$$\sigma_T^2 = \mathscr{E}(T - \hat{T})^2 .$$

The expected value $\mathscr{E}\, T$ depends on ϑ, so that we should write $\mathscr{E}_\vartheta T$ instead of $\mathscr{E}\, T$. We shall require that this expected value be equal to ϑ or at least be close to ϑ. The difference

$$\hat{T} - \vartheta = \mathscr{E}_\vartheta T - \vartheta = b(\vartheta) \tag{1}$$

is called the *bias*, or *systematic error*, of the estimate. The variance σ_T^2 will be required to be as small as possible. An unbiased estimate with the smallest possible variance is said to be a *minimum variance unbiased estimate.*

It is easy to give estimates with variance zero simply by setting T equal to any constant T_0 independent of the observed event, provided we are willing to accept in return a large bias $T_0 - \vartheta$, if T_0 differs considerably from ϑ. Bias and variance therefore can not both be made zero (except in trivial cases in which ϑ is known beforehand or can be determined with probability 1 from the observed event).

The foregoing considerations can be made precise by means of an inequality which yields a minimum value for the variance, given the bias. This inequality is an important result due independently to Fréchet, Rao, and Cramér[1]. In English literature it is known as the Cramér-Rao inequality, or more recently as the *information inequality.*

If y and z are chance variables and if y^2 and z^2 have finite expected value, then the *Schwarz inequality* yields

$$(\mathscr{E}\, y\, z)^2 \leqq (\mathscr{E}\, y^2)(\mathscr{E}\, z^2). \tag{2}$$

---

[1] M. Fréchet, Rev. Intern. de Stat. 1943, p. 182. C. R. Rao, Bull. Calcutta Math. Soc. 37, p. 81. H. Cramér, Skandinavisk Aktuarie-tidsk. 29, p. 85, or Math. Methods of Stat. p. 480. In addition, see J. Wolfowitz, Ann. of Math. Stat. 18 (1947) p. 215. For applications, see J. L. Hodges and E. L. Lehmann, Proceedings of the Second Berkeley Symposium on Mathematical Statistics (1951) p. 13.

The proof is quite simple. The quadratic form

$$\mathscr{E}(\lambda y+\mu z)^2=\lambda^2\,\mathscr{E}\,y^2+2\mu\lambda\,\mathscr{E}\,y\,z+\mu^2\,\mathscr{E}\,z^2 \tag{3}$$

is positive semi-definite and therefore its determinant is positive or zero:

$$(\mathscr{E}\,y^2)(\mathscr{E}\,z^2)-(\mathscr{E}\,y\,z)^2\geqq 0. \tag{4}$$

The Schwarz inequality (2) follows immediately from (4). In the case that one of the right-hand factors in (2) is infinite, inequality (2) is trivially true.

Now let $x_1,\ldots,x_n$ be observed quantities whose probability density

$$g(x\,|\,\vartheta)=g(x_1,\ldots,x_n\,|\,\vartheta) \tag{5}$$

depends on a single unknown parameter ϑ. (We shall no longer differentiate between the observed quantities $x_1,\ldots,x_n$ and the variables $t_1,\ldots,t_n$.) Let $T=T(x)$ be an estimate of this parameter. We shall derive a lower bound for the variance σ_T^2.

If $g(x\,|\,\vartheta)$ is zero on a portion of the x-space, this portion may be excluded from the integration. Therefore we shall integrate only over that part of the x-space on which $g(x\,|\,\vartheta)>0$. We assume that this part is independent of ϑ. Furthermore, we assume that $g(x\,|\,\vartheta)$ is differentiable with respect to ϑ. If the derivative with respect to ϑ is denoted by a prime, then the logarithm

$$L(x\,|\,\vartheta)=\ln g(x\,|\,\vartheta)$$

has the derivative

$$L'(x\,|\,\vartheta)=\frac{g'(x\,|\,\vartheta)}{g(x\,|\,\vartheta)}.$$

Furthermore

$$\vartheta+b(\vartheta)=\mathscr{E}_\vartheta\,T=\int T\,g(x\,|\,\vartheta)\,dx \tag{6}$$

and

$$1=\int g(x\,|\,\vartheta)\,dx. \tag{7}$$

We assume that (6) and (7) may be differentiated with respect to ϑ under the integral sign. Carrying out this differentiation, we have

$$1+b'(\vartheta)=\int T\,g'(x\,|\,\vartheta)\,dx=\mathscr{E}_\vartheta(T\,g'\,g^{-1})=\mathscr{E}_\vartheta(T\,L'), \tag{8}$$

$$0=\int g'\,dx=\mathscr{E}_\vartheta(g'\,g^{-1})=\mathscr{E}_\vartheta\,L'. \tag{9}$$

If (9) is multiplied by $\hat{T}$ and subtracted from (8), we get

$$1+b'(\vartheta)=\mathscr{E}_\vartheta[(T-\hat{T})\,L']. \tag{10}$$

On the right-hand side we have the expected value of a product to which the Schwarz inequality may be applied to yield

$$(1+b')^2 \leqq \sigma_T^2 \cdot \mathscr{E}_\vartheta L'^2 . \tag{11}$$

Assuming that $\mathscr{E}_\vartheta L'^2 \neq 0$ and replacing $\mathscr{E}_\vartheta L'^2$ by $I(\vartheta)$, we have from (11)

$$\sigma_T^2 \geqq \frac{[1+b'(\vartheta)]^2}{I(\vartheta)} . \tag{12}$$

This is the Fréchet inequality (information inequality). The hypotheses under which it was derived are the following:

1. The portion of the x-space on which $g(x\,|\,\vartheta) > 0$ is independent of ϑ.
2. (6) and (7) may be differentiated with respect to ϑ under the integral sign.
3. The denominator in (12) is strictly positive.

The denominator in (12) is the integral

$$I(\vartheta) = \mathscr{E}_\vartheta L'(x\,|\,\vartheta)^2 = \int (\ln g)' g' \, dx , \tag{13}$$

which, following R. A. Fisher, we have already called "information". Another expression for $I(\vartheta)$ obtained by integration by parts is

$$I(\vartheta) = -\mathscr{E}_\vartheta L''(x\,|\,\vartheta) . \tag{14}$$

If the estimate T is unbiased in a neighborhood of the true ϑ-value, then the numerator in (12) is equal to one and we have

$$\sigma_T^2 \geqq I(\vartheta)^{-1} . \tag{15}$$

The right-hand side does not depend on the estimate T. Therefore there exists a sure lower bound for the variance of an unbiased estimate, namely the reciprocal of the information, $I(\vartheta)^{-1}$.

The inequality of Fréchet and the derivations from it also hold true if $x_1, \ldots, x_n$ are discrete variables. In all the equations the integrals must be replaced by sums, and the sums (6) and (7) must be assumed differentiable term by term, which, for example, is always the case with finite sums.

§ 38. Sufficiency and Minimum Variance

When does the equality sign hold in the foregoing inequality?

Because the form (3) § 37 is a pure quadratic, the equality sign in the Schwarz inequality (2) § 37 obviously holds only in the case that there exists a λ and a μ, both not zero, such that $\lambda y + \mu z$ takes on the value zero with probability 1. In the case of inequality (12), this means that either $T = \hat{T}$ with probability one, or with probability one

$$L'(x|\vartheta) = K \cdot (T - \hat{T}), \tag{1}$$

where K does not depend on x.

The first case, that the estimate T has a constant value T_0 independent of the observations, may be dropped from consideration. In this case $b(\vartheta) = T_0 - \vartheta$ is very much dependent upon ϑ. It is the case of an extreme "bias" in the literal sense of a preconceived notion: the true value of ϑ is believed to be known beforehand and no consideration is given to the observations. This attitude can be entirely satisfactory in the case that the preconceived idea is well-founded and is not refuted by the observations. The problem of the "most exact estimate based on observations" does not exist in this case.

There remains case (1). Integration yields

$$L(x|\vartheta) = \ln g(x|\vartheta) = A(\vartheta) \cdot T + B(\vartheta) + C(x),$$

and hence

$$g(x|\vartheta) = e^{AT+B} h(x), \tag{2}$$

where A and B depend only on ϑ and h only on x.

Therefore, two conditions are required so that the equality sign holds in inequality (12) § 37, namely:

a) *The likelihood function* $g(x|\vartheta)$ *is a product of two factors*

$$g(x|\vartheta) = e(T|\vartheta)\, h(x), \tag{3}$$

of which the first depends only on ϑ *and* T *and the second only on* x.

b) *The first factor has the form*

$$e(T|\vartheta) = e^{AT+B}, \tag{4}$$

where A *and* B *depend only on* ϑ.

If condition a) is fulfilled, T is called a *sufficient estimate* or *sufficient statistic* for ϑ in the terminology of R.A. Fisher.

We shall now prove the following:

If conditions 1 *through* 3 (§ 37) *and, moreover,* a) *and* b) *are fulfilled, then the estimate* T *has the smallest variance among all estimates with the same bias* $b(\vartheta)$.

Proof. From (3) and (4) we have

$$L'(x|\vartheta) = A'T + B'. \tag{5}$$

From (9) § 37 it follows that

$$A'\mathscr{E}T+B'=\mathscr{E}(A'T+B')=\mathscr{E}L'=0,$$

and therefore

$$B'=-A'\mathscr{E}T=-A'\hat{T}. \tag{6}$$

If the expression (6) for B' is substituted into (5), we then have

$$L'(x|\vartheta)=A'(T-\hat{T}). \tag{7}$$

Therefore, since L' and $T-\hat{T}$ are proportional, the equality sign in Fréchet's inequality is valid:

$$\sigma_T^2=\frac{[1+b'(\vartheta)]^2}{I}.$$

For every other estimate the sign $\geqq$ holds true. Therefore T has the smallest possible variance σ_T^2 among all estimates with bias $b(\vartheta)$.

In order to establish the relation to the maximum likelihood method, we add to assumptions a) and b) a further hypothesis, namely:

c) *The estimate T is unbiased.*

Hypothesis c) states that

$$b(\vartheta)=\hat{T}-\vartheta=0,$$

or $\hat{T}=\vartheta$. If this is substituted into (7), we have

$$L'(x|\vartheta)=A'(T-\vartheta). \tag{8}$$

The equality (10) § 37 becomes now

$$\mathscr{E}_\vartheta[(T-\hat{T})L']=1.$$

Substituting the expression (7) for L' we have

$$\mathscr{E}_\vartheta[A'(T-\hat{T})^2]=1,$$

or

$$A'\sigma_T^2=1. \tag{9}$$

From here it follows that A' is always positive. Furthermore, it follows from (7) that

$$\begin{aligned}I&=\mathscr{E}L'(x|\vartheta)^2\\&=A'^2\mathscr{E}(T-\hat{T})^2\\&=A'^2\sigma_T^2,\end{aligned}$$

and therefore from (9) that

$$I=A', \tag{10}$$

or that *the information I is the derivative with respect to ϑ of the coefficient A occurring in* (4).

Setting up the likelihood equation, we have, using (8),

$$A'(T-\vartheta)=0, \tag{11}$$

which has the unique solution $\tilde{\vartheta}=T$, since A' is always positive. Since $L'(x|\vartheta)$ is positive for $\vartheta<T$ and negative for $\vartheta>T$, L assumes its maximum for $\vartheta=T$. Furthermore, for the same value of ϑ the likelihood function

$$g(x|\vartheta)=\exp L(x|\vartheta)$$

also assumes its maximum. Therefore:

Under hypotheses a), b), *and* c) *the maximum likelihood method yields a minimum variance unbiased estimate.*

If hypothesis c) is dropped and $f(\vartheta)$ defined by

$$\mathscr{E}T=f(\vartheta),$$

then T is a minimum variance unbiased estimate of $f(\vartheta)$.

§ 39. Examples

The conditions 1 through 3 (§ 37) and a) through c) (§ 38) are satisfied in certain important cases, the simplest of which is the following.

Example 25. Estimation of the mean of a normal distribution.

Let the independent observations $x_1, \ldots, x_n$ be from a normal distribution with unknown mean μ. It is immaterial whether or not the standard deviation σ is known. We shall assume for the sake of simplicity that $\sigma=1$. The likelihood function is then (cf. § 35, Example 22)

$$g(x|\mu)=\exp\{-\tfrac{1}{2}\sum(x-\mu)^2\}.$$

Equivalently, we can write

$$g(x|\mu)=\exp\left(-\frac{1}{2}\sum x^2+\sum x\,\mu-\frac{n}{2}\mu^2\right).$$

If the sample mean M is defined by

$$M=\frac{1}{n}\sum x,$$

then $g(x|\mu)$ can be written as the product

$$g(x|\mu)=\exp n(\mu M-\tfrac{1}{2}\mu^2)\cdot\exp(-\tfrac{1}{2}\sum x^2).$$

The first factor depends only on M and μ and the second only on the observed x's. Condition a) is therefore satisfied and *M is a sufficient estimate of μ.*

Conditions 1 through 3 (§ 37) give no difficulty. Conditions b) and c) (§ 38) are obviously satisfied as well. *Therefore the mean M is a minimum variance unbiased estimate of μ.*

Example 26. Estimation of the variance of a normal distribution with known mean.

Without loss of generality we may assume that the mean value μ is zero, for if μ is known, then the variables may be shifted by the amount μ so that their common expected value is zero. Hence the probability density is, except for a constant factor,

$$g(x|\sigma)=\sigma^{-n}\exp\left(-\frac{\sum x^2}{2\sigma^2}\right). \tag{1}$$

We wish to estimate $\vartheta=\sigma^2$. Setting $\sum x^2=n\,s^2$, we may write in place of (1)

$$g(x|\sigma)=\exp\left(-n\ln\sigma-\frac{n}{2}\frac{s^2}{\sigma^2}\right). \tag{2}$$

This function is already of the form $\exp(A\,s^2+B)$. Therefore the estimate $T=s^2$ fulfills conditions a) and b). Since the expected value of s^2 is σ^2, condition c) is also fulfilled. Conditions 1 through 3 (§ 37) are easily verified. *Therefore* $s^2=\frac{1}{n}\sum x^2$ *is a minimum variance unbiased estimate of* σ^2.

Example 27. The method of least squares.

Let the observations $x_1, \ldots, x_n$ be independent and normally distributed with known standard deviations $\sigma_1, \ldots, \sigma_n$. In § 26 the mean values $\xi_1, \ldots, \xi_n$ were assumed to be differentiable functions of the unknown parameters $\vartheta_1, \vartheta_2, \ldots$. We used linear approximations to these functions. Here, however, we shall assume that the ξ_i are *linear* functions of a *single* unknown parameter ϑ. The theory may be extended to the case of more parameters, but in the case of non-linear functions it is only an approximation.

By the transformation $x_i=\sigma_i\,x_i'$ the general case may be reduced to the case that all the x's have unit variance. The probability density is then, except for a constant factor,

$$g(x|\vartheta)=\exp\{-\tfrac{1}{2}\sum(x_i-\xi_i)^2\}. \tag{3}$$

If the ξ_i are replaced by linear functions of ϑ,

$$\xi_i=c_i+a_i\,\vartheta, \tag{4}$$

then $g(x|\vartheta)$ takes the form

$$g(x|\vartheta)=\exp\tfrac{1}{2}(-k\,\vartheta^2+2l\,\vartheta-m), \tag{5}$$

where $k=\sum a_i^2=\sum a\,a$ is a constant, $l=\sum(x_i-c_i)\,a_i$ a linear function of the x's, and m a quadratic function of the x's. If $k=0$, then $a_i=0$ for every i, $g(x|\vartheta)$ is independent of ϑ, and ϑ cannot be estimated. If $k\neq 0$, then (5) may be rewritten as

$$g(x|\vartheta)=\exp\{-\tfrac{1}{2}k(\vartheta-T)^2+h(x)\}, \tag{6}$$

where

$$T=\frac{\sum a\,x-\sum a\,c}{\sum a\,a}. \tag{7}$$

It is clear that the expression { } in (6) assumes its maximum when $\vartheta=T$. The expression { } is the same as that which has occurred already in (3) as the exponent and as that which is to be maximized in applying the method of least squares. Therefore the method of least squares gives rise to T as the estimate of ϑ. The expected value of T is $\hat{T}=\vartheta$, so that T is unbiased. Conditions a) and b) and 1 through 3 are satisfied, and hence we have the result: *T is a minimum variance unbiased estimate of ϑ.*

The statistic T is normally distributed with probability density

$$c\exp\{-\tfrac{1}{2}k(\vartheta-T)^2\},$$

and its variance is

$$\sigma_T^2=\frac{1}{k}=\frac{1}{\sum a\,a}. \tag{8}$$

Since the equality sign in the Fréchet inequality holds, the constant $k=\sum a\,a$ is just the information I.

Considering the fact that with the transformation $x_i=\sigma_i\,x_i'$ we have made all the standard deviations equal to one and correspondingly all the weights one also, we see that (8) agrees with the previous result

$$\sigma_u^2=h^{11}\,\sigma^2=\frac{\sigma^2}{[g\,a\,a]}.$$

Example 28. Estimation of a probability.

Suppose that an event with unknown probability p has occurred x times in n trials. What is the best estimate of p?

The fact that x is a discrete chance variable causes no difficulty. In the first example of § 35 we saw that the likelihood function is, except for a constant factor independent of p,

$$g(x\,|\,p)=p^x(1-p)^{n-x}. \tag{9}$$

Equivalently, we may write

$$g(x\,|\,p)=\exp\{x\ln p+(n-x)\ln(1-p)\}. \tag{10}$$

If the frequency

$$h=\frac{x}{n}$$

is introduced, we then have

$$g(x\,|\,p)=\exp\{h\,n\ln p+(1-h)\,n\ln(1-p)\}. \tag{11}$$

This expression has exactly the form required in a) and b). We have seen already that the estimate h has expected value p. Conditions 1 through 3 are also satisfied. *Therefore the frequency h is a minimum variance unbiased estimate; i.e., it has the smallest variance among all unbiased estimates.*

In all cases considered thus far, the minimal character of the unbiased estimates has been established by using the fact that the equality sign in the Fréchet inequality holds. If, however, conditions a) and b) § 38 are not satisfied, then it may happen that the equality sign does not hold at all. However, there exist other methods for finding minimum variance unbiased estimates. These methods have been developed by Rao and under more general conditions by Lehmann and Scheffé[2]. In order to discuss their methods we shall consider next the Kolmogorov concept of conditional expectation.

---

[2] E. L. Lehmann and H. Scheffé, Completeness, similar regions, and unbiased estimation I. Sankhyā 10 (1950) p. 305.

§ 40. Conditional Expectation

Let the bold-face letters $\mathbf{t}, \mathbf{u}, \mathbf{v}, \mathbf{x}, \ldots$ represent random variables. Assume that $\mathbf{x}_1, \ldots, \mathbf{x}_n$ are observable variables and that all the others are functions of the $\mathbf{x}_k$:

$$\mathbf{t}=T(\mathbf{x}); \quad \mathbf{u}=U(\mathbf{x}); \ldots.$$

Let the values which these functions take on in the case of the observed values x be

$$t=T(x); \quad u=U(x); \ldots.$$

We shall now define the concept of *the conditional expectation of* $\mathbf{u}$ *given a particular value* t *of* $\mathbf{t}$.

The definition is very simple in the case that $\mathbf{t}$ and $\mathbf{u}$ both may take on only finitely many values. If t is a value which $\mathbf{t}$ may take on with positive probability, $P(t) \neq 0$, then the conditional probability

$$P_t(u_k)=\frac{P(u_k, t)}{P(t)}=\frac{\mathscr{P}(\mathbf{u}=u_k \,\&\, \mathbf{t}=t)}{\mathscr{P}(\mathbf{t}=t)} \tag{1}$$

may be calculated for each of the finitely many values $u_1, \ldots, u_m$ and the conditional mean $\mathscr{E}_t \mathbf{u}$ may be defined as the sum of all values u_k multiplied by their conditional probabilities:

$$\mathscr{E}_t \mathbf{u}=\sum u_k P_t(u_k). \tag{2}$$

If (2) is multiplied by $P(t)$ and summed over all those t-values belonging to a set M, the result is, because of (1),

$$\sum_{t \text{ in } M} (\mathscr{E}_t \mathbf{u}) P(t)=\sum_k u_k \mathscr{P}(\mathbf{u}=u_k \,\&\, \mathbf{t} \text{ in } M). \tag{3}$$

Conversely, if (3) holds for every set M, then (3) holds for the set consisting of the single value t. Dividing by $P(t)$ we again have (2).

The hypothesis that $\mathbf{u}$ may take on only finitely many values is not essential. The finite sums on the right-hand side of (2) or of (3) may be replaced by infinite sums or integrals, as we have done in § 3 in defining unconditional expectation. If $F_t(u)$ is the conditional distribution function of $\mathbf{u}$ – i.e., $F_t(u)$ is the conditional probability of the event $\mathbf{u}<u$ given that $\mathbf{t}=t$ – then we may write in place of (2)

$$\mathscr{E}_t \mathbf{u}=\int_{-\infty}^{\infty} u \, dF_t(u), \tag{4}$$

and in place of (3)

$$\sum_{t \text{ in } M} (\mathscr{E}_t \mathbf{u}) P(t)=\int_{M'} \mathbf{u} \, d\mathscr{P}(E), \tag{5}$$

where M' is the event that $\mathbf{t}$ belongs to M. The integral on the right-hand side is the Lebesgue integral of the function $\mathbf{u}$ over the set M' with measure $\mathscr{P}(A)$ (§ 3 A).

The left-hand side of (5) may also be thought of as a Lebesgue integral over the set M by letting $H(t)$ be the distribution function of the variable $\mathbf{t}$ and expressing (5) as

$$\int_M (\mathscr{E}_t\,\mathbf{u})\,dH(t) = \int_{M'} \mathbf{u}\,d\mathscr{P}(E). \tag{6}$$

Thus far we have taken as the distribution function of $\mathbf{t}$ a step-function with finitely many steps of height different from zero. If $\mathbf{t}$ has a continuous distribution function, the definitions (1) and (2) are no longer applicable because the denominator of (1) is then zero. Formula (6) however is still meaningful and can be taken as the definition of conditional expectation. Kolmogorov proved (Foundations of Probability V § 4, translated by N. Morrison. New York: Chelsea 1950), using a theorem of Nikodym, that under the single hypothesis of the existence of $\mathscr{E}\mathbf{u}$ there always exists a measurable function $f(t) = \mathscr{E}_t\,\mathbf{u}$ such that (6) holds for all measurable sets M on the t-axis. The function $f(t) = \mathscr{E}_t\,\mathbf{u}$ is, to be sure, not uniquely defined by (6), but two solutions $f_1(t)$ and $f_2(t)$ of (6) differ from one another only on subsets of the t-axis with probability zero.

If $\mathbf{x}_1, \ldots, \mathbf{x}_n$ have a probability density function $g(x)$, then in place of (6) we may write

$$\int_M (\mathscr{E}_t\,\mathbf{u})\,dH(t) = \int_{M'} U(x)\,g(x)\,dx. \tag{7}$$

If $\mathbf{t}$ as well has a probability density function $h(t)$, then the Stieltjes integral on the left-hand side of (7) may be replaced by an ordinary integral:

$$\int_M (\mathscr{E}_t\,\mathbf{u})\,h(t)\,dt = \int_{M'} U(x)\,g(x)\,dx. \tag{8}$$

If (8) holds for every interval $-\infty < t < b$ on the t-axis, then (8) holds for every measurable set M. Hence instead of (8) we may require

$$\int_{-\infty}^{b} (\mathscr{E}_t\,\mathbf{u})\,h(t)\,dt = \int_{t<b} U(x)\,g(x)\,dx. \tag{9}$$

The function $\mathscr{E}_t\,\mathbf{u}$ can be determined, provided it is continuous and $h(t) \neq 0$, from (9) by means of differentiation with respect to the upper limit b.

We shall give a few simple examples of calculation of the conditional expectation.

First let $\mathbf{t}=\mathbf{x}_1$. The probability density $h(t)$ is found by integration of the probability density $g(t, x_2, \ldots, x_n)$ with respect to $x_2, \ldots, x_n$:

$$h(t)=\int g(t, x_2, \ldots, x_n)\, dx_2 \ldots dx_n. \tag{10}$$

Setting

$$\mathscr{E}_t \mathbf{u}=\frac{\int U(t, x_2, \ldots, x_n)\, g(t, x_2, \ldots, x_n)\, dx_2 \ldots dx_n}{\int g(t, x_2, \ldots, x_n)\, dx_2 \ldots dx_n}, \tag{11}$$

where the integration is over the whole space of the variables $x_2, \ldots, x_n$, we see immediately that (9) is satisfied.

Secondly let $\mathbf{t}=(\mathbf{x}_1^2+\cdots+\mathbf{x}_n^2)^{\frac{1}{2}}$. If we introduce polar coordinates r, $\varphi_1, \ldots, \varphi_{n-1}$, we then have the preceding case again, and

$$\mathscr{E}_r \mathbf{u}=\frac{\int U(x)\, g(x)\, d\omega}{\int g(x)\, d\omega}, \tag{12}$$

where $d\omega$ is a surface element of the unit sphere $r=1$ and the integration in the numerator and denominator extends over a sphere of radius r.

Whenever the conditional expectation is well-defined – that is, outside a set on the t-axis with probability zero – it has the following properties:

1. $\mathscr{E}_t(\mathbf{u}-\mathbf{v})=\mathscr{E}_t \mathbf{u}-\mathscr{E}_t \mathbf{v}$.
2. If $\mathbf{u}$ is a constant c, then $\mathscr{E}_t \mathbf{u}=c$.
3. If $\mathscr{E}_t \mathbf{u}$ is zero for all t, then $\mathscr{E}\mathbf{u}=0$.
4. If $\mathbf{v}=\varphi(\mathbf{t})$, then $\mathscr{E}_t(\mathbf{u}\,\mathbf{v})=(\mathscr{E}_t \mathbf{u}) \cdot \varphi(t)$.

The first three properties follow immediately from the definition. The last one is due to Kolmogorov (Foundations of Probability, translated by N. Morrison, p. 56).

§ 41. Sufficient Statistics

Returning to the problem of optimal estimation of an unknown parameter ϑ, let us assume once more that the probability density of the observed variables $\mathbf{x}_k$ has the form

$$g(x|\vartheta)=e(t|\vartheta)\, h(x), \tag{1}$$

where t is a function of the x's independent of ϑ:

$$t=T(x). \tag{2}$$

In the previous notation, $\mathbf{t}=T(\mathbf{x})$ would have been called a sufficient estimate of ϑ. In as much as $\mathbf{t}$ need not be an estimate of ϑ, it is preferable to refer to $\mathbf{t}$ as a *sufficient statistic*. We shall also say that $\mathbf{t}=T(\mathbf{x})$ is *sufficient for* ϑ.

The conditional expectation $\mathscr{E}_t \mathbf{u}$ of a statistic $\mathbf{u}=U(\mathbf{x})$ is defined as in §40 by

$$\int_{M'} U(x)\, g(x|\vartheta)\, dx = \int_{M} (\mathscr{E}_t \mathbf{u})\, dH(t), \tag{3}$$

where $H(t)$ is the distribution function of the variable $\mathbf{t}$. We shall now prove the following theorem:

If the probability density $g(x|\vartheta)$ *has the form* (1), *then the function* $\mathscr{E}_t \mathbf{u}$ *may be determined in such a way that it is independent of* ϑ.

In our proof we shall consider first the case that there exists a ϑ such that $e(t|\vartheta) \neq 0$ for every t. Then from (1) for arbitrary ϑ' we have

$$g(x|\vartheta') = \frac{e(t|\vartheta')}{e(t|\vartheta)}\, e(t|\vartheta)\, h(x) = \frac{e(t|\vartheta')}{e(t|\vartheta)}\, g(x|\vartheta),$$

or, if the fraction on the right is called $Q(t)$,

$$g(x|\vartheta') = Q(t)\, g(x|\vartheta). \tag{4}$$

If we let $\mathscr{E}$ denote expected value when the parameter value is ϑ and $\mathscr{E}'$ when it is ϑ', then from (3) we have

$$\int_{M'} U(x)\, g(x|\vartheta)\, dx = \int_{M} (\mathscr{E}_t \mathbf{u})\, dH(t) \tag{5}$$

and

$$\int_{M'} U(x)\, g(x|\vartheta')\, dx = \int_{M} (\mathscr{E}'_t \mathbf{u})\, dH'(t), \tag{6}$$

or, on account of (4),

$$\int_{M'} U(x)\, Q(t)\, g(x|\vartheta)\, dx = \int_{M} (\mathscr{E}'_t \mathbf{u})\, dH'(t). \tag{7}$$

We shall call the statistic $Q(\mathbf{t})$, whose value is equal to $Q(t)$ in every case, $V(\mathbf{x})=\mathbf{v}$:

$$Q(\mathbf{t}) = Q(T(\mathbf{x})) = V(\mathbf{x}) = \mathbf{v}.$$

Applying property 4 (§40) to the product $\mathbf{u}\,\mathbf{v} = U V$, we have

$$\mathscr{E}_t(\mathbf{u}\,\mathbf{v}) = (\mathscr{E}_t \mathbf{u}) \cdot Q(t),$$

and therefore from the definition of $\mathscr{E}_t(\mathbf{u}\,\mathbf{v})$, we have

$$\int_{M'} U(x)\, V(x)\, g(x|\vartheta)\, dx = \int_{M} (\mathscr{E}_t \mathbf{u})\, Q(t)\, dH(t). \tag{8}$$

Since $Q(t) = V(x)$ by definition of V, the left side of (8) and the left side of (7) are equal. Hence

$$\int_{M} (\mathscr{E}'_t \mathbf{u})\, dH'(t) = \int_{M} (\mathscr{E}_t \mathbf{u})\, Q(t)\, dH(t). \tag{9}$$

Applying (9) to the particular case $\mathbf{u}=1$, we have (for every measurable set M)

$$\int_M dH'(t) = \int_M Q(t)\,dH(t). \tag{10}$$

Hence for every piecewise constant function $f(t)$,

$$\int_M f(t)\,dH'(t) = \int_M f(t)\,Q(t)\,dH(t). \tag{11}$$

The proof of (11) is immediate once the set M is broken into subsets on which $f(t)$ is constant and (10) is applied on each of these subsets.

Since every measurable function can be approximated by piecewise constant functions such that their respective integrals differ by an arbitrarily small amount, then (11) must hold for every measurable function and, hence, for every function $f(t)$ for which the left-hand side is defined. If we set $f(t)=\mathscr{E}_t\,\mathbf{u}$, it follows that

$$\int_M (\mathscr{E}_t\,\mathbf{u})\,dH'(t) = \int_M (\mathscr{E}_t\,\mathbf{u})\,Q(t)\,dH(t). \tag{12}$$

Comparison of (9) and (12) yields

$$\int_M (\mathscr{E}'_t\,\mathbf{u})\,dH'(t) = \int_M (\mathscr{E}_t\,\mathbf{u})\,dH'(t). \tag{13}$$

Thus $\mathscr{E}'_t\,\mathbf{u}$ in (7) may be replaced by $\mathscr{E}_t\,\mathbf{u}$ without altering the equality; i.e., for every ϑ', $\mathscr{E}'_t\,\mathbf{u}$ may be chosen equal to $\mathscr{E}_t\,\mathbf{u}$, as was to be proved.

The proof becomes somewhat more difficult in the case that $e(t|\vartheta)$ is equal to zero for values of t depending on ϑ.

We shall assume that $e(t|\vartheta)$ is piecewise smooth, which suffices in applications. At the points of discontinuity we may set $e(t|\vartheta)=0$. Then the set of points on the t-axis for which $e(t|\vartheta)\neq 0$ is an open set for every ϑ.

There can exist points t for which all $e(t|\vartheta)$ are zero. These points form a set B_0 which has probability zero for all ϑ. Without loss of generality we may set $\mathscr{E}_t\,\mathbf{u}=0$, say, on B_0. The set of interest is the set C complementary to B_0.

For every point t belonging to C there exists a ϑ such that $e(t|\vartheta)\neq 0$. Hence there exists as well an interval $B(t)$ about t in which $e(t|\vartheta)\neq 0$. The open sets $B(t)$ cover the whole set C and therefore countably many of them, say $B_1, B_2, \ldots$, suffice to cover C. In B_1 let $e(t|\vartheta_1)\neq 0$, in B_2 let $e(t|\vartheta_2)\neq 0$, etc.

B_2 may be so modified to contain only points not in B_1, and likewise B_3 to contain only points not already in B_1 or B_2, etc. The disjoint sets $B_1, B_2, \ldots$ thus modified still cover the whole set C.

Applying the methods of the preceding case, we may modify all functions $\mathscr{E}_t\,\mathbf{u}$ in B_1 in such a way that they agree with the function $\mathscr{E}_{1t}\,\mathbf{u}$ defined for $\vartheta=\vartheta_1$. Likewise, on B_2 the $\mathscr{E}_t\,\mathbf{u}$ may be set equal to $\mathscr{E}_{2t}\,\mathbf{u}$

corresponding to $\vartheta=\vartheta_2$, and so forth. Hence we have a definition of $\mathscr{E}_t\mathbf{u}$ which does not depend on ϑ and which for all ϑ and M satisfies condition (3). That is, every set M can be broken into countably many sets $M_0, M_1, M_2, \ldots$ which are contained in $B_0, B_1, B_2, \ldots$ and if (3) holds for these sets, then (3) holds also for M.

Hence the assertion is proved in general.

§ 42. Application to the Problem of Unbiased Estimation

A. Improvement of an Estimate

Let $x_1, \ldots, x_n$ again be observed random variables with probability density $g(x|\vartheta)$ depending on ϑ, and let $\mathbf{t}=T(\mathbf{x})$ be a sufficient statistic. Hence

$$g(x|\vartheta)=e(t|\vartheta)\,h(x). \tag{1}$$

Let $\mathbf{u}=U(\mathbf{x})$ be an estimate of ϑ with finite expected value $\hat{\mathbf{u}}$ and finite variance $\sigma_{\mathbf{u}}^2$. These hypotheses may hold either for the true value of ϑ only or in some neighborhood of the true value. The following assertions hold for those values of ϑ for which the expected value and variance of $\mathbf{u}$ are finite.

We shall now define an *improved estimate* $\mathbf{v}$ which depends only on the sufficient statistic $\mathbf{t}$:

$$\mathbf{v}=W(\mathbf{t}), \tag{2}$$

where the value v of $\mathbf{v}$ corresponding to the value t of $\mathbf{t}$ is equal to the conditional expected value $\mathscr{E}_t\mathbf{u}$:

$$v=W(t)=\mathscr{E}_t\mathbf{u}. \tag{3}$$

Kolmogorov proved that $\mathbf{v}=W(\mathbf{t})$ is a random variable. In § 41 we saw that $W(t)=\mathscr{E}_t\mathbf{u}$ depends not on ϑ but only on t.

We shall now prove that $\mathbf{v}$ *has the same expected value as* $\mathbf{u}$ *and the variance of* $\mathbf{v}$ *is at most equal to the variance of* $\mathbf{u}$.

The proof rests entirely on properties 1 through 4 (§ 40). From 2 and 4 it follows that

$$\mathscr{E}_t\mathbf{v}=\mathscr{E}_t(1\cdot\mathbf{v})=(\mathscr{E}_t 1)\cdot W(t)=W(t). \tag{4}$$

Furthermore from 1 we have

$$\mathscr{E}_t(\mathbf{u}-\mathbf{v})=\mathscr{E}_t\mathbf{u}-\mathscr{E}_t\mathbf{v}=W(t)-W(t)=0. \tag{5}$$

Therefore from 3 it follows that

$$\mathscr{E}(\mathbf{u}-\mathbf{v})=0, \tag{6}$$

or $\mathscr{E}\mathbf{u}=\mathscr{E}\mathbf{v}$, and the first assertion is proved.

The variance of **u** is

$$\begin{aligned}\sigma_{\mathbf{u}}^2 &= \mathscr{E}(\mathbf{u}-\hat{\mathbf{u}})^2 = \mathscr{E}(\mathbf{u}-\hat{\mathbf{v}})^2 \\ &= \mathscr{E}(\mathbf{u}-\mathbf{v}+\mathbf{v}-\hat{\mathbf{v}})^2 \\ &= \mathscr{E}(\mathbf{u}-\mathbf{v})^2 + 2\mathscr{E}(\mathbf{u}-\mathbf{v})(\mathbf{v}-\hat{\mathbf{v}}) + \mathscr{E}(\mathbf{v}-\hat{\mathbf{v}})^2.\end{aligned} \tag{7}$$

Since $\mathbf{v}-\hat{\mathbf{v}}$ is a function of **t** only, let us call this function $\varphi(\mathbf{t})$, so that using 4 and (5) we have

$$\mathscr{E}_t(\mathbf{u}-\mathbf{v})(\mathbf{v}-\hat{\mathbf{v}}) = \mathscr{E}_t(\mathbf{u}-\mathbf{v}) \cdot \varphi(t) = 0. \tag{8}$$

From 3 we then have

$$\mathscr{E}(\mathbf{u}-\mathbf{v})(\mathbf{v}-\hat{\mathbf{v}}) = 0. \tag{9}$$

Hence (7) simplifies to

$$\sigma_{\mathbf{u}}^2 = \mathscr{E}(\mathbf{u}-\mathbf{v})^2 + \sigma_{\mathbf{v}}^2. \tag{10}$$

The second assertion follows immediately:

$$\sigma_{\mathbf{u}}^2 \geqq \sigma_{\mathbf{v}}^2. \tag{11}$$

If $\sigma_{\mathbf{v}}^2$ were infinite, then from (11) $\sigma_{\mathbf{u}}^2$ would necessarily be infinite, contradicting the hypothesis. Therefore $\sigma_{\mathbf{v}}^2$ is finite and at most equal to $\sigma_{\mathbf{u}}^2$.

The equality sign in (11) applies only in the case that $\mathbf{u}-\mathbf{v}$ differs from zero at most on a set with probability zero.

The hypothesis of finiteness of $\sigma_{\mathbf{u}}^2$ may also be dropped, for if $\sigma_{\mathbf{u}}^2$ is infinite, then (11) is trivially true.

Therefore there exists for every estimate $\mathbf{u}=U(\mathbf{x})$ a better estimate $\mathbf{v}=V(\mathbf{x})$ which has the same bias and at most the same variance as **u** and which depends only on the sufficient statistic $\mathbf{t}=T(\mathbf{x})$. If **u** is unbiased, then **v** is also unbiased.

From this point on we may discontinue the use of the bold-face letters and denote the observed random variables and their values by $x_1, \ldots, x_n$, the sufficient statistic by $T=T(x)$, and the estimates by $U(x)$ and $V(x)=W(T)$.

B. An Integral Equation of Unbiased Estimation

On the basis of the results just derived, in seeking a minimum variance unbiased estimate we may always limit ourselves to estimates $V=W(T)$, which depend only on the sufficient statistic T. We shall now assume that T has a probability density $q(t|\vartheta)$.

Let us generalize the problem to the case of estimating not ϑ itself but a function $\varphi(\vartheta)$. The condition that the estimate be unbiased leads immediately to the integral equation

$$\int W(t)\, q(t|\vartheta)\, dt = \varphi(\vartheta), \tag{12}$$

integration being over the entire space of possible values t of the statistic T.

If W and W_1 are two such solutions to the integral equation, then their difference $D(t)$ satisfies the integral equation

$$\int D(t)\, q(t|\vartheta)\, dt = 0. \tag{13}$$

If the functions $q(t|\vartheta)$ form a *complete family* of probability densities on the t-axis – that is, if there does not exist a non-zero function $D(t)$ orthogonal to every member of the family – then from (13) it follows that

$$D(t) = 0.$$

Hence the solution of (12) is uniquely determined, and we have the following:

Theorem. If $T = T(x)$ is a sufficient statistic for ϑ and if the densities $q(t|\vartheta)$ form a complete family, then every unbiased estimate of $\varphi(\vartheta)$ depending only on T is a minimum variance estimate.

§ 43. Applications

The method discussed in § 42 for finding optimal unbiased estimates lends itself to many examples. First of all, all the previous examples may be treated using the present methods. We also include some new examples, the first two of which are taken from a book by Rao[3].

Example 29. χ^2-distribution with factor α.

Let $x_1, \ldots, x_n$ be n independent observations of a random variable x with a distribution of the χ^2-type except for an unknown parameter α appearing in the exponent:

$$f(x|\alpha) = c\, \alpha^p\, e^{-\alpha x}\, x^{p-1} \qquad (x>0), \tag{1}$$

with $c = \Gamma(p)^{-1}$. The joint probability density of $x_1, \ldots, x_n$ is, if $\sum x = T(x) = T$,

$$g(x|\alpha) = c^n\, \alpha^{np}\, e^{-\alpha T} (x_1 \cdot \ldots \cdot x_n)^{p-1}. \tag{2}$$

From the form of the function g it is apparent that T is sufficient for α. If we make the transformation to new coordinates T, y_i defined by

$$x_i = T y_i, \tag{3}$$

then the y_i are subject to the condition

$$\sum y_i = 1, \tag{4}$$

[3] C. R. Rao, Adv. Stat. Meth. in Biom. Res. New York: John Wiley & Sons 1952.

so that only T and $y_1, \ldots, y_{n-1}$ may vary independently. If we integrate with respect to $y_1, \ldots, y_{n-1}$ over the region

$$y_1 > 0, \ldots, y_n > 0, \quad \sum y_i = 1, \tag{5}$$

we obtain as the probability density of T

$$q(T\,|\,\alpha) = c'\,\alpha^{np}\,e^{-\alpha T}\,T^{np-1}. \tag{6}$$

In order that the integral from 0 to ∞ be one, it must be true that

$$c' = \Gamma(n\,p)^{-1}.$$

The mean value of T^{-1} is

$$c'\,\alpha^{np} \int_0^\infty e^{-\alpha T}\,T^{np-2}\,dT = \frac{\alpha\,\Gamma(n\,p-1)}{\Gamma(n\,p)} = \frac{\alpha}{n\,p-1}.$$

Therefore

$$W(T) = (n\,p-1)\,T^{-1} \tag{7}$$

is an unbiased estimate of α.

If there were to exist a second unbiased estimate depending only on T, then there would be a solution to the integral equation

$$\int_0^\infty D(t)\,e^{-\alpha t}\,t^{np-1}\,dt = 0. \tag{8}$$

If we introduce the new variable $z = e^{-t}$ and put $D(t)\,t^{np-1} = G(z)$, then from (8) we have

$$\int_0^\infty z^{\alpha-1}\,G(z)\,dz = 0. \tag{9}$$

This holds for all real α, and hence in particular for $\alpha = 1, 2, 3, \ldots$. But the functions $1, z, z^2, \ldots$ form a complete family[4] on the interval from 0 to 1. Therefore from (9) we have $G(z) = 0$; i.e., the integral equation has only the null-solution. *The estimate* (7) *is therefore minimum variance unbiased.*

The maximum likelihood estimate

$$\tilde{\alpha} = n\,p\,T^{-1}$$

has a small bias, but agrees asymptotically as $n \to \infty$ with the unbiased estimate (7).

The variance of (7) is

$$\sigma_W^2 = \frac{\alpha^2}{n\,p-2}.$$

As we have seen above, for every unbiased estimate the inequality $\sigma^2 \geqq \sigma_W^2$ holds. The Fréchet inequality yields only

$$\sigma^2 \geqq I^{-1} = \frac{\alpha^2}{n\,p}.$$

Therefore the integral equation method gives a sharper lower bound.

Example 30. Rectangular distribution.

Let $x_1, \ldots, x_n$ be independent observations of a random variable having a rectangular distribution on the interval from 0 to ϑ, where ϑ is to be estimated.

---

[4] See, for example, R. Courant, and D. Hilbert, Methods of Mathematical Physics I, translated and revised from the German original, New York: Interscience 1953, Chapter 2, §4.

The probability that all n observations are less than t is

$$(t/\vartheta)^n = t^n \vartheta^{-n}.$$

The probability density of the largest observation T is therefore

$$q(t\,|\,\vartheta) = n\,\vartheta^{-n}\,t^{n-1}. \tag{10}$$

Let the remaining observations x_j (ordered according, say, to increasing index j) be $y_1, \ldots, y_{n-1}$. We wish to determine the joint probability density of $T, y_1, \ldots, y_{n-1}$.

Let G be a subset of the event set of the variables $T, y_1, \ldots, y_{n-1}$. We may restrict attention to that portion of G defined by the inequalities

$$y_1 < T, \ldots, y_{n-1} < T,$$

because, say, $y_1 > T$ is impossible and $y_1 = T$ has probability zero. A single point P in G corresponds to n points $P_1, \ldots, P_n$ in x-space, for if T and $y_1, \ldots, y_{n-1}$ are given, then either $x_1 = T$ and the remaining x_j's equal $y_1, \ldots, y_{n-1}$, or $x_2 = T, \ldots$, or $x_n = T$. Therefore n different subsets $G_1, \ldots, G_n$ in x-space correspond to the subset G. All these subsets have equal volume V, since each is obtained from the other by a permutation of the variables. The probability that P belongs to G is the sum of the probabilities of the subsets $G_1, \ldots, G_n$ and hence is equal to n times the volume of G_1 divided by ϑ^n:

$$\mathscr{P}(G) = n\,V\,\vartheta^{-n}.$$

Therefore the probability density of the entire system $T, y_1, \ldots, y_{n-1}$ is

$$g(t, y\,|\,\vartheta) = n\,\vartheta^{-n}\,h(t, y), \tag{11}$$

where $h(t, y)$ is equal to one if $0 < y_i < t < \vartheta$ for all i and equal to zero otherwise.

From the form of the density (11), we see that T is sufficient for ϑ, and from (10), that the expected value of T is

$$\mathscr{E}\,T = n\,\vartheta^{-n} \int_0^{\vartheta} t^n\,dt = \frac{n}{n+1}\,\vartheta. \tag{12}$$

Therefore an unbiased estimate of ϑ is

$$W(T) = \frac{n+1}{n}\,T, \tag{13}$$

which has variance

$$\sigma^2 = \frac{\vartheta^2}{n(n+2)}.$$

If there were to exist another unbiased estimate depending only on T, then there would be a solution to the integral equation

$$\int_0^{\vartheta} D(t)\,n\,\vartheta^{-n}\,t^{n-1}\,dt = 0, \tag{14}$$

or

$$\int_0^{\vartheta} D(t)\,t^{n-1}\,dt = 0. \tag{15}$$

But if (15) holds for all ϑ, we must have $D(t) = 0$. Therefore (13) is the unique minimum variance unbiased estimate.

The maximum likelihood estimate $\tilde{\vartheta} = T$ yields a systematic under-estimation of ϑ.

The following example was kindly communicated to me by E. L. Lehmann. It is particularly interesting in that it involves a direct application of the method discussed in § 42 A for improving an unbiased estimate.

Example 31. A manufacturer supplies a product in lots. A customer takes a sample of n items out of each lot and tests them. If the sample contains more than two defectives, the lot is refused, which involves a loss to the manufacturer. The manufacturer is informed of the number of defective items found in each sample. He assumes that each item has the same probability of being defective, say p, in as much as all items are produced in the same way. Let the observed number of defective items be $x_1, \ldots, x_r$. The minimum variance unbiased estimate of p is obviously

$$h = \frac{x_1 + \cdots + x_r}{r\,n}.$$

The manufacturer desires an unbiased estimate of his expected loss. Improving his product (perhaps by testing the items more carefully before they leave the factory) costs money, which he will spend only in the case that it is justified.

The probability that a lot will not be returned is

$$\vartheta = q^n + n\,p\,q^{n-1} + \binom{n}{2} p^2\, q^{n-2}.$$

The expected loss is proportional to $1 - \vartheta$, so that it is enough to find an unbiased estimate of ϑ.

The observations $x_1, \ldots, x_r$ are independent binomial variables. The probability that the values $x_1, \ldots, x_r$ occur is

$$P(x) = \binom{n}{x_1} \cdots \binom{n}{x_r} p^{x_1} q^{n-x_1} \ldots p^{x_r} q^{n-x_r}.$$

Letting $x_1 + \cdots + x_r = T$, we may write $P(x)$ as

$$P(x) = \binom{n}{x_1} \cdots \binom{n}{x_r} p^T q^{nr-T}.$$

From the form of this function it follows immediately that T is a sufficient statistic for p, or for ϑ. Hence a minimum variance unbiased estimate can be determined as a function of T only.

A poor, but simple unbiased estimate is

$$U = 1 \quad \text{if the first lot is accepted,}$$
$$U = 0 \quad \text{if it is refused.}$$

We shall now take the conditional expected value of the statistic U, given that $T = t$. The first lot is accepted if the sample from it produces zero, one, or two defectives. Multiplying the conditional probability of each of these events by $U = 1$ and adding yields

$$\mathscr{E}_t\, U = \frac{\mathscr{P}(x_1 = 0\ \&\ x_2 + \cdots + x_r = t)}{\mathscr{P}(x_1 + \cdots + x_r = t)} + \frac{\mathscr{P}(x_1 = 1\ \&\ x_2 + \cdots + x_r = t-1)}{\mathscr{P}(x_1 + \cdots + x_r = t)}$$
$$+ \frac{\mathscr{P}(x_1 = 2\ \&\ x_2 + \cdots + x_r = t-2)}{\mathscr{P}(x_1 + \cdots + x_r = t)}.$$

The factor $p^t q^{rn-t}$ occurs in each term in both the numerator and the denominator and hence cancels. Therefore we have

$$\mathscr{E}_t U = \frac{\binom{rn-n}{t}+\binom{n}{1}\binom{rn-n}{t-1}+\binom{n}{2}\binom{rn-n}{t-2}}{\binom{rn}{t}}.$$

This yields

$$V=\binom{rn}{T}^{-1}\left[\binom{rn-n}{T}+\binom{n}{1}\binom{rn-n}{T-1}+\binom{n}{2}\binom{rn-n}{T-2}\right].$$

as the improved estimate of ϑ.

In order to show that we have at hand a minimum variance unbiased estimate, we need only to verify that V is the unique unbiased estimate depending only on T; i.e., that the equation

$$\sum_{t=0}^{nr}\binom{nr}{t} p^t q^{nr-t}\, W(t)=\vartheta$$

has but one solution. If W and W_1 are two solutions, then their difference $D(t)$ satisfies the homogeneous equation

$$\sum \binom{nr}{t} p^t q^{nr-t}\, D(t)=0.$$

But a polynomial in p can be zero for every $0 \leqq p \leqq 1$ only if all the coefficients vanish. Therefore $D=0$ is the only solution.

§ 44. Estimation of the Variance of a Normal Distribution

Both Rao and Lehmann and Scheffé have extended their theories of optimal estimation to the case of several unknown parameters. We shall not discuss the general theory here but shall instead restrict attention to an example which is particularly important in applications.

Let $x_1, \ldots, x_n$ be independent normally distributed random variables with unknown mean μ and unknown variance ϑ. The probability density function is therefore

$$\begin{aligned} g(x_1, \ldots, x_n | \vartheta, \mu) &= c\, \vartheta^{-\frac{n}{2}} \exp\left(-\frac{\sum (x-\mu)^2}{2\vartheta}\right) \\ &= c\, \vartheta^{-\frac{n}{2}} \exp\left(-\frac{\sum x^2 - 2\mu \sum x + n\mu^2}{2\vartheta}\right). \end{aligned} \tag{1}$$

From the form of the density function we see immediately that $\sum x^2$ and $\sum x$ are sufficient for ϑ and μ. As we already have seen, the optimal estimate of μ is

$$\bar{x} = \frac{1}{n} \sum x. \tag{2}$$

We want a minimum variance estimate of ϑ which is unbiased for all values of ϑ and μ.

We begin by introducing an orthogonal transformation to new coordinates $y_1, \ldots, y_n$ with

$$y_1 = \bar{x}\sqrt{n} = n^{-\frac{1}{2}}(x_1 + \cdots + x_n). \tag{3}$$

On account of the orthogonality of the transformation, we have

$$\sum x^2 = \sum y^2,$$

which yields

$$\begin{aligned}\sum x^2 - 2\mu \sum x + n\mu^2 &= \sum y^2 - 2\mu\, y_1 \sqrt{n} + n\mu^2 \\ &= (y_1 - \mu\sqrt{n})^2 + y_2^2 + \cdots + y_n^2.\end{aligned}$$

The probability density hence becomes

$$f(y_1, \ldots, y_n | \vartheta, \mu) = c\, \vartheta^{-\frac{n}{2}} \exp\left\{-\frac{(y_1 - \mu\sqrt{n})^2 + y_2^2 + \cdots + y_n^2}{2\vartheta}\right\}. \tag{4}$$

If we transform the variables $y_2, \ldots, y_n$ to polar coordinates $r, \varphi_1, \ldots, \varphi_{n-2}$, the probability density becomes

$$f(y_1, r, \varphi | \vartheta, \mu) = c\, \vartheta^{-\frac{n}{2}} \exp\left\{-\frac{(y_1 - \mu\sqrt{n})^2 + r^2}{2\vartheta}\right\} r^{n-2}\, h(\varphi_1, \ldots, \varphi_{n-2}). \tag{5}$$

The variables r and y_1 which occur in (5) instead of the previous $\sum x^2$ and $\sum x$ are likewise sufficient for ϑ and μ.

As in § 40 the conditional expected value $\mathscr{E}_{r, y_1} \mathbf{u}$ may be defined for every random variable $\mathbf{u}$. The general theory of Kolmogorov is not necessary here: the conditional expected value can be defined simply by integration with respect to the angular coordinates $\varphi_1, \ldots, \varphi_{n-2}$ (cf. conclusion of § 40).

By taking conditional expected values, we can, as in § 42A, derive from every estimate $\mathbf{u}$ of ϑ a better estimate $\mathbf{v}$, which has the same bias and at most the same variance and which depends only on r and y_1. Therefore we may restrict attention to functions of r and y_1.

One such function is

$$s^2 = \frac{\sum (x - \bar{x})^2}{n-1} = \frac{\sum x^2 - n\bar{x}^2}{n-1} = \frac{\sum y^2 - y_1^2}{n-1} = \frac{r^2}{n-1}. \tag{6}$$

We know already that s^2 is an unbiased estimate of $\sigma^2 = \vartheta$. If a second unbiased estimate were to exist, then the integral equation

$$\iint D(y,r)\exp\left\{-\frac{(y-\mu\sqrt{n})^2+r^2}{2\vartheta}\right\} r^{n-2}\,dr\,dy=0 \tag{7}$$

would have a non-zero solution.

Setting

$$\int_0^\infty D(y,r)\exp\left(-\frac{r^2}{2\vartheta}\right) r^{n-2}\,dr = F(y\,|\,\vartheta), \tag{8}$$

we have the integral equation

$$\int_{-\infty}^{\infty} F(y\,|\,\vartheta)\exp\left(-\frac{y^2-2y\mu\sqrt{n}+\mu^2 n}{2\vartheta}\right)dy=0,$$

or, if the constant factor $\exp\dfrac{-\mu^2 n}{2\vartheta}$ is factored out and if $\alpha=\dfrac{\mu\sqrt{n}}{\vartheta}$ is introduced,

$$\int_{-\infty}^{\infty} F(y\,|\,\vartheta)\exp\left(-\frac{y^2}{2\vartheta}\right)e^{\alpha y}\,dy=0. \tag{9}$$

This equation must hold for arbitrary α and ϑ. The left side of (9) is a holomorphic function of α, which is defined[5] for all complex α. If such a function is zero on a small section of the imaginary axis, then it is identically zero. Therefore in (9) we may replace α by it and obtain a Fourier transform which is zero. This implies that the function itself is zero, which in turn implies that

$$F(y\,|\,\vartheta)=0. \tag{10}$$

Substituting this result into (8), the integral equation for $D(y,r)$ becomes

$$\int_0^\infty D(y,r)\exp(-\beta r^2)\,r^{n-2}\,dr=0. \tag{11}$$

If r^2 is taken to be the variable of integration, the integral equation then has the same form as (8) § 43. Therefore, as before, it follows that

$$D(y,r)=0.$$

In the above proof certain weak regularity conditions on the function $D(y,r)$ are required. It is sufficient, for example, to assume that the integrals (7) and (8) are absolutely convergent for all μ and ϑ in a finite

---

[5] *Proof.* In (9) the limits of integration $-\infty$ and ∞ may be replaced by $-M$ and $+M$ such that in any circle $|\alpha|<R$ the resulting error is $<\varepsilon$. Expanding $e^{\alpha y}$ into a power series with respect to α and integrating term by term, we obtain a power series in α for the integral from $-M$ to $+M$. Letting M tend to ∞, the result follows because the uniform limit of a holomorphic function is again a holomorphic function in the circle $|\alpha|<R$.

rectangle

$$a < \mu < b,$$

$$0 < \vartheta < c$$

and that convergence is uniform on every closed subset.

The case of several observed sequences $x_1, \ldots, x_m; y_1, \ldots, y_n; \ldots$ of normally distributed random variables with common variance ϑ but, possibly, with unequal means $\mu, \nu, \ldots$ can be handled with exactly the same methods of proof. The result is the same: *the empirical variance*

$$s^2 = \frac{\sum (x-\bar{x})^2 + \sum (y-\bar{y})^2 + \cdots}{(m-1)+(n-1)+\cdots} \tag{12}$$

is a minimum variance unbiased estimate of ϑ.

If each sequence consists of two observations only, then we have formula (10) § 35 as a special case of (12). Hence, Example 23 in § 35 serves as an example of the practical application of formula (12).

§ 45. Asymptotic Properties

All the results considered thus far hold for small samples as well as large, which is especially important in applications. In conclusion we wish to present briefly and without proofs the most important asymptotic properties of estimates based on large samples.

A. Consistency of Maximum Likelihood Estimates

Returning to the case of a single unknown parameter ϑ, let $x_1, \ldots, x_n$ be independent, identically distributed random variables having probability density $f(x|\vartheta)$. The joint probability density function is then

$$g(x|\vartheta) = f(x_1|\vartheta) \ldots f(x_n|\vartheta). \tag{1}$$

An estimate T of ϑ is said to be *consistent* if as $n \to \infty$ the probability that T differs from ϑ by less than ε, $|T-\vartheta| < \varepsilon$, tends to one. It can be shown that under certain regularity conditions the method of maximum likelihood yields a consistent estimate of ϑ.

A relatively simple proof due to A. Wald and J. Wolfowitz[6] requires quite weak regularity assumptions. We shall not reproduce it here but rather refer the reader to the original paper.

If as $n \to \infty$ the number of unknown parameters also increases, the consistency property need not hold, as we have already seen in Example 23 (§ 35).

[6] A. Wald and J. Wolfowitz, Ann. of Math. Stat. 20 (1949) pp. 595 and 601.

B. Asymptotic Normality, Expected Value, and Variance

The consistency of the maximum likelihood estimate $\tilde{\vartheta}$ was proved prior to the Wald-Wolfowitz paper, but under stronger hypotheses, by Hotelling[7] and by Doob[8]. The results of these authors, however, are stronger, in that they proved that the estimate $\tilde{\vartheta}$ is asymptotically normally distributed with mean ϑ and standard deviation $c/\sqrt{n}$. That is to say, if

$$U = (\tilde{\vartheta} - \vartheta)\sqrt{n} \tag{2}$$

is taken as a new variable, then the distribution function of U tends (as $n \to \infty$) to a normal distribution function with mean zero and standard deviation c, where c is defined by

$$\frac{1}{c^2} = \mathscr{E}\left(\frac{\partial \ln f}{\partial \vartheta}\right)^2. \tag{3}$$

We have denoted the right-hand side of this defining equation in § 36 by $j(\vartheta)$. If we multiply it by n, we obtain the "information" $I = I(\vartheta)$ introduced there:

$$I = \frac{n}{c^2} = \mathscr{E}\left(\frac{\partial \ln g}{\partial \vartheta}\right)^2. \tag{4}$$

The variance of the asymptotic normal distribution is therefore exactly the inverse of the information:

$$\frac{c^2}{n} = I^{-1}. \tag{5}$$

The student should note carefully the definitions of the concepts "asymptotic mean" and "asymptotic variance". It may well happen that the exact distribution of $\tilde{\vartheta}$ has an infinite variance for every n but that nevertheless $\tilde{\vartheta}$ is asymptotically normally distributed with finite mean ϑ and finite variance c^2/n. Hence we are not permitted to calculate first the variance and then its limit as $n \to \infty$, but rather we must determine first the distribution of U, then the limiting distribution as $n \to \infty$, and finally the variance of this limiting distribution. It is in this sense that the expressions "asymptotic mean" and "asymptotic variance" are always to be understood in the following.

If T is an estimate of ϑ and if in the above sense the asymptotic mean of $T - \vartheta$ is small compared with $n^{-\frac{1}{2}}$ – that is, if the quantity

$$U = (T - \vartheta)\sqrt{n} \tag{6}$$

[7] H. Hotelling, Trans. Amer. Math. Soc. 32 (1930) p. 847.

[8] J. L. Doob, Trans. Amer. Math. Soc. 36 (1934) p. 766 and 39 (1936) p. 410.

has an asymptotic distribution with mean zero – then the estimate T is said to be *asymptotically unbiased*. According to the above-mentioned theorems of Hotelling and Doob, this is always the case with maximum likelihood estimates.

C. Efficiency

R. A. Fisher conjectured that the estimate $\tilde{\vartheta}$ is *asymptotically efficient*, in the sense that among all asymptotically unbiased estimates it has the smallest possible variance. Later research (see the comprehensive publication by L. LeCam[9]) has established that this conjecture is true only under strong regularity conditions.

By relaxing the unbiasedness restriction to asymptotic unbiasedness, we can construct "super-efficient" estimates which have a smaller asymptotic variance for certain parameter values than the estimate $\tilde{\vartheta}$. J. L. Hodges has given an example of such an estimate. Let $f(x\,|\,\vartheta)$ be a normal probability density function with unit standard deviation and with mean ϑ, to be estimated:

$$f(x\,|\,\vartheta)=(2\pi)^{-\frac{1}{2}}\,e^{-\frac{1}{2}(x-\vartheta)^2}. \tag{7}$$

Given n observations, consider first of all their mean $\bar{x}$, an estimate which has variance n^{-1} and no bias. If we then define an estimate T by

$$T=\bar{x} \qquad \text{if } |\bar{x}|\geqq n^{-\frac{1}{4}}$$

$$T=\tfrac{1}{2}\bar{x} \qquad \text{if } |\bar{x}|<n^{-\frac{1}{4}},$$

we can prove the following:

1. T is asymptotically normal for every ϑ;
2. T is asymptotically unbiased for every ϑ;
3. T has asymptotic variance n^{-1} if $\vartheta\neq 0$, but asymptotic variance $\frac{1}{4}n^{-1}$ if $\vartheta=0$.

Proof. If $\vartheta\neq 0$, then for n sufficiently large the event $|\bar{x}|<n^{-\frac{1}{4}}$ has an arbitrarily small probability, so that with probability arbitrarily close to one we have $T=\bar{x}$ and therefore the asymptotic distribution of T is the same as that of $\bar{x}$. On the other hand, if $\vartheta=0$, then the event $|\bar{x}|\geqq n^{-\frac{1}{4}}$ has an asymptotically negligible probability and T has asymptotically the same distribution as $\frac{1}{2}\bar{x}$.

For a more thorough investigation of the questions discussed here in brief, the reader is referred to the publication by Le Cam cited above.

[9] L. Le Cam, On some asymptotic properties of maximum likelihood estimates, Univ. of Calif. Publ. in Stat. 1, No. 11 (1953) p. 277.

Chapter Nine

Inferences Based on Observed Frequencies

In this chapter we are concerned with the same type of problem as in the preceding chapter: the estimation of a parameter ϑ on the basis of observations. In the present chapter we shall restrict attention to observed *frequencies* $h=x/n$, where n denotes the number of trials and x the number of times a particular event has occurred. The frequency h is a random variable whose numerator x has a binomial distribution (§ 5). Other than on the known number n, the sample *size*, the binomial distribution depends only on a single parameter p, the *probability* of the event. In the case of several events with corresponding observed frequencies h_i and probabilities p_i, these p_i may be either completely unknown or known functions of an unknown parameter ϑ. The problem is to estimate whatever is unknown and to study the reliability of the estimate.

This chapter assumes knowledge of the most important points in Chapters 7 and 8.

§ 46. The Maximum Likelihood Method

To be precise, let us assume that n independent trials have been performed and that on each trial there were three mutually exclusive outcomes possible. Suppose that the first event occurred x_1 times, the second x_2 times, and the third x_3 times. Then we have

$$x_1+x_2+x_3=n.$$

The observed frequencies are $h_i=x_i/n$ with

$$h_1+h_2+h_3=1.$$

Let the probabilities be p_1, p_2, and p_3 with

$$p_1+p_2+p_3=1.$$

The expected value of x_i is $n\,p_i$ and therefore that of h_i is p_i. We shall derive the expected value of x_i^2 and of $x_i x_k$ $(i \neq k)$, which we shall need later.

The variance of x_i is $n\,p_i(1-p_i)$. Therefore

$$\begin{aligned}\mathscr{E}\,x_i^2 &= (\mathscr{E}\,x_i)^2 + n\,p_i(1-p_i)\\ &= n^2\,p_i^2 + n\,p_i - n\,p_i^2 \\ &= n(n-1)\,p_i^2 + n\,p_i\,.\end{aligned} \tag{1}$$

The same holds true for $x_i + x_k$:

$$\mathscr{E}(x_i+x_k)^2 = n(n-1)(p_i+p_k)^2 + n(p_i+p_k)\,.$$

Subtracting $\mathscr{E}\,x_i^2 + \mathscr{E}\,x_k^2$ from both sides and dividing by 2, we get

$$\mathscr{E}\,x_i\,x_k = n(n-1)\,p_i\,p_k \qquad (i \neq k)\,. \tag{2}$$

After these introductory remarks we come to the actual problem: let the probabilities p_1, p_2, p_3 be functions of an unknown parameter ϑ and estimate ϑ.

The probability that in n trials we observe the first possible outcome x_1 times, the second x_2 times, and the third x_3 times is

$$\frac{n!}{x_1!\,x_2!\,x_3!}\,p_1^{x_1}\,p_2^{x_2}\,p_3^{x_3}\,.$$

The maximum likelihood method involves maximizing this probability as a function of ϑ. Since the combinatorial factor does not depend on ϑ, we may drop it and consider the likelihood function simply as

$$g(x\,|\,\vartheta) = p_1^{x_1}\,p_2^{x_2}\,p_3^{x_3}\,.$$

Instead of $g(x\,|\,\vartheta)$ we may just as well maximize its logarithm

$$L(x\,|\,\vartheta) = x_1 \ln p_1 + x_2 \ln p_2 + x_3 \ln p_3\,. \tag{3}$$

If the maximum occurs at an interior point of the interval of permissible ϑ-values, then the derivative there must be zero. Hence we have the *likelihood equation*

$$L'(x\,|\,\vartheta) = 0\,. \tag{4}$$

In simple cases the likelihood equation may be solved directly, but in others the method of successive approximation discussed in § 36 must be applied.

If we have to determine just one probability p, the method leads immediately, as we have already seen in § 35, to the estimate

$$\tilde{p} = h = \frac{x}{n}\,,$$

which is unbiased and consistent. According to Example 28 (§ 39), this estimate is an optimal unbiased estimate as well.

In the first of the following two examples, the maximum likelihood method leads to a very good estimate, but in the second we shall see that there exist cases in which the maximum likelihood method fails.

Example 32. This is a famous example discussed in detail by R.A. Fisher in Chapter 9 of his Statistical Methods for Research Workers.

W. A. Carver studied the genetic relation between two hereditary factors in corn. The characteristics in question are starchy versus sugary and green versus white base leaf. After self-fertilization of the heterozygotes, Carver observed the following numbers in the next generation:

	Starchy	Sugary	Totals
Green	1,997	904	2,901
White	906	32	938
Totals	2,903	936	3,839

The overall ratios starchy: sugary and green: white are very close to the 3:1 predicted by Mendelian law. Within the sugary-class, however, the ratio green: white is far from 3:1. The deviation lies well outside the limits of chance; thus we are bound to conclude that the two factors are linked.

Let $\frac{1}{2}p$ be the probability of the formation of a female gamete with the two recessive characteristics sugary and white, and let $\frac{1}{2}p'$ be the probability of the same event for the male gamete. Then the probability that the offspring possesses both recessive factors is $\frac{1}{4}pp'$. If $\vartheta = pp'$, then the probabilities of the four phenomena are

$$p_1 = \tfrac{1}{4}(2+\vartheta), \qquad p_2 = p_3 = \tfrac{1}{4}(1-\vartheta), \qquad p_4 = \tfrac{1}{4}\vartheta.$$

It is obvious that only the product $pp' = \vartheta$ can be estimated from the above observations. If we assume that $p = p'$, then we can also estimate the "recombination rate" $p = \sqrt{\vartheta}$. We shall use the maximum likelihood method to find an estimate of ϑ.

If x_1, x_2, x_3, x_4 are the four observed numbers, the likelihood function is

$$g(x|\vartheta) = p_1^{x_1} p_2^{x_2} p_3^{x_3} p_4^{x_4} = \left(\frac{2+\vartheta}{4}\right)^{x_1} \left(\frac{1-\vartheta}{4}\right)^{x_2+x_3} \left(\frac{\vartheta}{4}\right)^{x_4}.$$

The logarithm is, except for an additive constant,

$$L(x|\vartheta) = x_1 \ln(2+\vartheta) + (x_2+x_3)\ln(1-\vartheta) + x_4 \ln\vartheta.$$

Hence the likelihood equation becomes

$$\frac{x_1}{2+\vartheta} - \frac{x_2+x_3}{1-\vartheta} + \frac{x_4}{\vartheta} = 0,$$

or

$$n\vartheta^2 - (x_1 - 2x_2 - 2x_3 - x_4)\vartheta - 2x_4 = 0. \tag{5}$$

The positive root of this equation is the maximum likelihood estimate $\tilde{\vartheta}$.

It is easy to find other estimates of ϑ, including other unbiased estimates. For example,

$$T_0 = 4h_4 = \frac{4x_4}{n}$$

and

$$T_1 = h_1 - h_2 - h_3 + h_4 = \frac{x_1 - x_2 - x_3 + x_4}{n}$$

are obviously unbiased estimates of ϑ. However, if we consider not only the bias but also the variance, the maximum likelihood estimate $\tilde{\vartheta}$ is far superior to the others. The bias of $\tilde{\vartheta}$ is of the order of n^{-1} and the variance is asymptotically (for large n) equal to the minimum variance as given by the Fréchet inequality applied to unbiased estimates. To be sure, the estimates T_0 and T_1 are unbiased, but their variance is considerably larger than that of $\tilde{\vartheta}$; i.e., they are not efficient.

Fisher compared five different estimates $T_1, \ldots, T_5$. His T_1 is our T_1 and his T_4 is our $\tilde{\vartheta}$. T_5 is the estimate minimizing

$$\chi^2 = \sum \frac{(x_j - n\,p_j)^2}{n\,p_j}.$$

Fisher's research shows that only the last three estimates, T_3, T_4, T_5, are efficient in the sense that their bias is small compared with $n^{-\frac{1}{2}}$ and their variance is asymptotically equal to the minimum variance

$$V_{\min} = \frac{1}{I(\vartheta)} = \frac{c}{n},$$

which according to the Fréchet inequality is the smallest variance possible for unbiased estimates.

We shall discuss the efficiency of the maximum likelihood estimate later.

Example 33. A cannon is fired n times without changing the sight adjustment at a small target (to be idealized as a geometrical point in our analysis). k shots overshoot the target, the remaining l fall short ($k+l=n$). When we speak of the "height of a shot", we shall mean the height of the point of impact in a vertical plane through the target and perpendicular to the correct line of fire. Let the standard deviation of the height be known and the distribution of such heights be normal. How much of a correction must be made in the sight adjustment for the average height to agree as closely as possible with the target?

The probability density of the height of the shot is, if height is measured so that the target is at zero and the variance is unity,

$$g(x) = \frac{1}{\sqrt{2\pi}}\, e^{-\frac{1}{2}(x-\mu)^2}.$$

The mean value of this distribution is μ, and therefore the correction sought is $-\mu$. The probability of overshooting the target is

$$\int_0^\infty g(x)\,dx = \Phi(\mu)$$

and the probability of overshooting it k times is

$$\binom{n}{k} \Phi(\mu)^k \,[1-\Phi(\mu)]^{n-k}.$$

Hence the logarithm of the likelihood function is

$$L(k|\mu) = k \ln \Phi(\mu) + (n-k) \ln [1-\Phi(\mu)],$$

which assumes its maximum whenever

$$\Phi(\mu) = \frac{k}{n}.$$

Hence the most likely value $\tilde{\mu}$ is defined in terms of the inverse function:

$$\tilde{\mu} = \Psi\left(\frac{k}{n}\right).$$

If $k=0$, then $\tilde{\mu}=-\infty$ and if $k=n$, then $\tilde{\mu}=+\infty$. Hence, strictly speaking, $\tilde{\mu}$ has neither a finite mean nor a finite variance.

In practice this problem is easily corrected in the two extreme cases $k=0$ and $k=n$ by replacing the estimate $\tilde{\mu}$ by a finite value which appears reasonable, in as much as we certainly know something about the position of the target. The fact remains that a straightforward application of the maximizing technique leads to an estimate with an infinite variance even for large n.

The estimate $\tilde{\mu}$ is nevertheless consistent: as $n\to\infty$ it converges in probability to the true value μ.

§ 47. Consistency of the Maximum Likelihood Estimate

The maximum likelihood estimate is consistent under quite general conditions. In § 45 we referred to the general proof of Wald and Wolfowitz. We shall now examine in more detail the case that the observed quantities are frequencies.

Let there again be three observed frequencies

$$h_i=\frac{x_i}{n} \qquad (i=1,2,3),$$

and let the probabilities p_1, p_2, p_3 of the three mutually exclusive events be functions of a parameter ϑ.

We shall assume that the mapping $\vartheta\to(p_1,p_2,p_3)$ and its inverse mapping are both continuous. Hence we shall assume that to each (p_1,p_2,p_3) corresponds a unique value of ϑ and that to any neighborhood of (p_1,p_2,p_3) corresponds a neighborhood of ϑ. If we were not to make this assumption, it would be impossible to obtain approximate values for ϑ from the observed material, which gives only frequencies and therefore approximate values for the probabilities p_i only.

As in § 46 the likelihood function is

$$g(x\,|\,\vartheta)=p_1^{x_1}\,p_2^{x_2}\,p_3^{x_3}.$$

If we multiply it by the factor

$$n^n\,x_1^{-x_1}\,x_2^{-x_2}\,x_3^{-x_3},$$

which does not depend on ϑ, we obtain the function

$$G(x\,|\,\vartheta)=\left(\frac{n p_1}{x_1}\right)^{x_1}\left(\frac{n p_2}{x_2}\right)^{x_2}\left(\frac{n p_3}{x_3}\right)^{x_3}, \tag{1}$$

which for our purposes is equivalent to the likelihood function and which has logarithm

$$L(x\,|\,\vartheta)=x_1\ln\left(\frac{n p_1}{x_1}\right)+x_2\ln\left(\frac{n p_2}{x_2}\right)+x_3\ln\left(\frac{n p_3}{x_3}\right). \tag{2}$$

This formula holds for all ϑ and, in particular, for the (unknown) true value to be denoted by ϑ^*. If ϑ equals its true value ϑ^*, the numbers x_i differ from their expected values $n p_i$ in probability at most by a term of the order of magnitude $\sqrt{n p_i}$ (§ 5): i.e., the contributions of the differences

$$z_i = x_i - n p_i \tag{3}$$

are, with probability arbitrarily close to one, not larger than a constant multiple of $\sqrt{n p_i}$. Here we have

$$\sum z = z_1 + z_2 + z_3 = 0. \tag{4}$$

If we express the $n p$ in (2) in terms of x and z and apply (4), we have

$$\begin{aligned} L(x\,|\,\vartheta) &= \sum x \ln \frac{x-z}{x} = \sum z + \sum x \ln\left(1 - \frac{z}{x}\right) \\ &= \sum x \left[\frac{z}{x} + \ln\left(1 - \frac{z}{x}\right)\right] = \sum x\, \varphi\left(\frac{z}{x}\right). \end{aligned} \tag{5}$$

Obviously this formula holds quite generally for arbitrarily many observed frequencies and for arbitrarily many parameters $\vartheta_1, \ldots, \vartheta_r$. Furthermore it holds not only for the true ϑ^* but also for all ϑ.

The function

$$\varphi(t) = t + \ln(1-t)$$

occurring in (5) attains its maximum value, zero, at $t=0$, and decreases monotonically as we move away from zero in either direction, since the derivative

$$\varphi'(t) = 1 - \frac{1}{1-t}$$

is positive for negative values of t and negative for positive values. Therefore in (5) each term is either strictly negative or zero.

For $|t|<1$, $\varphi(t)$ may be expanded into a power series:

$$\varphi(t) = -\tfrac{1}{2}t^2 - \tfrac{1}{3}t^3 - \cdots. \tag{6}$$

In particular, if we set $t = z/x$ and assume that z is at most of order $\sqrt{n p}$, then $t = \frac{z}{x} = \frac{z}{n p + z}$ is at most of order $1/\sqrt{n p}$ and $-\varphi(t)$ is, from (6), at most of order $1/n p$. Hence every term in (5) is at most of order 1, and the same conclusion holds for the entire sum; i.e., the inequality

$$L(x \mid \vartheta) \geqq -g \tag{7}$$

holds for suitable choice of g with probability arbitrarily close to one.

If, on the other hand, one z is large compared with $\sqrt{x}$, then $t=z/x$ is large compared with $1/\sqrt{x}$, hence $-\varphi(t)$ is large compared with $1/x$, and from (5) it follows that $-L(x \mid \vartheta)$ is large compared with 1, yielding the reverse inequality

$$L(x \mid \vartheta) < -g. \tag{8}$$

For the true value ϑ^* and the corresponding true p^* it follows from the preceding discussion that inequality (7) has probability close to one. For $\vartheta=\tilde{\vartheta}$, which maximizes $G(x \mid \vartheta)$ and hence $L(x \mid \vartheta)$, (7) is even more probable. Therefore the corresponding $\tilde{z}=x-n\tilde{p}$ can not be large compared with $\sqrt{x}$; i.e., they have at most the order

$$\sqrt{x} \sim \sqrt{n p^*}.$$

Also, since the $z=x-np^*$ formed using the true p^* have at most the order $\sqrt{np^*}$ in probability, the same is true for the differences $n\tilde{p}-np^*$. Hence:

The differences $\tilde{p}-p^*$ have with arbitrarily high probability at most the order $n^{-\frac{1}{2}}$.

Because of the continuity of ϑ as a function of the p, the differences $\tilde{\vartheta}-\vartheta^*$ are also small. Hence we have proved the consistency of maximum likelihood estimates.

Let us now make the further assumption that the p_i are differentiable functions of ϑ, and conversely. Hence:

The differences $\tilde{\vartheta}-\vartheta^*$ have with arbitrarily high probability the order $n^{-\frac{1}{2}}$.

If we substitute (6) into (5), we obtain the useful series expansion for $L(x \mid \vartheta)$ given by

$$\begin{aligned} L(x \mid \vartheta) &= -\frac{1}{2}\sum\frac{z^2}{x}-\frac{1}{3}\sum\frac{z^3}{x^2}-\cdots \\ &= -\frac{1}{2}\sum\frac{(x-np)^2}{x}-\frac{1}{3}\sum\frac{(x-np)^3}{x^2}-\cdots. \end{aligned} \tag{9}$$

By and large, the approximation

$$L(x \mid \vartheta) \sim -\frac{1}{2}\sum\frac{(x-np)^2}{x} \tag{10}$$

suffices.

§ 48. Maximum Likelihood, Minimum χ^2, and Least Squares

The last approximation in § 47 to the likelihood function yields a first approximation to $\tilde{\vartheta}$ quickly and with a minimum of computation: instead of maximizing $L(x|\vartheta)$, we minimize the quadratic form

$$\chi_x^2 = \sum \frac{(x-np)^2}{x}. \tag{1}$$

The quadratic form χ_x^2 differs only slightly from the well-known expression

$$\chi^2 = \sum \frac{(x-np)^2}{np}. \tag{2}$$

Minimizing χ^2 also yields an estimate of ϑ, but this method of estimation frequently leads to complicated computations, in as much as the denominator must be differentiated as well.

To avoid differentiating the denominator in (2), one alternative procedure is to replace the p in the denominator by an estimated value $p^{(0)}$ and minimize the expression

$$\chi_0^2 = \sum \frac{(x-np)^2}{np^{(0)}}. \tag{3}$$

We shall now examine somewhat more closely the calculation of these various estimates and prove that they differ from one another only in the order of magnitude n^{-1}. We shall restrict ourselves to the following three estimates:

A. Minimum χ_0^2,

B. Minimum χ_x^2,

C. Maximum Likelihood.

To simplify the calculations we shall assume in addition that the p_i are *linear* functions of the unknown parameters ϑ. The results can be applied to nonlinear differentiable functions, for these functions may be approximated in a neighborhood of the true ϑ-value by linear functions. In the examples the p_i are often nonlinear functions, but throughout the chapter the theory will be developed for linear functions only.

A. Minimum χ_0^2

The quadratic form

$$\chi_0^2 = \sum \frac{(x-x')^2}{np^{(0)}} \tag{4}$$

defines a Euclidean metric in X-space: χ_0^2 is the square of the distance between the points X and X'. If we set

$$x_i' = n\, p_i(\vartheta),$$

where the $p_i(\vartheta)$ are by hypothesis linear functions of the parameters ϑ_α, then the point X' lies in a linear subspace G. To minimize χ_0^2 is to determine the nearest point X' in this subspace to the observed point X. Therefore X' is the point of intersection with G of the perpendicular line from X to the subspace G.

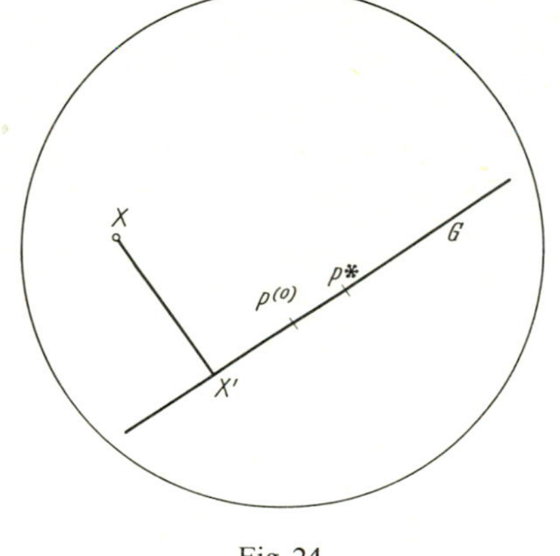

Fig. 24

Computation yields the same result as the geometric argument. Differentiating (3) with respect to ϑ_α and setting the derivatives equal to zero, we have the conditions

$$\sum \frac{[x_i - n\, p_i(\vartheta)]\, q_{i\alpha}}{p_i^{(0)}} = 0, \tag{5}$$

where

$$q_{i\alpha} = \frac{\partial p_i(\vartheta)}{\partial \vartheta_\alpha}. \tag{6}$$

The Eqs. (5) *imply that the vector from the point X with coordinates x_i/n to the point X' with coordinates $p_i(\vartheta)$ is perpendicular to all vectors tangential to the manifold G through the point X'.*

In this formulation the assertions are also valid if G is a non-linear sub-manifold of X-space. We shall, however, retain the assumption that the $p_i(\vartheta)$ are linear functions of the parameters ϑ_α. In this case the $q_{i\alpha}$ are constants and we may write

$$p_i(\vartheta) = p_i(0) + \sum_\beta q_{i\beta}\, \vartheta_\beta. \tag{7}$$

By relocating the origin in the parameter space so that the initial approximation $p^{(0)}$ corresponds to all $\vartheta_\beta = 0$ – i.e., so that $p_i(0) = p_i^{(0)}$ – we can simplify the computation somewhat.

If we substitute (7) into (5), we get r linear equations in the r unknowns $\vartheta_1, \ldots, \vartheta_r$:

$$\sum_\beta h_{\alpha\beta}\,\vartheta_\beta = \sum_i \frac{[x_i - n\,p_i(0)]\,q_{i\alpha}}{p_i^{(0)}}, \tag{8}$$

where

$$h_{\alpha\beta} = \sum_i \frac{n\,q_{i\alpha}\,q_{i\beta}}{p_i^{(0)}}. \tag{9}$$

The coefficients $h_{\alpha\beta}$ have exactly the same form as the sums $[g\,a\,a]$, $[g\,a\,b], \ldots$ in the theory of least squares. In fact, the problem of minimizing the quadratic form χ_0^2 is exactly the same minimization problem as the least squares problem, the weights of the observed frequencies x_i/n being $n/p_i^{(0)}$.

We are now in a position to examine the effect on the point X' of a change in the arbitrarily chosen $p_i^{(0)}$.

Let us take as the coordinates of X not the x_i but the frequencies $h_i = x_i/n$ and as the definition of distance

$$r^2 = \frac{1}{n}\,\chi_0^2 = \sum \frac{(h_i - h_i')^2}{p_i^{(0)}}. \tag{10}$$

We shall assume that the point X with coordinates h_i, as well as the point P_0 with coordinates $p_i^{(0)}$, lies in a neighborhood of the order of magnitude $n^{-\frac{1}{2}}$ of the "true point" P^* with coordinates p_i^* and that in this neighborhood all the p_i are bounded away from zero: $p_i \geqq \delta > 0$.

If the $p_i^{(0)}$ are replaced by $s_i^{(0)}$ in the same neighborhood, then the order of magnitude of all the differences $p_i^{(0)} - s_i^{(0)}$ is $\varepsilon = n^{-\frac{1}{2}}$. There is a change neither in the point X nor in the linear space G, but only in the metric defined by (10), and this is only by an amount of order ε. The perpendicular XX' can change direction, but the angle between the old direction and the new has the order of magnitude ε. Since the length of the perpendicular has the order ε, the change in the coordinates of X' is only of the order of magnitude $\varepsilon^2 = n^{-1}$.

In general, if the differences in coordinates $p_i^{(0)} - s_i^{(0)}$ are of the order η, then the coordinates of X' change only in the order $\varepsilon\,\eta$; or, more precisely, if all $|p_i^{(0)} - s_i^{(0)}|$ are less than η, then the differences of the coordinates of the two X' corresponding to $p_i^{(0)}$ and to $s_i^{(0)}$ respectively are less than $c\,\varepsilon\,\eta$, for some constant factor c.

There is little point in determining the constant c theoretically, since such computations are clumsy and the differences occurring are for the most part small. For practical purposes it suffices to assert that the point X' depends very little on the choice of the initial approximation $p_i^{(0)}$.

B. Minimum χ_x^2

If the $p_i^{(0)}$ are chosen equal to the observed frequencies $h_i = x_i/n$, the resulting χ_0^2 equals χ_x^2. Hence: *The minimum value of χ_x^2 is calculated in exactly the same manner as that of χ_0^2 – namely, by the method of least squares – and the resulting point P_x differs from the previous X' in probability only in the order of magnitude n^{-1}.*

C. Maximum Likelihood

If the logarithm of the likelihood function

$$L(x|\vartheta) = \sum x_i \ln p_i(\vartheta)$$

assumes its maximum at an interior point of the set of possible ϑ-values, then the derivative must be zero at that point. This gives rise to the conditions

$$\sum_i \frac{x_i q_{i\alpha}}{p_i} = 0, \tag{11}$$

where the $q_{i\alpha}$ are again the derivatives of the p_i. Since the sum of the p_i is one, the sum of their derivatives must be zero:

$$\sum_i q_{i\alpha} = 0. \tag{12}$$

Multiplying (12) by n and subtracting the result from (11), we have

$$\sum_i \frac{(x_i - n p_i) q_{i\alpha}}{p_i} = 0. \tag{13}$$

Here the p_i in the numerator and denominator are linear functions of the ϑ. The solution of (13) is the maximum likelihood estimate $\tilde{\vartheta}$, so that, explicitly, we have

$$\sum_i \frac{[x_i - n p_i(\tilde{\vartheta})] q_{i\alpha}}{p_i(\tilde{\vartheta})} = 0. \tag{14}$$

The Eqs. (14) may be solved easily by a method of successive approximation in which the ϑ in the denominator are first replaced by an arbitrary approximate value $\vartheta^{(0)}$. The resulting equations

$$\sum_i \frac{[x_i - n p_i(\vartheta)] q_{i\alpha}}{p_i(\vartheta^{(0)})} = 0 \tag{15}$$

are identical with the Eq. (5) of the minimum χ_0^2 method and therefore may be solved easily by the method of least squares. If the solution $\vartheta^{(1)}$ is substituted into the denominator in (14), repetition of the same process yields an improved solution $\vartheta^{(2)}$, and so on.

The process converges and the limit is independent of the choice of the initial approximation $p_i^{(0)}=p_i(\vartheta^{(0)})$. In particular, if $p_i^{(0)}$ and $s_i^{(0)}$ are two initial approximations whose differences are of the order $\varepsilon=n^{-\frac{1}{2}}$, then according to the previous proofs the differences of the first approximations $p_i^{(1)}$ and $s_i^{(1)}$ are only of the order ε^2, those of the second approximations only of the order ε^3, and so on. Suppose $s_i^{(0)}$ is a solution to the likelihood equation, so that $s_i^{(0)}=s_i^{(1)}=s_i^{(2)}=\cdots$; then the sequence $p_i^{(0)}, p_i^{(1)}, p_i^{(2)}, \ldots$ converges to this solution $s_i^{(0)}$ independently of the choice of $p_i^{(0)}$ in the ε-neighborhood of $s_i^{(0)}$.

In the above proof we assumed that there exists a solution to the maximum likelihood equation. This is always the case, provided all the x_i are positive. The conditions $p_i \geqq 0$ define a closed, bounded set in p-space and the part of the linear subspace G which belongs to this set is likewise closed and bounded. The likelihood function

$$g(x|\vartheta)=\prod_i p_i(\vartheta)^{x_i} \tag{16}$$

is continuous and therefore assumes a maximum value on this closed set. This maximum can not occur on the boundary because the function $g(x|\vartheta)$ is zero there.

From the above proof it also follows that so long as X lies sufficiently close to the subspace G there can exist only one solution to the likelihood equation. If $p_i^{(0)}$ and $s_i^{(0)}$ were two different solutions, then we would have $p_i^{(0)}=p_i^{(1)}=p_i^{(2)}=\cdots$ and $s_i^{(0)}=s_i^{(1)}=s_i^{(2)}=\cdots$; but, the first series must converge to $s_i^{(0)}$, contradicting the hypothesis $p_i^{(0)} \neq s_i^{(0)}$.

With slight modification this proof also holds in the case that the p_i are no longer linear functions, so long as either the manifold consisting of all points $P=P(\vartheta)$ is closed or its boundary lies on the boundary of our domain $p_i \geqq 0$. If neither of the above two cases apply, complications can arise.

In applying this method, the fact that it converges quickly – the differences $p_i^{(k)}-s_i^{(0)}$ tend to zero as the powers ε^k – is particularly important. If the initial approximation is chosen well, one can be satisfied with the first approximation $p^{(1)}$. Further approximations differ from $p^{(1)}$ only in the order $\varepsilon^2=n^{-1}$, but since the unavoidable random deviations in the $\tilde{\vartheta}$ are of the order $\varepsilon=n^{-\frac{1}{2}}$, there is little reason for demanding greater accuracy.

From the proofs we have the following: *The differences of the estimates A and B from the maximum likelihood estimate C are of the order* n^{-1}.

All these results apply equally well to the case that the observations consist of not one but several sequences of frequencies each summing to one; *e.g.*,

$$h_{11}+h_{12}=1, \quad \text{or} \quad x_{11}+x_{12}=n_1,$$
$$h_{21}+h_{22}=1, \quad \text{or} \quad x_{21}+x_{22}=n_2, \ldots.$$

The expressions $\chi^2, \chi_0^2, \ldots$ remain unchanged, except that each p_{ij} must be multiplied by the corresponding n_i; *e.g.*,

$$\chi^2 = \sum_i \sum_j \frac{(x_{ij} - n_i p_{ij})^2}{n_i p_{ij}}.$$

§ 49. Asymptotic Distributions of χ^2 and $\tilde{\vartheta}$

For the sake of simplicity we shall limit ourselves to the case of a single parameter ϑ and under this assumption discuss the distribution functions of χ^2 and of the estimate $\tilde{\vartheta}$.

In the method of least squares, $\tilde{\vartheta}$ was a linear function of the observations x_i, which were assumed to be normally distributed, resulting in $\tilde{\vartheta}$ being normally distributed as well. Now, however, the observations x_i are natural numbers and, hence, discrete quantities, which are only approximately normally distributed, and also the $\tilde{\vartheta}$ are only approximately linear functions of the x_i. Hence at best we can now expect $\tilde{\vartheta}$ to be normally distributed only asymptotically (as $n \to \infty$).

The probability that the observed point X belongs to a subset B in X-space is the sum of the probabilities of the individual points in B:

$$\mathscr{P}(B) = \sum_{X \text{ in } B} \mathscr{P}(X), \tag{1}$$

where

$$\mathscr{P}(X) = \frac{n!}{x_1! \ldots x_m!} p_1^{x_1} \ldots p_m^{x_m}. \tag{2}$$

Here the $p_i = p_i(\vartheta)$ are the *true* probabilities $p_1^*, \ldots, p_m^*$, but we shall omit the asteriks.

For large n we may apply Stirling's formula to (2), which yields

$$\mathscr{P}(X) \sim x_1^{-(x_1+\frac{1}{2})} \ldots x_m^{-(x_m+\frac{1}{2})} n^{n+\frac{1}{2}} (2\pi)^{\frac{1-m}{2}} p_1^{x_1} \ldots p_m^{x_m}, \tag{3}$$

or

$$\gamma \mathscr{P}(X) \sim \left(\frac{n p_1}{x_1}\right)^{x_1+\frac{1}{2}} \ldots \left(\frac{n p_m}{x_m}\right)^{x_m+\frac{1}{2}}$$

where

$$\gamma = [(2\pi n)^{m-1} p_1 \ldots p_m]^{\frac{1}{2}}. \tag{4}$$

The logarithm of $\gamma \mathscr{P}(X)$ is therefore

$$\ln \gamma \mathscr{P}(X) = \sum (x + \tfrac{1}{2}) \ln \frac{n p}{x} + \cdots, \tag{5}$$

where the terms omitted are of the order n^{-1}. Again let

$$x_i = n p_i + z_i, \tag{6}$$

where the z_i have with arbitrarily high probability the order of magnitude $(n p_i)^{\frac{1}{2}}$, yielding

$$\begin{aligned} \ln \gamma \mathscr{P}(X) &= -\sum (n p + z + \tfrac{1}{2}) \ln \frac{n p + z}{n p} + \cdots \\ &= -\sum (n p + z + \tfrac{1}{2}) \left(\frac{z}{n p} - \frac{z^2}{2 n^2 p^2} + \frac{z^3}{3 n^3 p^3} \right) + \cdots \\ &= -\frac{1}{2} \sum \frac{z^2}{n p} - \frac{1}{2} \sum \frac{z}{n p} + \frac{1}{6} \sum \frac{z^3}{n^2 p^2} + \cdots , \end{aligned}$$

where the terms indicated by ... are of the order n^{-1} in probability. If we put

$$\sum \frac{z^2}{n p} = \chi^2,$$

we get

$$\begin{aligned} \gamma \mathscr{P}(X) &\sim e^{-\frac{1}{2}\chi^2} \exp\left(-\frac{1}{2} \sum \frac{z}{n p} + \frac{1}{6} \sum \frac{z^3}{n^2 p^2} \right) \\ &\sim e^{-\frac{1}{2}\chi^2} \left(1 - \frac{1}{2} \sum \frac{z}{n p} + \frac{1}{6} \sum \frac{z^3}{n^2 p^2} \right). \end{aligned} \tag{7}$$

We shall take up the last two terms, which are of order $n^{-\frac{1}{2}}$, but first let us consider the dominating term

$$\gamma \mathscr{P}(X) \sim e^{-\frac{1}{2}\chi^2}, \tag{8}$$

where

$$\chi^2 = \sum \frac{(x - n p)^2}{n p}. \tag{9}$$

From relation (8) we see that the probability of a point X is a decreasing function of the distances of its coordinates x_i from their expected values $n p_i$; that is, the probability decreases as the function χ^2 increases.

As we have seen above, the function χ^2 defines a Euclidean metric in X-space, in that it represents, up to a factor n, the square of the distance of the variable points X with coordinates x_i/n from the fixed point P with coordinates p_i. The more distant X is from P the smaller is its probability as given by (8). In this metric the surfaces on which χ is constant are concentric spheres about the point P. If these spherical surfaces are cut (in the case $m=3$) by the plane $x_1 + x_2 + x_3 = n$, which passes through the common center P, the result is concentric circles about the point P.

Since the $x - n p$ have with arbitrarily high probability the order of magnitude $\sqrt{n}$, the χ^2 have with arbitrarily high probability the order 1; i.e., for every η there exists an upper bound R^2 such that the inequality

$\chi^2 < R^2$ holds with probability $> 1-\eta$. Hence in asymptotic calculations of probabilities we can always restrict attention to a neighborhood of P of the form $\chi^2 < R^2$.

In particular, we want to calculate the following probabilities:

a) the distribution function of χ^2; i.e., the probability that the event $\chi^2 < u$ occurs, and

b) the distribution function of $\mathfrak{F}$.

If we introduce in the neighborhood of the point P new coordinates

$$y_i = \frac{x_i - n p_i}{\sqrt{n p_i}}, \tag{10}$$

which are of the order of one, the distance function χ^2, which is the square of the distance of the point X with coordinates x_i/n from P, becomes simply

$$\chi^2 = \sum y^2. \tag{11}$$

The y_i therefore are ordinary rectangular coordinates in the metric defined by χ^2.

The points X, which correspond to the sets of m non-negative integers x_i whose sum $\sum x_i$ is n, define a lattice in the hyperplane $\sum y_i \sqrt{n p_i} = 0$, which has dimension $m-1$. The basis-vectors of this lattice are all of the order $n^{-\frac{1}{2}}$ in the metric defined by (11); therefore the volume of the lattice is of the order $n^{-(m-1)/2}$. The probability that the point X belongs to some set B is the sum of the probabilities of the lattice points contained in B.

The asymptotic evaluation of these probabilities is most simple in case a), in which the set B is a sphere defined by $\chi^2 < u$. K. Pearson, who introduced the formula for χ^2 into mathematical statistics, evaluated the sum (1) over the lattice points inside such a sphere by the following method.

First replace the probability $\mathscr{P}(X)$ by the approximation given by (8):

$$\mathscr{P}(X) \sim \gamma^{-1} e^{-\frac{1}{2}\chi^2}. \tag{12}$$

The error of this approximation is of the order $n^{-\frac{1}{2}}$. Then replace the sum over all lattice points by an integral over the set B, divided by the volume of the lattice. The error of this approximation occurs chiefly on the boundary. The order of magnitude of this boundary error is again at most $n^{-\frac{1}{2}}$. The resulting integral is

$$(2\pi)^{-(m-1)/2} \int \cdots \int e^{-\frac{1}{2}\chi^2} \, dV_{m-1}, \tag{13}$$

integrated over the set $\chi^2 < u$. Evaluation of the integral yields in the end, as we have already seen in § 27, the familiar χ^2-distribution with $m-1$ degrees of freedom. The reader should refer to Pearson's[1] work for the details.

We have thus far ignored the two correction terms of order $n^{-\frac{1}{2}}$ in (7). If we do take them into consideration, however, it becomes apparent that they have no influence at all on the result. The function

$$e^{-\frac{1}{2}\chi^2}\left(-\frac{1}{2}\sum\frac{z}{np}+\frac{1}{6}\sum\frac{z^3}{n^2p^2}\right) \tag{14}$$

is an odd function of $z_1, \ldots, z_m$, in as much as it changes its sign if all the z_i are replaced by $-z_i$. But the integral of an odd function over the symmetric region $\chi^2 < u$ is zero. The asymmetry of the distribution therefore has virtually no influence on the result. The only error is the error arising on the boundary, and it can attain at most the order $n^{-\frac{1}{2}}$.

To attack the problem of the distribution of the maximum likelihood estimate $\tilde{\vartheta}$, we shall first consider instead of $\tilde{\vartheta}$ the estimate ϑ' defined by minimizing χ_0^2, where

$$\chi_0^2=\sum\frac{[x_i-n\,p_i(\vartheta)]^2}{n\,p_i}. \tag{15}$$

Unlike § 48, the values substituted in the denominator here are the true values p_i. It is quite true that in practice the p_i are unknown, but here we are concerned with a purely theoretical investigation of the distribution function. If χ_0^2 is minimized, the resulting estimate ϑ' is a quantity which differs from $\tilde{\vartheta}$, as we have seen in § 48, only in the order of magnitude n^{-1}.

This ϑ' is found by the method of least squares, and hence by dropping a perpendicular from the point X to the linear subspace G defined by the parameter representation $p_i(\vartheta)$. ϑ' is the parameter value corresponding to the foot X' of this perpendicular.

The equations for determining ϑ' have already been given in § 48 in the general case of r parameters $\vartheta_1, \ldots, \vartheta_r$. If the origin in the parameter space is chosen so that the probabilities $p_i(0)$ corresponding to it are the true probabilities p_i, then from (8) § 48 the equations for ϑ' read

$$\sum_\beta h_{\alpha\beta}\,\vartheta'_\beta=\sum_i\frac{(x_i-n\,p_i)\,q_{i\alpha}}{p_i}, \tag{16}$$

where

$$h_{\alpha\beta}=\sum_i\frac{n\,q_{i\alpha}\,q_{i\beta}}{p_i}. \tag{17}$$

[1] K. Pearson, On the criterion that a given system of deviations ... is such that it can be reasonably supposed to have arisen from random sampling. Philosophical Magazine 50 (1900) p. 157.

From (16) we see first of all that the ϑ'_β are linear functions of the observed frequencies $h_i = x_i/n$. Secondly, we see that their expected values are zero, since the expected values of $x_i - n p_i$ are zero. *Therefore the estimates ϑ'_β are unbiased.*

Furthermore, since the $x_i - n p_i$ and the $h_{\alpha\beta}$ have the orders of magnitude $\sqrt{n}$ and n, respectively, in probability, it follows that the ϑ'_α have in probability at most the order of magnitude $n^{-\frac{1}{2}}$.

Finally, since the x_i are approximately normally distributed, it is to be expected that the ϑ' are also approximately normally distributed. In the proof we shall again restrict attention to the case of a single parameter ϑ. The problem at hand is to calculate asymptotically for $n \to \infty$ the probability that $\vartheta' < t n^{-\frac{1}{2}}$.

ϑ' was the parameter value corresponding to the point X', which was the projection of the observed point X onto the line G (Fig. 24). If this projection on the line G is to lie to the left of a point $P_t = P(t n^{-\frac{1}{2}})$, then X must lie in a half-space bounded by a hyperplane H_t which is perpendicular to G through the point P_t.

The calculation is easiest if the rectangular coordinates $y_1, \ldots, y_m$ are transformed orthogonally so that the y_1-axis coincides with the line G. The point P_t then has the coordinate $y_1 = a t$ and the half-space bounded by H_t is defined by $y_1 < a t$.

The desired probability is now the sum of the probabilities $\mathscr{P}(X)$ of the lattice points in the half-space. Again the $\mathscr{P}(X)$ can be approximated by (8), or by (12), and the sum by an integral. Hence we obtain an integral of the form

$$\begin{aligned}(2\pi)^{-(m-1)/2} \int \cdots \int e^{-\frac{1}{2}\chi^2}\, dV_{m-1} \\ = (2\pi)^{-(m-1)/2} \int \cdots \int e^{-\frac{1}{2}(y_1^2 + \cdots + y_m^2)}\, dV_{m-1}\end{aligned} \tag{18}$$

integrated over that part of the hyperplane $\sum x_i = n$ which lies in the half-space $y_1 < a t$. Whether we integrate over the whole space or only over that part of the space contained in the sphere $\chi^2 < R^2$ is immaterial, as long as R is chosen sufficiently large. After integration with respect to $y_2, \ldots, y_{m-1}$ there remains the integral

$$(2\pi)^{-\frac{1}{2}} \int_{-\infty}^{at} e^{-\frac{1}{2}y_1^2}\, dy_1 = \Phi(a t). \tag{19}$$

Therefore the distribution of ϑ' is asymptotically normal.

To determine the normal distribution completely we have yet to determine the factor a. As is evident from Fig. 24, the coordinate $y_1 = a t$ is the distance $P P_t$ in the metric defined by χ^2. Computation yields

$$a^2 = \sum_i \frac{q_{i1}^2}{p_i} = \frac{h_{11}}{n}. \tag{20}$$

In (19) we had the probability of the event $\vartheta' < t\,n^{-\frac{1}{2}}$. If we set $t\,n^{-\frac{1}{2}} = t'$ and $a\,n^{\frac{1}{2}} = a'$, then we have $a\,t = a'\,t'$ and the asymptotic distribution function of ϑ' becomes

$$\Phi(a\,t) = \Phi(a'\,t'),$$

with

$$a' = a\sqrt{n} = \sqrt{h_{11}}. \tag{21}$$

The asymptotic standard deviation of ϑ' therefore becomes

$$\sigma_{\vartheta'} = \frac{1}{a'} = \sqrt{\frac{1}{h_{11}}} = \sqrt{h^{11}}. \tag{22}$$

The case of r parameters $\vartheta_1, \ldots, \vartheta_r$ is analogous and the standard deviation of ϑ'_α is

$$\sigma_\alpha = \sqrt{h^{\alpha\alpha}}, \tag{23}$$

where $(h^{\alpha\beta})$ is the inverse matrix of the matrix $(h_{\alpha\beta})$. The formulas are completely analogous to those in § 31. The derivation is the same in that we solve (16) for the ϑ' by means of the inverse matrix, square, and take the expected values. On the right-hand side only the x_i depend on chance. The mean values of x_i^2 and $x_i\,x_k$ are, however, known exactly, so that formulas (22) and (23) are valid, therefore, not only asymptotically as $n \to \infty$ but also exactly. From (17) we see that the $h_{\alpha\beta}$ are proportional to n and therefore that the elements of the inverse matrix $h^{\alpha\beta}$ are proportional to n^{-1}. The standard deviations as given by (23) are hence of the form $c\,n^{-\frac{1}{2}}$.

The transition from ϑ' to $\tilde{\vartheta}$ is now no longer difficult. If we put

$$\tilde{\vartheta} = \vartheta' + \eta, \tag{24}$$

then in probability η has the order n^{-1}. If we multiply both sides of (24) by $\sqrt{n}$ we get

$$\tilde{\vartheta}\sqrt{n} = \vartheta'\sqrt{n} + \eta\sqrt{n}. \tag{25}$$

The first term on the right has an asymptotically normal distribution with mean value zero and standard deviation independent of n, whereas the second tends to zero in probability as $n \to \infty$. Therefore we may apply the elementary limit law § 24 G to the sum (25). *Hence $\tilde{\vartheta}$ is asymptotically normally distributed with the same mean value and the same standard deviation as ϑ'.*

The same conclusion holds for all estimates which differ from ϑ' or $\tilde{\vartheta}$ in probability only by terms of the order n^{-1}; therefore, for example, the minimum χ_x^2 estimate and any minimum χ_0^2 estimate are asymptotically normal.

§ 50. Efficiency

We shall consider again the case of a single unknown parameter ϑ. The variance of the asymptotic distribution of $\tilde{\vartheta}$ is, as we have seen, given by

$$\sigma_{\tilde{\vartheta}}^2 = \frac{1}{h_{11}} = h^{11} = \frac{c^2}{n}. \tag{1}$$

This does not mean, however, that the standard deviation of $\tilde{\vartheta}$ tends to zero as $n \to \infty$. As illustrated in Example 33 (§ 46), it may even happen that the standard deviation of $\tilde{\vartheta}$ is infinite for every finite n, and therefore that the limit of these standard deviations is also infinite. Formula (1) is not an asymptotic formula for the exact variance, but it represents the *asymptotic variance* of the estimate $\tilde{\vartheta}$ in the sense of § 45 B. Furthermore, $\tilde{\vartheta}$ is *asymptotically unbiased*, also in the sense of § 45 B.

We shall now compare the asymptotic variance (1) with the minimum variance for an unbiased estimate as given by the Fréchet inequality. To this end we must first calculate the information $I(\vartheta)$ defined in § 37 as

$$I(\vartheta) = \mathscr{E}\{L'(x \mid \vartheta)^2\}. \tag{2}$$

In the present situation

$$L(x \mid \vartheta) = \sum x_i \ln p_i(\vartheta),$$

so that, if the derivative of p_i is denoted by q_i (the previous $q_{i\alpha}$),

$$L'(x \mid \vartheta) = \sum \frac{q_i}{p_i} x_i$$

and

$$\begin{aligned} I(\vartheta) &= \mathscr{E}\left(\sum \frac{q_i}{p_i} x_i\right)^2 \\ &= \sum_i \sum_k \frac{q_i q_k}{p_i p_k} \mathscr{E}(x_i x_k). \end{aligned} \tag{3}$$

The expected values of $x_i x_k$ were found in § 46 to be

$$\mathscr{E}(x_i x_k) = n(n-1) p_i p_k \qquad \text{for } i \neq k, \tag{4}$$

$$\mathscr{E}(x_i^2) = n(n-1) p_i^2 + n p_i. \tag{5}$$

Hence $I(\vartheta)$ may be written as

$$I(\vartheta) = \sum \sum n(n-1) q_i q_k + \sum n \frac{q_i^2}{p_i}. \tag{6}$$

The first term is zero because $(\sum q_i)^2 = 0$. The second term is simply h_{11}. Therefore, as in the theory of least squares, we have

$$I(\vartheta) = h_{11}. \tag{7}$$

The Fréchet inequality for the case of unbiased estimates now reads

$$\sigma^2 \geqq I(\vartheta)^{-1} = \frac{1}{h_{11}} = h^{11} = \frac{c^2}{n}. \tag{8}$$

The estimate $\tilde{\vartheta}$, which is asymptotically unbiased, has the asymptotic variance c^2/n, as we have seen in (1). In this sense it is asymptotically efficient.

This does not mean that the estimate $\tilde{\vartheta}$ has the smallest asymptotic variance among all asymptotically unbiased estimates. We can construct examples (as in § 45 D) of estimates which are asymptotically unbiased but which have a smaller asymptotic variance for some ϑ-values. A minimality property for the estimate $\tilde{\vartheta}$ can be proved only if the competing estimates satisfy regularity conditions which we shall now discuss.

If the likelihood Eq. (14) § 48 is divided by n, the result is, if the index α in $q_{i\alpha}$ is dropped,

$$\sum_i \frac{[h_i - p_i(\tilde{\vartheta})]\, q_i}{p_i(\tilde{\vartheta})} = 0. \tag{9}$$

The equation no longer contains the x_i and n explicitly but only the frequencies h_i. The maximum likelihood estimate $\tilde{\vartheta}$ is therefore a function of the h_i alone, and, in fact, so long as the h_i are not improbably far from the p_i and also not too near zero, it is a *differentiable* function of the h_i.

We shall now allow only those estimates T which are differentiable functions of the h_i. Such estimates shall be called *regular*. Therefore let T be a regular, asymptotically unbiased estimate. We wish to compare the asymptotic variance of T with that of $\tilde{\vartheta}$.

We know that with high probability the h_i lie in a neighborhood of the p_i with a diameter which is small relative to one. In such a neighborhood every differentiable function can be approximated by a linear function. A useful linear approximation for $\tilde{\vartheta}$ is our ϑ' defined earlier. Let a linear approximation for T be

$$T' = \sum c_i\, h_i. \tag{10}$$

Just as $\tilde{\vartheta}$ has asymptotically the same distribution as ϑ', so the differentiable function T has asymptotically the same distribution as the linear function T'. The lines of reasoning in the two proofs are exactly the same. Also the asymptotic derivation of the distribution of T' runs exactly as that of ϑ'. We have again the problem of summing probabilities

$\mathscr{P}(X)$ over all points in a half-space. Summation is replaced by integration and the probability $\mathscr{P}(X)$ by a normal probability density

$$C e^{-\frac{1}{2}\chi^2}, \tag{11}$$

where $\chi^2 = y_1^2 + \cdots + y_m^2$. The lattice points over which we must sum all lie in a hyperplane $\sum h_i = 1$ over which, therefore, we must integrate. By making a suitable orthogonal transformation of the y_i we may take this hyperplane to have the equation $y_m = 0$, in which case the probability density becomes

$$f(y_1, \ldots, y_{m-1}) = C \exp - \tfrac{1}{2}(y_1^2 + \cdots + y_{m-1}^2). \tag{12}$$

This formula is true only if the point whose coordinates are $p_i(\vartheta)$ in the space of the h_i is taken as the origin for the rectangular coordinates $y_1, \ldots, y_m$. If, however, a fixed origin independent of ϑ is desired, then (12) must be replaced by

$$f(y_1, \ldots, y_{m-1}) = C \exp - \tfrac{1}{2}[(y_1 - \hat{y}_1)^2 + \cdots + (y_{m-1} - \hat{y}_{m-1})^2], \tag{13}$$

where the $\hat{y}_i$ are the expected values of the y_i.

Hence it follows that T' has asymptotically a normal distribution. The asymptotic mean and the asymptotic variance of the estimate T' are by definition equal to the mean and the variance of this asymptotic normal distribution; i.e., they are equal to the mean and the variance of the linear function

$$T' = \sum c_i h_i = b_0 + \sum b_k y_k \tag{14}$$

obtained from the normal distribution (13) by integration. Here the coefficients c_i do not depend on n. Also, the mean value and variance of T' may be calculated exactly from these coefficients so that no longer is there any difference between the limit of the variance and the asymptotic variance. The same is true for ϑ'.

We now have exactly the same situation as in the theory of least squares. The $y_1, \ldots, y_{m-1}$ are independent, normally distributed variables whose standard deviation is unity and whose expected values $\hat{y}_i$ are linear functions of a parameter ϑ. The method of least squares yields an optimal unbiased estimate ϑ'. The estimate T' is unbiased as well and, therefore, its variance is at least equal to the variance of ϑ'. Equality of the variances occurs only in the case that the coefficients occurring in the linear function T' equal those in ϑ'. Therefore:

Among all regular, asymptotically unbiased estimates T, the maximum likelihood estimate $\tilde{\vartheta}$ has asymptotically the smallest variance. If T and $\tilde{\vartheta}$ have the same variance and if both are expanded in powers of $h_i - p_i$ in a neighborhood of the point $p_i = p_i(\vartheta)$, they must agree at least in the linear terms. Hence the difference $T - \tilde{\vartheta}$ is small compared with $n^{-\frac{1}{2}}$ in probability.

The statement of the above theorem can be made more concise if we define the concepts of *efficiency* and *asymptotic equivalence*. An asymptotically unbiased, regular estimate T is said to be *efficient* if it has the smallest variance asymptotically among all such estimates. Two estimates T_1 and T_2 are said to be *asymptotically equivalent* if their difference $D = T_1 - T_2$ is small in probability compared with $n^{-\frac{1}{2}}$ – that is, if $D n^{\frac{1}{2}}$ becomes arbitrarily small in probability. We may now state the theorem in the following manner:

The maximum likelihood estimate $\tilde{\vartheta}$ *is efficient and every efficient regular estimate is asymptotically equivalent to it. The variance of such an estimate is asymptotically equal to the inverse of the information,* $I(\vartheta)^{-1}$.

J. Neyman[2] proved this theorem under more general conditions. In particular, his proof does not require the hypothesis of linearity of the functions $p_i(\vartheta)$.

Example 34. According to the hypothesis of F. Bernstein[3] generally accepted today, there exist three genes A, B, and O, of which A and B are dominant over O, which are responsible for the human blood types O, A, B, and AB. An individual possessing the pair of genes OO belongs to blood type O, one with AO or AA to type A, one with BO or BB to B, and, finally, one with AB belongs to blood type AB. Suppose that the frequencies

$$h_1 = \frac{x_1}{n}, \ldots, h_4 = \frac{x_4}{n}$$

of the blood types in a sample of n individuals are given. Efficient estimates of the frequencies p, q, and r of the genes A, B, and O in the population are desired.

We shall assume that the population is homogeneous in the sense that it can not be broken into nearly exclusive groups with different gene frequencies. Under this assumption the probability that the genotype OO occurs is r^2, that of AO is $p\,r$, etc. The probabilities of the four blood types O, A, B, and AB are therefore

$$\begin{aligned} p_1 &= r^2 \\ p_2 &= 2pr + p^2 = (p+r)^2 - r^2 \\ p_3 &= 2qr + q^2 = (q+r)^2 - r^2 \\ p_4 &= 2pq. \end{aligned} \tag{15}$$

Solving these equations for p and q, we have

$$\begin{aligned} p &= 1-(q+r) = 1-\sqrt{p_1+p_3} \\ q &= 1-(p+r) = 1-\sqrt{p_1+p_2}. \end{aligned} \tag{16}$$

In order to obtain a preliminary estimate for p and q, we may replace the probabilities p_1, p_2, and p_3 in (16) by the observed frequencies h_1, h_2, and h_3. Hence we get

$$\begin{aligned} p_0 &= 1-\sqrt{h_1+h_3} \\ q_0 &= 1-\sqrt{h_1+h_2}. \end{aligned} \tag{17}$$

[2] J. Neyman, Contribution to the theory of the χ^2-test. Proc. Berkeley Symposium on Mathematical Statistics and Probability (1949) pp. 239–273.

[3] F. Bernstein, Z. f. induktive Abstammungs- und Vererbungslehre 37 (1925) p. 236.

That this estimate is not efficient can be seen from the following argument. In the observation space, let h_1, h_2, h_3 be introduced as coordinates with $h_4 = 1 - h_1 - h_2 - h_3$ as the remaining (dependent) coordinate. According to (17), the points H with coordinates h_i giving rise to equal values of the estimates lie on a straight line

$$\begin{aligned} h_1 + h_3 &= (1 - p_0)^2 \\ h_1 + h_2 &= (1 - q_0)^2. \end{aligned} \tag{18}$$

This line intersects the surface represented by (15) in a point P_0 with coordinates $p_i(0)$, which correspond to the parameter values (p_0, q_0). If the estimate were efficient, the line (18) would be perpendicular (or at least asymptotically perpendicular) to this surface in the metric defined by the quadratic form

$$\chi_0^2 = \sum \frac{[h_i - p_i(0)]^2}{p_i(0)}. \tag{19}$$

The orthogonality condition reads

$$\sum \frac{u_i v_i}{p_i(0)} = 0, \tag{20}$$

where $u = (1, -1, -1, 1)$ is a vector parallel to the line (18) and v is any vector tangent to the surface. Differentiation of (15) with respect to p and to q yields two such vectors v and v'. If they are substituted into (20) it is then apparent that orthogonality is not even approximately satisfied.

An efficient estimate can be obtained by maximizing the logarithm of the likelihood function

$$L(x|p, q) = x_1 \ln r^2 + x_2 \ln(2pr + p^2) + x_3 \ln(2qr + q^2) + x_4 \ln 2pq. \tag{21}$$

Differentiation of L with respect to p and q, in which it is to be noted that $r = 1 - p - q$, leads to the equations

$$\frac{x_2 + x_4}{p} + \frac{x_2}{2r + p} = \frac{x_3 + x_4}{q} + \frac{x_3}{2r + q} = \frac{2x_1}{r} + \frac{2x_2}{2r + p} + \frac{2x_3}{2r + q}. \tag{22}$$

They may be solved by starting with

$$\begin{aligned} p &= p_0 + u \\ q &= q_0 + v, \end{aligned} \tag{23}$$

expanding the fractions in (22) in powers of u and v, and retaining only the linear terms. The result is two linear equations in u and v to be solved.

Alternatively to the above estimation procedure, χ_0^2 or χ_x^2 may be minimized as in § 48.

§ 51. The χ^2-Test

In § 49 we calculated the distribution function of

$$\chi^2 = \sum \frac{(x - np)^2}{np} \tag{1}$$

under the assumption that the p_i were the true probabilities

$$p_i^* = p_i(\vartheta^*). \tag{2}$$

In practice the true p_i^* are unknown and must be replaced by the estimated values

$$\tilde{p}_i = p_i(\tilde{\vartheta}). \tag{3}$$

If we use these $\tilde{p}_i$ in forming the expression

$$\tilde{\chi}^2 = \sum \frac{(x - n\tilde{p})^2}{n\tilde{p}}, \tag{4}$$

the resulting $\tilde{\chi}^2$ is in general smaller than χ^2 and has a different distribution function. Whereas χ^2 defined by (1) has, according to § 49, asymptotically a χ^2-distribution with $m-1$ degrees of freedom, we shall see that $\tilde{\chi}^2$ has asymptotically a χ^2-distribution with $m-1-r$ degrees of freedom, where r is the number of estimated parameters $\vartheta_1, \ldots, \vartheta_r$.

The numerator and the denominator of every term of the sum (4) are each of the order of magnitude n. In the denominator we may replace the $\tilde{p}$ by the true p^*, from which $\tilde{p}$ differs in probability only in the order of magnitude $n^{-\frac{1}{2}}$, and effect only an arbitrarily small change in the distribution function of $\tilde{\chi}^2$. Hence we obtain a modified expression

$$\chi_1^2 = \sum \frac{(x - n\tilde{p})^2}{np^*}, \tag{5}$$

whose distribution is somewhat easier to determine.

We shall now apply the theory of § 48, choosing the then arbitrary approximation $p^{(0)}$ equal to p^*. There

$$x_i' = n\,p_i(\vartheta') \tag{6}$$

was defined by the requirement that the quadratic form

$$\chi_0^2 = \sum \frac{(x - x')^2}{np^*} \tag{7}$$

be a minimum. The minimum χ_0^2-estimate so obtained,

$$p_i' = p_i(\vartheta') = \frac{x_i'}{n},$$

differs from the maximum likelihood estimate $\tilde{p}$ only in the order of magnitude n^{-1}, as was proved in § 48. Hence we can replace the $n\tilde{p}$ in the right-hand side of (5) by x', thus changing χ_1^2 into χ_0^2, without altering the asymptotic distribution.

Therefore there remains only the problem of determining the asymptotic distribution of χ_0^2. The probability of the event $\chi_0^2 < u$ is again a sum of probabilities $\mathscr{P}(X)$, summed over all points X in the region $\chi_0^2 < u$, and as in § 49, the sum can be replaced by an integral. Hence we obtain the desired asymptotic distribution function

$$F(u)=(2\pi)^{-\frac{m-1}{2}}\int\cdots\int e^{-\frac{1}{2}\chi^2}\,dV_{m-1}, \tag{8}$$

where integration is over the region $\chi_0^2<u$.

As in (11) § 49 let $y_1,\ldots,y_m$ be rectangular coordinates such that

$$\chi^2=y_1^2+\cdots+y_m^2.$$

These coordinates may be transformed orthogonally so that the hyperplane $\sum x_i=n$ has the equation $y_m=0$; the variables of integration are then $y_1,\ldots,y_{m-1}$ and we have

$$F(u)=(2\pi)^{-\frac{m-1}{2}}\int\cdots\int e^{-\frac{1}{2}(y_1^2+\cdots+y_{m-1}^2)}\,dy_1\ldots dy_{m-1}. \tag{9}$$

We can now make yet another orthogonal transformation such that the y_1- through y_r-axes lie in the linear space G defined by the parameter representation $p_i=p_i(\vartheta)$, the remaining axes then being perpendicular to these axes. In these new coordinates the point defined by the condition that it minimize (7) is especially easy to determine. The point X' lies in G, so that only $y_1',\ldots,y_r'$ can differ from zero and the remaining $y_{r+1}',\ldots,y_{m-1}'$ must be zero. The form χ_0^2 becomes

$$\chi_0^2=(y_1-y_1')^2+\cdots+(y_r-y_r')^2+y_{r+1}^2+\cdots+y_{m-1}^2 \tag{10}$$

in the new coordinates, and the minimum is attained when all $y_1-y_1',\ldots,$ y_r-y_r' are zero. The first r members of the right-hand side of (10) then drop out and we have

$$\chi_0^2=y_{r+1}^2+\cdots+y_{m-1}^2. \tag{11}$$

The limits of integration $\chi_0^2<u$ therefore restrict only $y_{r+1},\ldots,y_{m-1}$. If the integration with respect to $y_1,\ldots,y_r$ is carried out, we have

$$F(u)=(2\pi)^{-\frac{m-r-1}{2}}\int\cdots\int e^{-\frac{1}{2}(y_{r+1}^2+\cdots+y_{m-1}^2)}\,dy_{r+1}\ldots dy_{m-1}, \tag{12}$$

integrated over

$$y_{r+1}^2+\cdots+y_{m-1}^2<u. \tag{13}$$

The result is a χ^2-distribution with $m-1-r$ degrees of freedom.

Thus we may formulate the *generalized χ^2-test:*

The hypothesis that the true p^* are of the form $p(\vartheta)$ for some ϑ (not specified by this hypothesis) is rejected whenever the expression $\tilde{\chi}^2$ exceeds the upper limit obtained from Table 6 *for $m-1-r$ degrees of freedom.*

The tilde over $\tilde{\chi}^2$ was introduced only to distinguish it clearly from the true χ^2. In applications it is usually omitted.

In applications the important question arises whether another estimate $p(T)$ may be used in the right-hand side of (4) instead of $\tilde{p}=p(\tilde{\vartheta})$. We shall see that the following answer is correct:

If T is an efficient estimate (§ 50) *and furthermore if $|T-\tilde{\vartheta}|$ has in probability a smaller order than $n^{-\frac{1}{2}}$, then $\tilde{p}$ may be replaced by $p(T)$ in the right-hand side of* (4) *and the χ^2-test may be applied.*

The proof is clear. If the $n\tilde{p}$ in the numerator of (4) are changed by an additive term of smaller order than $n^{\frac{1}{2}}$, the change in χ^2 tends to zero as $n\to\infty$ and the asymptotic distribution is unchanged.

If the changes in the $\tilde{p}$ are by terms of order at least $n^{-\frac{1}{2}}$, then the above proof breaks down, and it can easily happen that the resulting χ^2 is much too large. *Hence it is important to use only efficient estimates.*

Example 35. To test the hypothesis of Bernstein discussed in § 50 concerning the blood types O, A, B, and AB, we may as before estimate the parameters p, q, and $r=1-p-q$ by the method of least squares, and then using

$$\begin{aligned} p_1 &= r^2 \\ p_2 &= 2pr+p^2 \\ p_3 &= 2qr+q^2 \\ p_4 &= 2pq, \end{aligned} \tag{14}$$

compute the quantity

$$\chi^2=\frac{(x_1-np_1)^2}{np_1}+\frac{(x_2-np_2)^2}{np_2}+\frac{(x_3-np_3)^2}{np_3}+\frac{(x_4-np_4)^2}{np_4}. \tag{15}$$

Two parameters p and q have been estimated, so that the number of degrees of freedom is

$$f=4-1-2=1.$$

The estimation of p and q is quite laborious, but there exists a simpler method of testing which yields practically the same result. This is apparent once we observe that a surface F is defined in the observation space by (14) and that the point $\tilde{P}$ is that point on the surface which is closest to the observed point H in the sense of the metric defined by χ^2. χ^2 is therefore the square of the distance from H to the surface F.

Instead of using the parametric equations, we may represent this surface by its equation, which reads[4]

$$\sqrt{p_1+p_2}+\sqrt{p_1+p_3}-\sqrt{p_1}-1=0. \tag{16}$$

If the point H does not lie on the surface, then its distance from the surface is proportional to the magnitude of

$$D=\sqrt{h_1+h_2}+\sqrt{h_1+h_3}-\sqrt{h_1}-1. \tag{17}$$

The quantity D can be approximated in a neighborhood of a point P of the surface by a linear function of the coordinates h_1, h_2, and h_3.

If we set $h_i=p_i+u_i$, after a little computation we have

$$D\sim\frac{1}{2}\frac{u_1+u_2}{\sqrt{p_1+p_2}}+\frac{1}{2}\frac{u_1+u_3}{\sqrt{p_1+p_3}}-\frac{1}{2}\frac{u_1}{\sqrt{p_1}}=a_1u_1+a_2u_2+a_3u_3. \tag{18}$$

---

[4] F. Bernstein, Z. f. induktive Abstammungs- und Vererbungslehre 37 (1925) p. 245.

The $u_i = h_i - p_i$ are approximately normally distributed with mean 0 and variance

$$\mathscr{E}u_i^2 = p_i(1-p_i)/n. \tag{19}$$

Also the expected values of $u_i\, u_k$ are known from (4) § 50 to be

$$\mathscr{E}u_i\, u_k = -p_i\, p_k/n. \tag{20}$$

Hence the sum $a_1 u_1 + a_2 u_2 + a_3 u_3$ and likewise D are approximately normally distributed with zero means and variance

$$\sigma^2 = a_1^2\, \mathscr{E}u_1^2 + a_2^2\, \mathscr{E}u_2^2 + a_3^2\, \mathscr{E}u_3^2 + 2a_1\, a_2\, \mathscr{E}u_1\, u_2 + 2a_1\, a_3\, \mathscr{E}u_1\, u_3 + 2a_2\, a_3\, \mathscr{E}u_2\, u_3. \tag{21}$$

Hence the quantity

$$\chi_D^2 = \frac{D^2}{\sigma^2} \tag{22}$$

has approximately a χ^2-distribution with a single degree of freedom. This χ_D^2 is nearly equal to the previous χ^2 and may be used as a test statistic.

In the calculation of σ^2 the p_i which occur in (18) through (20) may be replaced by their approximate values h_i. For large n, as is usually the case in such investigations, this approximation is quite inconsequential, especially since χ^2 is not strongly affected by variation of the p_i.

Chapter Ten

Bio-Assay

This chapter is concerned with the biological assessment of poisons and other stimuli, a branch of science known as bio-assay.

Poisons, or other stimuli, are administered to animals in varying doses. With each such dose there is observed a corresponding mortality, or other response-rate, among the subjects treated with that dose. The graph of the response-rate versus the dose is called the *dose-response diagram.* We shall discuss here various methods of drawing inference from the observed response curve. In most of our discussion, we shall assume knowledge only of the contents of Chapters 1 and 2.

§ 52. Response Curves and Logarithmic Response Curves

There exist stimuli whose effects can be determined only by administering them in various doses to a number of animals and observing a specified response, such as death. For each dose there exists a corresponding probability of reaction, which can be approximated by the empirical frequency. In general, this probability p increases with increasing dose. If we consider p as a function of the dose we get the *response curve* of the preparation.

Frequently the abscissa is taken not to be the dose itself but to be the logarithm l of the dose and is called the *log dose*, or *dosage*. The resulting graph of p vs. l is called the *logarithmic response curve*. The use of the logarithm has the advantage, among others, that the response curve of one preparation which differs from another only in the con-

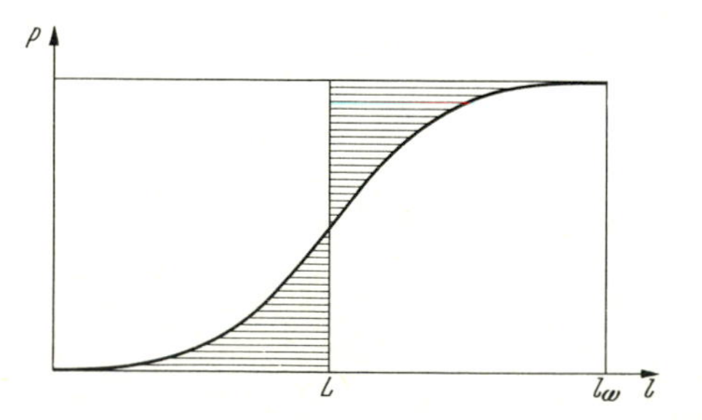

Fig. 25. Logarithmic Response Diagram

centration of the compound can be obtained from that of the other preparation by a horizontal shift. The magnitude of this horizontal shift is obviously equal to the logarithm of the ratio of the two concentrations.

Likewise, for comparing different preparations causing similar effects it is usually assumed that their logarithmic response curves differ from one another by a shift. Only under this assumption is it meaningful to speak of the ratio of effectiveness. The logarithm of this ratio is again equal to the magnitude of the horizontal shift.

In practice it is customary to compare every preparation with a standard preparation and to attempt to estimate the ratio of effectiveness of an arbitrary preparation relative to the standard preparation.

Many response curves increase from zero to one, which is to say that all animals react to very strong doses. There exist, however, other cases in which a certain percentage of the animals are insensitive to the preparation and withstand the strongest doses. In such cases the curve rises at most as far as a maximum ordinate $p_\infty < 1$. The assay is much more difficult with curves of this type. We shall be concerned here primarily with curves of the first type, those which increase from zero to one.

The logarithmic response curves observed in nature very frequently have the form of a normal distribution function:

$$p = \Phi\left(\frac{l-L}{\sigma}\right), \tag{1}$$

where l is the logarithm of the dose, L the logarithm of the 50%-dose, σ the standard deviation of a normal distribution, and, as throughout this book,

$$\Phi(u) = \frac{1}{\sqrt{2\pi}} \int_{-\infty}^{u} e^{-\frac{1}{2}t^2}\, dt. \tag{2}$$

There exist methods of assay which are based entirely on the assumption of such a "normal" response curve. These are called "probit methods". There exist, however, only very few instances in which this assumption has been justified on the basis of a sufficiently large amount of observed data. With this in mind we shall consider first a method which is independent of the normality assumption and which requires only the weaker assumption that the probability p increase from zero to one with increasing dosage.

For the sake of simplicity we shall always refer to p as the "mortality" in that which follows. These methods are applicable not only to lethal poisons but also to other biological stimuli. Also we shall frequently refer to the "dosage l" when in fact we mean the dose with logarithm l. The letters l or L refer clearly enough to the logarithm.

§ 53. Integral-Approximation Method of Behrens and Kärber[1]

This method is based on the concept of the *mean lethal dose*. If in Fig. 25 we consider a perpendicular line drawn so that the resulting two figures bounded by it and the response curve are equal in area, the corresponding abscissa L is the logarithm of the mean lethal dose. In the case of symmetric response curves, which are mapped into themselves by half a complete rotation about the 50%-point, the mean lethal dose is equal to the 50%-dose.

The expression "mean lethal dose", or in more general situations, "mean effective dose", can be justified as follows: If one assumes that for each individual animal there is a "lethal dose" which just kills it, then the logarithm of this lethal dose is dependent on the chance selection of the animal and therefore is a random quantity in the sense of § 2. The distribution function $F(l)$ of this random variable $\mathbf{x}$ is the probability that an animal dies from the dosage l and hence is the mortality p corresponding to dosage l. The graphical representation of this distribution function is therefore just our logarithmic response curve. The mean value of $\mathbf{x}$ is from § 3 the integral

$$\mathscr{E}\mathbf{x} = \int_{-\infty}^{\infty} l \, dF(l).$$

If we introduce $p = F(l)$ into this integral as a new variable, we get

$$\mathscr{E}\mathbf{x} = \int_{0}^{1} l \, dp = L. \tag{1}$$

Therefore L is exactly the mean value of the logarithms of the doses lethal to the individual animals.

The standard deviation σ of $\mathbf{x}$ is likewise defined by

$$\sigma^2 = \mathscr{E}(\mathbf{x} - L)^2 = \int_{-\infty}^{\infty} (l - L)^2 \, dF(l) = \int_{0}^{1} (l - L)^2 \, dp. \tag{2}$$

There is also the following alternative definition of the dosage L. If in Fig. 25 we draw a line $l = l_0$ so far to the left that the corresponding mortality p_0 is practically zero and a second line $l = l_\omega$ so far to the right that the corresponding mortality p_ω is practically one, then the area of the rectangle between L and l_ω is equal (except for a small error) to the

[1] G. Kärber, Archiv f. exp. Pathol. 162 (1931) p. 480. Behrens and Kärber, Archiv f. exp. Pathol. 177 (1935) p. 637.

area under the curve:

$$l_\omega - L = \int_{l_0}^{l_\omega} p\,dl.$$

From this expression we can get a solution for L:

$$L = l_\omega - \int_{l_0}^{l_\omega} p\,dl. \tag{3}$$

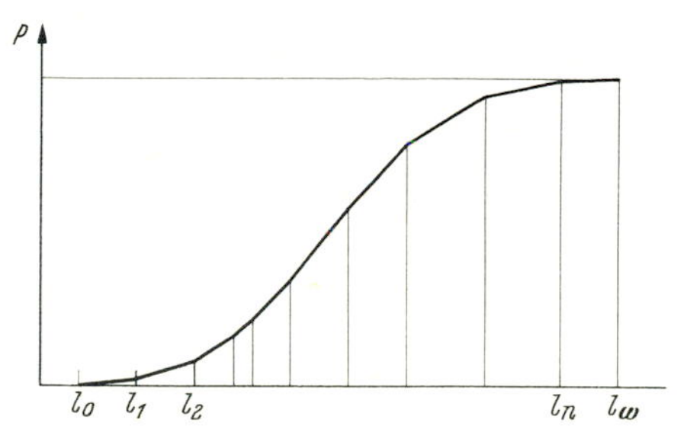

Fig. 26. The Trapezoidal Approximation

If we partition the interval from l_0 to l_ω into $n+1$ intervals with division points $l_1, l_2, \ldots, l_n$ and set $l_{n+1} = l_\omega$, the integral in (3) can be approximated by a sum of areas of trapezoids:

$$L \sim l_\omega - \tfrac{1}{2}\{(p_0+p_1)(l_1-l_0)+(p_1+p_2)(l_2-l_1)+\cdots \\ +(p_n+p_\omega)(l_{n+1}-l_n)\}. \tag{4}$$

If we rearrange the terms of the sum on the right-hand side, collecting coefficients of $p_0, p_1, \ldots, p_n, p_\omega$ and recalling that $p_0=0$, $p_\omega=1$, and $l_\omega = l_{n+1}$, we have

$$L \sim \tfrac{1}{2}(l_n + l_{n+1}) - \tfrac{1}{2}\{p_1(l_2-l_0)+p_2(l_3-l_1)+\cdots+p_n(l_{n+1}-l_{n-1})\}. \tag{5}$$

This expression is the starting point for the empirical determination of L by Kärber's method.

Each of the dosages $l_1, \ldots, l_n$ is administered to a number of animals and the frequencies of death $h_1, \ldots, h_n$ are observed. The number of animals used with each dosage need not be equal, but *the dosages must be given in sufficiently small increments*, so that we may replace the integral by a sum. *At the two ends of the sequence the frequencies must be so near* 0 *and* 1 *respectively that we may be practically certain that with the next smaller dosage* l_0 *the mortality is almost zero and that with the next larger dosage* $l_{n+1} = l_\omega$ *the mortality is almost one.*

The sequence must therefore be extended sufficiently far in both directions. It is not essential that the first frequency h_1 be zero and the last h_n be one; it is enough if, for example, h_1 and h_2 are both nearly zero. The experimentalist will be able to judge on the basis of intuition and experience whether or not p_0 is nearly zero and p_{n+1} nearly one, once he has observed the sequence of frequencies $h_1, \dots, h_n$.

Hence we obtain an approximate value for L by replacing the probabilities $p_1, \dots, p_n$ in (5) by the corresponding frequencies:

$$M = \tfrac{1}{2}(l_n + l_{n+1}) - \tfrac{1}{2}\{h_1(l_2 - l_0) + h_2(l_3 - l_1) + \cdots + h_n(l_{n+1} - l_{n-1})\}. \quad (6)$$

In practice the mean lethal dosage is calculated using this formula.

It is not necessary that the numbers of animals $N_1, \dots, N_n$ to which the dosages are administered be equal. The accuracy is even increased if the numbers in the middle are chosen to be larger than those at the ends of the sequence. Too much emphasis on the strong doses, with which almost all animals die, and on the weak doses, with which almost none die, is a waste of time and available animals, as Gaddum[2] has already pointed out.

The quantity M is a random variable whose mean value is immediately seen to be L. We shall now derive its standard deviation σ_M. From (6) M is the sum of a constant term, which has no influence on the standard deviation, plus n additional independent random terms. Therefore from (15) § 3

$$\sigma_M^2 = \sigma_1^2 + \sigma_2^2 + \cdots + \sigma_n^2,$$

where σ_1, for example, is the standard deviation of the term $-\frac{1}{2}h_1(l_2 - l_0)$:

$$\sigma_1^2 = \frac{p_1(1-p_1)}{4N_1}(l_2 - l_0)^2.$$

Hence we have

$$\sigma_M^2 = \sum_{k=1}^{n} \frac{p_k(1-p_k)}{4N_k}(l_{k+1} - l_{k-1})^2. \quad (7)$$

Since the p_k are unknown, according to the previously discussed procedure (§ 5, end), we shall replace the p_k by the h_k and the N by $N-1$. Hence we obtain the estimated variance:

$$s_M^2 = \sum_{k=1}^{n} \frac{h_k(1-h_k)}{4(N_k - 1)}(l_{k+1} - l_{k-1})^2. \quad (8)$$

The mean value of s_M^2 is σ_M^2, and for large numbers of animals s_M^2 does not differ much from σ_M^2. Therefore the accuracy of M can be judged from the value of s_M.

[2] J. H. Gaddum, Med. Res. Council Rep. on Biol. Standards 3 (1933) p. 27.

Example 36. Chen, Anderson, and Robbins[3] observed in groups of ten New Zealand red rabbits the following frequencies of death corresponding to the indicated doses of gelsemicine in the form of hydrochloride:

Dose	6	7	8	9	10	11	12	13
Frequency of death	0.0	0.1	0.3	0.6	0.8	0.5	0.9	1.0

In the computations based on formula (6), three-place logarithms were used:

$$l_1 = \log 6 = 0.778, \quad etc.$$

Judging from the nature of these observations, it is reasonable to assume that corresponding to the next smaller dosage (say $l_0 = \log 5$) the mortality would be zero for all practical purposes and that likewise corresponding to the next larger dosage (say $l_{n+1} = \log 14$) would be a mortality essentially one. Moreover, the choice of l_0 and l_{n+1} is immaterial since in our case $h_1 = 0$ and $h_n = 1$, so that l_0 and l_{n+1} drop out of formulas (6) and (8).

Formulas (6) and (8) yield[4]

$$M = 0.957,$$

$$s_M = 0.016.$$

The logarithm of the mean lethal dose is therefore estimated to be

$$L = 0.957 \pm 0.016.$$

We see from this result that there is no point in working with more than three-place logarithms. Even the second decimal place in M is not guaranteed, since the standard error s_M is larger than unity in the second decimal place.

Bliss[5] applied another method, which involved much more computation, to the same experimental material and obtained

$$L = 0.961 \pm 0.0166.$$

Hence we see that the integral-approximation method is hardly inferior to the more complicated procedure, which is, furthermore, based on the assumption of a normal curve. A more detailed study at the end of my paper quoted below shows that in the case of equal numbers of animals the standard error of the integral-approximation method is only about 1% larger than the standard error of the maximum likelihood method, which uses the assumption of normality of the response curve.

§ 54. Methods Based on the Normality Assumption

If we assume that the logarithmic response curve has the form of a normal distribution function, we then have at our disposal a whole series of methods of assay.

---

[3] Quarterly Journal Pharmacy and Pharmacol. 11 (1938) p. 84.

[4] Van der Waerden, Archiv f. exp. Pathol. 195 (1940) p. 389.

[5] C. I. Bliss, Quarterly Journal Pharmacy and Pharmacol. 11 (1938) p. 202.

A. Graphical Assay

The graphical method is based on the fact that a straight-line transformation of the dosage-response curve can be induced by a transformation of the p-axis.

Instead of denoting the abscissa by $l=\log$ dose, we shall now, as a matter of preference, write x. The normal dosage-response curve is then, according to (1) § 52, given by

$$p=\Phi\left(\frac{x-L}{\sigma}\right). \tag{1}$$

If we now introduce a new dependent variable y defined by

$$p=\Phi(y), \quad \text{or} \quad y=\Psi(p), \tag{2}$$

the equation of the dosage-response curve in the new coordinates x and y reads

$$y=\frac{x-L}{\sigma}, \tag{3}$$

an equation which represents a straight line.

To avoid working with negative numbers, one sometimes adds 5 to the y-values and then calls them *probits*. However, here we shall work with the y-values themselves, and not with $y+5$, in order to keep the formulas as simple as possible.

To estimate σ and L, the probits $y=\Psi(h)$ corresponding to the (observed) frequencies h are plotted vertically against the logarithms of the administered doses on the x-axis. A straight line is then drawn by eye to fit these points as satisfactorily as possible. There exists paper which is ruled so that the logarithms and the probits need not be computed first.

A difficulty arises from the fact that for $h=0$ or $h=1$ the probit y takes on the values $-\infty$ or $+\infty$. This difficulty may be avoided, as Prigge and Schäfer[6] have done, by first determining confidence bounds p_1 and p_2 corresponding to the frequencies h. For this purpose the formulas of § 6 should be used with $g=1$, so that the limits are not too wide. Corresponding to these p_1 and p_2 are bounds y_1 and y_2 on the probits, so that corresponding to each dosage we have an interval parallel to the y-axis extending from y_1 to y_2. If $h=0$ or $h=1$, then this interval is semi-infinite. The estimated response-line is then drawn so that it passes through all, or at least most, of these intervals. If this can be done in more than one way, then that line is drawn which is as close as possible to the points $y=\Psi(h)$ corresponding to the non-zero observed frequencies h.

[6] R. Prigge and W. Schäfer, Arch. f. exp. Pathol. 191 (1939) p. 303.

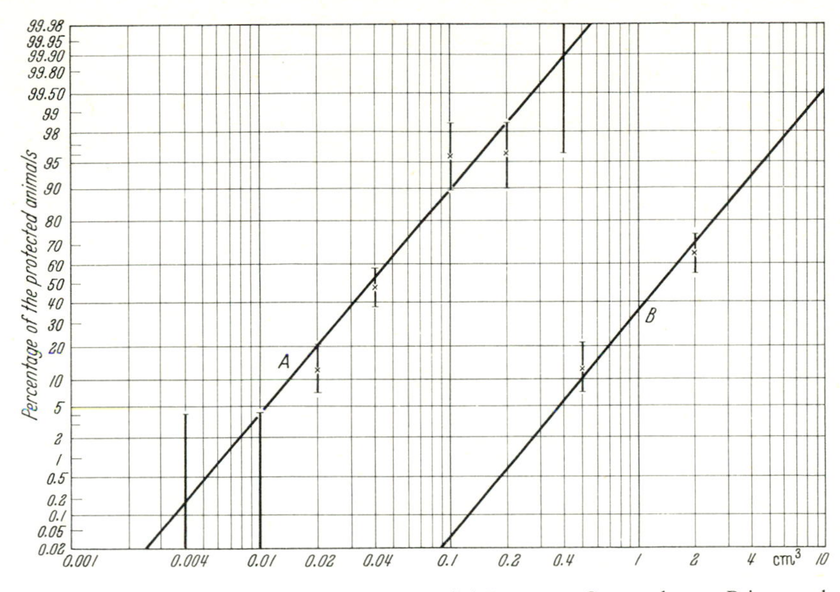

Fig. 27. Graphical Determination of Two Parallel Response Curves due to Prigge and Schäfer

Fig. 27, which is taken from the above-quoted paper of Prigge and Schäfer, illustrates this method.

B. Maximum Likelihood

Bliss and Fisher[7] have shown that the graphical assay can be improved by employing the method of maximum likelihood. The method, however, requires a great amount of computation, which in my mind is never worthwhile. The apparent precision in probit analysis is only illusionary, since everything depends on the highly uncertain normality assumption concerning the response curve. If one wishes to skate on such thin ice, then a rough graphical assay is quite sufficient. If, however, one desires to have a reliable assay whose precision can be estimated, then one should use the integral-approximation method, which is independent of the normality assumption.

There exists only one case in which it is essential to employ an efficient method of estimation: the case of testing the hypothesis of normality. In applying the χ^2-test (§ 51), the estimated expected values $n\tilde{p}$ which occur in χ^2 must be calculated using *efficient* estimates, lest χ^2 be too

[7] C. I. Bliss, Annals of Applied Biol. 22 (1935) p. 134, with an appendix by R. A. Fisher p. 149. The method is discussed fully in D. J. Finney, Probit Analysis, Cambridge University Press 1947.

large. Such efficient estimates can be found by the method of maximum likelihood (§ 50), but it is somewhat more convenient to use the equally efficient method of minimum χ_0^2 (§ 48), choosing the $n p^{(0)}$ in the denominator by using the graphical method, for example.

C. The Two-Point Method

The *two-point method* involves administering only two doses and drawing a line through the resulting points plotted on probability paper. The x-coordinate of its intersection with the horizontal axis is the estimated 50%-dosage.

The method can be used only if there are large numbers of animals and if approximate values of the slope and the intercept of the response-curve are known beforehand. The one dose must be considerably smaller than the 50%-dose, and the other considerably larger, to ensure that with high probability the two points lie on opposite sides of the 50%-line. On the other hand, the doses must not be too distant from the 50%-dose, lest the empirical mortality be 0 or 100%, in which cases no point could be plotted. In addition to all these disadvantages[8], the facts remain that a normal response-curve is quite an essential assumption in using the method and that its precision can be estimated only with great difficulty, if at all. It seems to me to be far better to design the experiment in such a way that the integral-approximation method can be used.

D. The One-Point Method

If the slope of the line is known approximately from previous experimentation, then the most exact method of assay is the *one-point method* of administering only *one* dose as close as possible to the median-dose. A line drawn through the resulting point (x_0, y_0) with the known slope $1/\sigma$ yields as x-intercept the median-dosage

$$M = x_0 - \sigma y_0 . \tag{4}$$

In practice σ is naturally replaced by an estimate s. The method is reliable only if the administered dosage is close to the median effective dosage L. Hence the approximate location of the median effective dosage must first be determined from preliminary experimentation (using, say, the integral-approximation method).

For calculation of the standard error of M refer to my previously quoted paper in Archiv für experimentelle Pathologie und Pharmakologie 195.

---

[8] With regard to this, see Behrens and Kärber, Archiv f. exp. Pathol. 177 (1935) p. 637.

The one-point method uses only the middle, steepest portion of the response-curve. It is therefore practically independent of the assumption that the response-curve be normal.

E. The Logistic Curve

Rather than assume a normal response-curve, one can also, as did Berkson[9], assume a logistic curve

$$p=\frac{1}{e^{-z}+1}, \tag{5}$$

where z is a linear function of x. The function (5) varies, just as the Φ-function, from 0 to $\frac{1}{2}$ to 1 as z varies from $-\infty$ to 0 to $+\infty$. The function (5) differs very little from a normal distribution function. The curve (5) is also symmetric with respect to the point $z=0$, $p=\frac{1}{2}$:

$$p(z)+p(-z)=1.$$

Eq. (5) can be solved for z easily:

$$z=-\ln\left(\frac{1}{p}-1\right). \tag{6}$$

Eq. (6) defines *logits* z, with which it is easier to work than with probits $y=\Psi(p)$. All the methods of probit analysis (graphical methods, maximum likelihood, two-point and one-point methods) are just as applicable, if not more so, to logits.

§ 55. "Up and Down" Methods

A. The Method of Dixon and Mood

Dixon and Mood[10] have proposed a method which yields just as accurate an estimate of L as the methods already discussed, but with fewer animals. They select an initial dose with logarithm l and administer it to one animal. Then, depending upon whether or not the animal reacts in a specified way (*e.g.*, dies), the log-dose is lowered or raised by an amount d and this new dose is given to a second animal. And so on, the log-dose for any test is always one step below or one step above that of the preceding test. Doses other than those with logarithms $l, l\pm d$, $l\pm 2d, \ldots$ are not used.

[9] J. Berkson, Journal Amer. Statist. Assoc., Vols. 39, 41, and 48.

[10] W. J. Dixon and A. M. Mood, Journal Amer. Statist. Assoc. 43 (1948) pp. 109 – 126.

With this procedure most of the points observed lie along the steepest part of the response-curve, since once the response-rate gets close to one, chances are that the dose is reduced again, and likewise it is increased once near zero. This is very fortunate, since it is just the part of the curve near the 50%-response-rate that is of interest. The difference d should be chosen between $\frac{1}{2}\sigma$ and 2σ in order that the method function properly. Hence a rough estimate of σ must be known beforehand.

For the assay one could apply Kärber's method. Dixon and Mood, however, use another, extremely simple method. They first count the total number of "successes" (occurrences of the reaction) and of "failures" and call the smaller of the two numbers N. Let the dosages corresponding to the less frequent event (success or failure) be, in increasing order, $l_0, l_1, l_2, \ldots$, and let the event occur $n_0, n_1, n_2, \ldots$ times with these dosages. Then let

$$A = \sum k\, n_k,$$

$$B = \sum k^2\, n_k.$$

As an estimate of L, the logarithm of the 50%-dose, we then take

$$M = l_0 + d\left(\frac{A}{N} \pm \frac{1}{2}\right). \tag{1}$$

The $+$-sign should be used when working with failures and the $-$-sign when using successes. Furthermore, as an estimate of σ we have

$$s = 1.62\, d\left(\frac{NB - A^2}{N^2} + 0.03\right). \tag{2}$$

For a justification of these formulas based upon the method of maximum likelihood, the reader is referred to the original paper. Brownlee, Hodges, and Rosenblatt have shown [Journal of the American Statistical Association 48 (1953) p. 262] that these formulas are quite good even with small numbers, and they have suggested some modifications that make the method even more efficient.

In the derivations of (1) and (2), the normality assumption is made, but it is not really a fundamental hypothesis. It suffices if the part of the curve near the median effective dose agrees approximately with a normal curve. It matters little about the outliers near the lines $p=0$ and $p=1$, since the method itself implies that the very large and the very small doses are seldom administered and therefore have almost no effect on the mean value and standard deviation of M.

A disadvantage of the method is that one can not perform any test after the first without observing first the success or failure of the preceding test. This disadvantage can be overcome in part by running, say, four sequences of tests simultaneously.

B. The Method of Stochastic Approximation

Robbins and Monro[11] have shown that the "up-and-down" method can be improved still more by increasing or decreasing the dosages not by a fixed amount d but by a variable, constantly decreasing amount. A decreasing sequence of positive numbers $a_1, a_2, \ldots$ and an initial dosage l_1 are chosen in advance. If the dosage l_n produces a reaction, then the next dosage is taken to be

$$l_{n+1} = l_n - \tfrac{1}{2} a_n, \tag{3}$$

but if it produces no reaction, then the next dosage is

$$l_{n+1} = l_n + \tfrac{1}{2} a_n. \tag{4}$$

A variation of this procedure is to test several animals at each dose simultaneously. If h_n is the frequency of success at the n-th stage, then the next dosage is taken to be

$$l_{n+1} = l_n + (\tfrac{1}{2} - h_n)\, a_n. \tag{5}$$

The last dosage l_{N+1} calculated is taken to be the estimate of the 50%-dosage. Robbins and Monro have proved that under certain regularity conditions the estimate l_{N+1} converges in probability as $N \to \infty$ to the 50%-dosage L; i.e., for N sufficiently large

$$|l_{N+1} - L| < \varepsilon$$

with arbitrarily high probability.

With regard to the asymptotic distribution of the estimate l_{N+1}, see in particular K. L. Chung, On a stochastic approximation method, Annals of Mathematical Statistics 25 (1954) pp. 463–483.

What is the best choice of the coefficients a_n? Whatever other conditions there may be, the sequence must converge to zero, for otherwise, (3) and (4) would imply that the sequence l_n could not possibly tend to L. If, on the other hand, we were to let the a_n tend to zero sufficiently fast for the series $\sum a_n$ to converge, then from some point n onwards success or failure would have almost no influence on the choice of dosages $l_{n+1}, l_{n+2}, \ldots$, since the correction terms $\pm\frac{1}{2}a_n, \pm\frac{1}{2}a_{n+1}, \ldots$ all taken together would amount to less than ε. Hence the a_n must be chosen so that the series $\sum a_n$ diverges.

Chung recommends the choice

$$a_n = c\, n^{-1+\varepsilon} \qquad (0 < \varepsilon < \tfrac{1}{2}). \tag{6}$$

[11] H. Robbins and S. Monro, A stochastic approximation method. Ann. Math. Stat. 22 (1951) pp. 400–407.

Robbins and Monro choose

$$a_n = \frac{c}{n}. \tag{7}$$

The coefficients (6) go to zero so slowly that the consistency of the procedure is assured under rather general assumptions about the form of the response curve. Use of the choice (7) requires greater caution. If a is the slope of the tangent to the response curve in a neighborhood of the 50%-dosage L, then the constant c in (7) must be chosen larger than $\frac{1}{2a}$. If the response curve is replaced by a straight line in a neighborhood of the 50%-dosage, Chung shows that under certain additional assumptions the standard deviation of the estimate l_{N+1} is asymptotically equal to

$$\frac{c}{\sqrt{2ac-1}} \frac{\sigma_h}{\sqrt{N}}, \tag{8}$$

where σ_h is the standard deviation of the frequencies h for dosages in a neighborhood of L. If n' is the number of animals at each dosage, then

$$\sigma_h^2 = \frac{pq}{n'} \sim \frac{1}{4n'}. \tag{9}$$

Expression (8) takes on its minimum value when we choose

$$c = \frac{1}{a}. \tag{10}$$

If, however, the slope a is not known exactly, it is wise to choose c somewhat larger. This induces only a slight increase in the standard deviation (8), since the function (8) changes only slowly near its minimum. In all cases, as was pointed out, c must be chosen larger than $\frac{1}{2a}$. If c is near $\frac{1}{2a}$, then (8) becomes quite large.

Chapter Eleven

Hypothesis Testing

Statistical tests are of the greatest importance in all applications. This chapter is a compendium of some of the most important statistical tests and includes many tests already mentioned, as well as a discussion of the general theory due to J. Neyman and E. S. Pearson.

With regard to examples, it is hoped that they will enable the reader to understand the basic principles upon which the choice of a suitable test for a given problem depends, even if he has not studied all the preceding chapters of this book. It goes without saying that a thorough knowledge of the basic concepts of Chapters 1 and 2 has been assumed. In the proofs, references will be made now and again to later chapters (8 and 9 in particular) as well as to the literature.

§ 56. Applications of the χ^2-Test

The general χ^2-test, which we deduced in § 51, embraces various special cases, some of which we have already considered. In each case the problem is to start with the observed frequencies of certain events and to test hypotheses which specify the probabilities of these events.

A. Hypotheses Specifying a Single Probability

If we wish to test the hypothesis that the probability of some event is a specified number p, and if in n independent trials this event occurs x_1 times and does not occur on the remaining $x_2=n-x_1$ trials, we can make our decision in the following manner. Let $q=1-p$, so that the expected values of x_1 and x_2 under this hypothesis are np and nq. Then compute

$$\chi^2=\frac{(x_1-np)^2}{np}+\frac{(x_2-nq)^2}{nq} \tag{1}$$

and reject the hypothesis that the probability is p if this observed value of χ^2 exceeds the cut-off value obtained from Table 6. Since there are two observed quantities x_1 and x_2 between which the linear relation $x_1+x_2=n$ holds, the number of degrees of freedom is

$$f=2-1=1.$$

But now

$$(x_1 - np) + (x_2 - nq) = 0,$$

and therefore

$$(x_1 - np)^2 = (x_2 - nq)^2.$$

Hence we may write (1) more simply as

$$\chi^2 = \frac{(x_1 - np)^2 (q + p)}{npq} = \frac{(x_1 - np)^2}{npq}, \tag{2}$$

from which we see that this test is equivalent to our previous test of this hypothesis.

B. Hypotheses Specifying Several Probabilities

Suppose that we have a sample of n items, which may be divided into m classes, or cells, according to certain characteristics. Let the observed numbers falling into these cells be $x_1, \ldots, x_m$, where $x_1 + \cdots + x_m = n$. The hypothesis to be tested gives rise to the m completely specified cell probabilities $p_1, \ldots, p_m$ (*e.g.*, $\frac{1}{16}, \frac{3}{16}, \frac{3}{16}, \frac{9}{16}$ in the case of two non-linked hereditary factors). Let

$$\chi^2 = \sum \frac{(x_i - np_i)^2}{np_i} \qquad (m-1 \text{ degrees of freedom}) \tag{3}$$

and reject the hypothesis whenever χ^2 exceeds the cut-off point obtained from Table 6.

Strictly speaking, the χ^2-distribution is valid only asymptotically as $n \to \infty$. The exact distribution of the statistic χ^2 is a discrete distribution, since for given n and p_i, χ^2 may assume only finitely many values.

In order to investigate how accurately the continuous χ^2-distribution approximates the exact distribution, I have computed the exact distribution of the statistic

$$Y = 1 - e^{-\frac{1}{2}\chi^2} \tag{4}$$

for the case

$$n = 10; \quad p_1 = 0.5, \quad p_2 = 0.3, \quad p_3 = 0.2$$

and have compared it with the asymptotic distribution.

We have in this case

$$\chi^2 = \frac{(x_1 - 5)^2}{5} + \frac{(x_2 - 3)^2}{3} + \frac{(x_3 - 2)^2}{2}. \tag{5}$$

There are 66 possible triples (x_1, x_2, x_3) such that $x_1 + x_2 + x_3 = 10$. The probability of any one such triple is

$$\frac{10!}{x_1!\, x_2!\, x_3!} 0.5^{x_1}\, 0.3^{x_2}\, 0.2^{x_3}. \tag{6}$$

Eqs. (4) and (5) yield a well-defined value of Y corresponding to each triple. These values and their probabilities obtained from (6) define a step-function (Fig. 28) as the distribution function of Y.

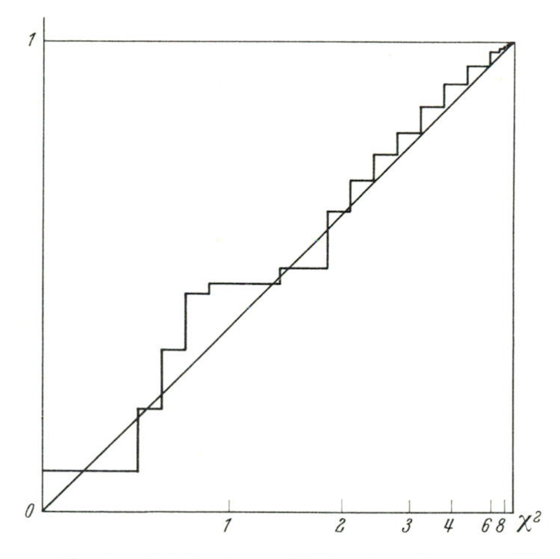

Fig. 28. Exact and approximate distribution of Y for two degrees of freedom, $n=10$, and probabilities 0.5, 0.3, and 0.2

The asymptotic distribution function of $\frac{1}{2}\chi^2$ with two degrees of freedom is

$$G(u)=\int_0^u e^{-t}\,dt=1-e^{-u}.$$

Hence the asymptotic distribution function of Y is a straight line

$$F(v)=v, \tag{7}$$

which is also shown in Fig. 28. As we see, the deviation between the two curves is small, particularly in the region between 0.95 and 1 which is most important for applications. The probability that χ^2 exceeds 9.21 is actually 0.0096, whereas the asymptotic distribution yields 0.01. The exact probability that $\chi^2>5.99$ is 0.0502, while the approximate probability is 0.05. The actual error probability in the application of the χ^2-criterion is in most cases even smaller than the nominal significance level based on the asymptotic distribution; i.e., we are on the safe side when using the asymptotic distribution.

It is often stated in the literature that neither the x_i nor their expected values $n\,p_i$ shall be too small if the asymptotic χ^2-distribution is to be applied. *The example here illustrates that we may allow the expected values to be as low as* 2 *or* 3 *and still employ the asymptotic distribution.*

Other examples substantiate this. *If there are many cells, the expected values in individual cells may be as low as one.* I once computed an example with 10 cells and

$$n\,p_1 = n\,p_2 = \cdots = n\,p_{10} = 1 \qquad (n = 10)$$

and still found satisfactory agreement between the exact distribution function and the asymptotic χ^2-distribution. Therefore one need not be all that careful.

C. Comparison of Two Probabilities

Let there be n_1 independent repetitions of an experiment and suppose that a particular event occurs on x_1 of these trials and does not occur on the remaining y_1 trials. Let there then be an additional n_2 independent repetitions with x_2 and y_2 denoting respectively the number of trials on which the event occurs and the number on which it does not occur. We wish to decide whether or not there is a change in the value of the probability of the event from the first sequence of trials to the second.

The hypothesis which we can test is the hypothesis of no change: the probability of the event on any one trial has the same, but unknown, value p for the two sequences of trials. To compute χ^2 we must replace p by an estimated value, which must be an efficient estimate to avoid obtaining a value of χ^2 which in certain circumstances is too large (§ 51).

To obtain such an estimate we shall use the method of maximum likelihood under the assumptions that the probability of the event is p for all $n_1 + n_2$ trials and that the probability of the observed numbers is

$$\frac{n_1!}{x_1!\,y_1!}\,\frac{n_2!}{x_2!\,y_2!}\,p^{x_1+x_2}(1-p)^{y_1+y_2}.$$

The maximum occurs when

$$\tilde{p} = \frac{x_1 + x_2}{n_1 + n_2}. \tag{8}$$

With this value $\tilde{p}$ we may compute $\tilde{q} = 1 - \tilde{p}$ and

$$\chi^2 = \frac{(x_1 - n_1\tilde{p})^2}{n_1\tilde{p}} + \frac{(y_1 - n_1\tilde{q})^2}{n_1\tilde{q}} + \frac{(x_2 - n_2\tilde{p})^2}{n_2\tilde{p}} + \frac{(y_2 - n_2\tilde{q})^2}{n_2\tilde{q}}, \tag{9}$$

or, more briefly,

$$\chi^2 = \frac{(x_1 - n_1\tilde{p})^2}{n_1\tilde{p}\tilde{q}} + \frac{(x_2 - n_2\tilde{p})^2}{n_2\tilde{p}\tilde{q}}. \tag{10}$$

But we have

$$(x_1 - n_1\tilde{p}) + (x_2 - n_2\tilde{p}) = 0,$$

so that we may write χ^2 even more briefly as

$$\chi^2 = \frac{(x_1 - n_1 \tilde{p})^2 (n_1 + n_2)}{n_1 n_2 \tilde{p} \tilde{q}}. \tag{11}$$

We have seen previously that with small numbers n_1 and n_2 we do well to replace the factor $N = n_1 + n_2$ in the numerator of (11) by $N-1$, in order that the error probability be not larger, or only slightly larger, than allowable (see § 9). Therefore instead of χ^2 we shall take as our test statistic

$$\chi_1^2 = \frac{(x_1 - n_1 \tilde{p})^2 (n_1 + n_2 - 1)}{n_1 n_2 \tilde{p} \tilde{q}} = \frac{(x_1 n_2 - x_2 n_1)^2 (n_1 + n_2 - 1)}{n_1 n_2 (x_1 + x_2)(y_1 + y_2)}. \tag{12}$$

The number of degrees of freedom is

$$f = 4 - 2 - 1 = 1,$$

because there are four observed numbers x_1, y_1, x_2, y_2 restricted by two linear relations and there is one unknown parameter p estimated by (8).

D. Testing the Independence of Two Events

Let N items be classified according to the occurrence or non-occurrence of each of two characteristics to yield four classes (a so-called 2×2 table), and let the numbers of items in the four classes be $x_{11}, x_{12}, x_{21}, x_{22}$. We wish to decide whether or not the two characteristics are independent. Let the probabilities be p_1 that the first characteristic appears and p_2 that it does not, $p_1 + p_2 = 1$, and let q_1 and q_2, $q_1 + q_2 = 1$, refer in like manner to the second. Then in the case of independence, the probabilities of the four classes are $p_1 q_1, p_1 q_2, p_2 q_1, p_2 q_2$.

In general, the p_i and q_k are unknown. If we once again employ the method of maximum likelihood, we obtain the estimates

$$\begin{aligned} \tilde{p}_i &= \frac{x_{i1} + x_{i2}}{N} \qquad (i = 1, 2) \\ \tilde{q}_k &= \frac{x_{1k} + x_{2k}}{N} \qquad (k = 1, 2). \end{aligned} \tag{13}$$

Calculation of χ^2 with these p_i and q_k yields

$$\begin{aligned} \chi^2 = {} & \frac{(x_{11} - N \tilde{p}_1 \tilde{q}_1)^2}{N \tilde{p}_1 \tilde{q}_1} + \frac{(x_{12} - N \tilde{p}_1 \tilde{q}_2)^2}{N \tilde{p}_1 \tilde{q}_2} + \frac{(x_{21} - N \tilde{p}_2 \tilde{q}_1)^2}{N \tilde{p}_2 \tilde{q}_1} \\ & + \frac{(x_{22} - N \tilde{p}_2 \tilde{q}_2)^2}{N \tilde{p}_2 \tilde{q}_2}. \end{aligned} \tag{14}$$

If we replace the $N\tilde{p}_i$ in the numerator and denominator by $n_i = x_{i1} + x_{i2}$, we obtain an expression analogous to (9), which may then be re-expressed as in (10) or (11) as

$$\chi^2 = \frac{(x_{11}x_{22} - x_{12}x_{21})^2 N}{(x_{11}+x_{12})(x_{21}+x_{22})(x_{11}+x_{21})(x_{12}+x_{22})}. \tag{15}$$

The number of degrees of freedom is

$$f = 4 - 1 - 2 = 1,$$

since the four observed numbers are restricted by one linear relation

$$x_{11} + x_{12} + x_{21} + x_{22} = N$$

and there are two estimated parameters p_1 and q_1.

For small N it is, as before, advisable to replace the N in the numerator of (15) by $N-1$.

The question whether the test of independence on the basis of a 2×2 table should be made on the basis of one or of three degrees of freedom led to a great debate among English statisticians. Karl Pearson first proposed the χ^2-criterion only for the case of completely specified probabilities p_i and q_k. Under this assumption, for large N the asymptotic distribution of the χ^2-criterion is the theoretical χ^2-distribution with $4-1=3$ degrees of freedom. If the true p_i and q_k are then replaced by the approximate values (13), the value of χ^2 is at most decreased. Hence we know that the true value of χ^2 is at least equal to the approximate value (15). Therefore, if the approximation (15) exceeds the value u, then the true value of χ^2 necessarily exceeds u also. Let the probability that the true value of χ^2 exceeds u be α (*e.g.*, $\alpha = 0.01$ or $\alpha = 0.05$), where u is obtained from the table for three degrees of freedom. Pearson's conclusion was that in order to be certain that the error probability in the application of the χ^2-test be at most α we must use the table with three degrees of freedom.

Fisher's counter-argument was that the χ^2 computed according to (15), say $\tilde{\chi}^2$, is nearly always smaller than the true χ^2. The probability that $\tilde{\chi}^2 > u$ is therefore considerably smaller than the probability that $\chi^2 > u$. The error probability of the χ^2-test using three degrees of freedom is therefore considerably smaller than α; that is, it is unnecessarily small. If, however, we take $f=1$, then the asymptotic error probability is exactly α.

Since Fisher's argument for his point of view was only heuristic and had not yielded an exact proof, Yule and Brownlee attempted to determine by means of extensive random experimentation whether the distribution function of $\tilde{\chi}^2$ is a χ^2-distribution with $f=3$ or with $f=1$. The research appeared to support Fisher's claim, but the conclusiveness of the method of experimentation was criticized by the opponents. Finally J. Neyman and E. S. Pearson[1] settled the matter by giving a mathematical proof that Fisher's intuitive notion had been correct.

E. Comparison of More Than Two Probabilities

Let an event occur on x_1 out of n_1 independent trials and not occur on the remaining y_1 trials, let it then occur on x_2 out of n_2 additional trials and not occur on y_2 trials, and so on. We want to decide whether or not the probability of the event is the same for all sequences of trials.

To consider an even more general problem, let n_1 items be divided according to some characteristic into h classes and let the numbers in the classes be $x_1, y_1, \ldots, z_1$. Likewise, let n_2 additional items yield the numbers $x_2, y_2, \ldots, z_2$, and so on up to $n_k, x_k, y_k, \ldots, z_k$. Hence we have hk observations, which we may display in a rectangular array:

$$\begin{array}{cccc|c} x_1 & y_1 & \cdots & z_1 & n_1 \\ x_2 & y_2 & \cdots & z_2 & n_2 \\ \vdots & \vdots & & \vdots & \vdots \\ x_k & y_k & \cdots & z_k & n_k \\ \hline \sum x & \sum y & \ldots & \sum z & N \end{array}$$

The column on the far right gives the row sums, the bottom row the column sums, and N the total number of observations. The problem is to decide whether or not the probabilities $p, q, \ldots, r$ associated with the h classes are the same for all rows.

The best estimates of these common values $p, q, \ldots, r$ for the probabilities are the marginal frequencies in the classes:

$$\tilde{p} = \frac{\sum x}{N}, \tilde{q} = \frac{\sum y}{N}, \ldots, \tilde{r} = \frac{\sum z}{N}. \tag{16}$$

If we now estimate the expected values by

$$n_i \tilde{p}, n_i \tilde{q}, \ldots, n_i \tilde{r}$$

and subtract them from the observed numbers $x_i, y_i, \ldots, z_i$, we can again display the resulting differences in a rectangular array:

$$\begin{array}{ccc} x_1 - n_1 \tilde{p} & y_1 - n_1 \tilde{q} & \ldots\ z_1 - n_1 \tilde{r} \\ x_2 - n_2 \tilde{p} & y_2 - n_2 \tilde{q} & \ldots\ z_2 - n_2 \tilde{r} \\ \ldots & \ldots & \ldots \end{array}$$

[1] J. Neyman and E. S. Pearson. On the use and interpretation of test criteria. Biometrika 20 A (1928) pp. 175 and 263.

In this array both the row sums and the column sums must be zero, a fact which provides a check on the computation of $\tilde{p}, \tilde{q}, \ldots, \tilde{r}$.

If we now divide the squares of these hk differences by their respective expected values and sum the resulting quotients we get

$$\chi^2 = \sum \frac{(x_i - n_i \tilde{p})^2}{n_i \tilde{p}} + \sum \frac{(y_i - n_i \tilde{q})^2}{n_i \tilde{q}} + \cdots. \tag{17}$$

The degrees of freedom are

$$f = hk - k - (h-1) = (h-1)(k-1). \tag{18}$$

There are hk observed numbers restricted by k linear equations

$$x_i + y_i + \cdots + z_i = n_i.$$

In addition, there are h parameters estimated from these observations using (16). However, there is one linear equation

$$p + q + \cdots + r = 1,$$

so that the estimates of any $h-1$ of the probabilities determine the estimate of the remaining probability. Hence only $h-1$ has been subtracted in (18).

F. Infrequent Events

We have already introduced the terminology *infrequent event* for an event whose probability p is so small that in all formulas $q = 1 - p$ may be replaced by 1. The Bernoulli distribution then becomes a Poisson distribution: the probability that an event occurs exactly x times in n trials becomes

$$P_x = \frac{(np)^x}{x!} e^{-np} = \frac{\lambda^x}{x!} e^{-\lambda}. \tag{19}$$

In the second expression for P_x, n and p no longer occur explicitly, but rather occur only through the expected value $\lambda = np$ of x. There is a corresponding simplification in the formula for χ^2. The term with nq in the denominator is negligible in comparison with that with np in the denominator. Hence formula (1) becomes

$$\chi^2 = \frac{(x - np)^2}{np} = \frac{(x - \lambda)^2}{\lambda}. \tag{20}$$

Any hypothesis which assigns a specific value to λ is then rejected whenever the expression given by (20) exceeds the critical value of χ^2 with one degree of freedom. Likewise in the case of two independent infrequent events, for which x denotes the number of occurrences of the first and y that of the second, a hypothesis assigning specific values to λ and μ, the expected values of x and of y, is rejected whenever the expression

$$\chi^2 = \frac{(x-\lambda)^2}{\lambda} + \frac{(y-\mu)^2}{\mu} \tag{21}$$

exceeds the critical value for two degrees of freedom.

G. Comparison of Two Infrequent Events

This problem has already been considered in detail (§ 10 B). We shall now briefly show that the test derived there can be obtained directly from the general χ^2-theory.

Given two independent infrequent events, let one event be observed x_1 times in time t_1 and let the other be observed x_2 times in time t_2. Let the expected values of x_1 and of x_2 be

$$\lambda_1 = \vartheta_1 t_1, \qquad \lambda_2 = \vartheta_2 t_2.$$

The hypothesis to be tested is that $\vartheta_1 = \vartheta_2$. If we assume that $\vartheta_1 = \vartheta_2 = \vartheta$, then we have

$$\lambda_1 = \vartheta t_1, \qquad \lambda_2 = \vartheta t_2, \tag{22}$$

where the parameter ϑ must be estimated in order to compute χ^2. The Poisson distribution (19) yields the likelihood function

$$(\vartheta t_1)^{x_1} e^{-\vartheta t_1} (\vartheta t_2)^{x_2} e^{-\vartheta t_2}.$$

Dropping the factors not dependent upon ϑ and taking the logarithm, we have

$$L(x_1, x_2 | \vartheta) = (x_1 + x_2) \ln \vartheta - (t_1 + t_2) \vartheta, \tag{23}$$

the maximum of which occurs when

$$\tilde{\vartheta} = \frac{x_1 + x_2}{t_1 + t_2}. \tag{24}$$

Therefore

$$\chi^2 = \frac{(x_1 - \tilde{\vartheta} t_1)^2}{\tilde{\vartheta} t_1} + \frac{(x_2 - \tilde{\vartheta} t_2)^2}{\tilde{\vartheta} t_2}, \tag{25}$$

with

$$f = 2 - 1 = 1 \tag{26}$$

degree of freedom, since there are two observations x_1, x_2 and a single parameter ϑ estimated by (24).

H. Testing for Normality

Given n independent, real-valued random observations $z_1, \ldots, z_n$, we wish to test the hypothesis that they have a common normal distribution.

To this end, it is possible to calculate the empirical distribution function and apply the Kolmogorov test (§16). In our previous discussion, however, it was pointed out that the "tails" of the empirical distribution, i.e., the very large and the very small z-values, have relatively little influence on this test. Yet it is precisely these tails which can, in certain circumstances, be decisive in judging a deviation from normality. Furthermore, the fact that in general the hypothesis of normality does not specify the mean and variance creates a difficulty in the application of the Kolmogorov test.

A good method which is somewhat more sensitive to the tails is the *method of moments*. We present here only a brief summary and refer the reader to Cramér, Mathematical Methods of Statistics, 27.1 through 28.4 and 29.3, for the foundations.

The *central moments* of an empirical distribution are defined by

$$m_k = \frac{1}{n} \sum (z - \bar{z})^k, \qquad k = 1, 2, \ldots .$$

The first moment m_1 is zero by definition. The empirical *coefficients of skewness* and *of excess* are defined in terms of m_2, m_3, m_4 to be

$$g_1 = m_3\, m_2^{-\frac{3}{2}},$$
$$g_2 = m_4\, m_2^{-2} - 3.$$

For large n, all m_k, as well as g_1 and g_2 are asymptotically normally distributed. They can be used as estimates of the true moments μ_k as well as of the true, or theoretical, coefficients of skewness and of excess

$$\gamma_1 = \mu_3\, \mu_2^{-\frac{3}{2}},$$
$$\gamma_2 = \mu_4\, \mu_2^{-2} - 3.$$

Both γ_1 and γ_2 are zero if the true distribution is normal.

For moderate n, it is advisable to replace g_1 and g_2 by

$$G_1 = \frac{\sqrt{n(n-1)}}{n-2}\, g_1$$

and

$$G_2 = \frac{n-1}{(n-2)(n-3)} \left[(n+1)\, g_2 + 6\right].$$

Under the assumption of normality, the expected values of G_1 and of G_2 are zero. Their variances are

$$\sigma_1^2 = \frac{6n(n-1)}{(n-2)(n+1)(n+3)},$$

$$\sigma_2^2 = \frac{24n(n-1)^2}{(n-3)(n-2)(n+3)(n+5)}.$$

Hence either G_1/σ_1 or G_2/σ_2 can be used as a test statistic for testing the hypothesis of normality, both quantities being asymptotically normally distributed with zero mean and unit standard deviation.

The χ^2-*method*, which may also be applied to test for distributions other than those of the normal family, involves partitioning the entire z-interval by means of $r-1$ endpoints $t_1, \ldots, t_{r-1}$ into r disjoint sub-intervals and counting the number of z_j in each such sub-interval. Let $x_1, \ldots, x_r$ denote these numbers.

In order to calculate χ^2 we must know the expected numbers $n p_i$. Hence we must obtain estimates m and s of the mean value and standard deviation of the normal distribution, in order to set

$$p_i = \Phi\left(\frac{t_i - m}{s}\right) - \Phi\left(\frac{t_{i-1} - m}{s}\right). \tag{27}$$

If we are to apply the theory of § 51, we must use efficient estimates m and s which depend only on the numbers x_i. As a first approximation we can use the familiar estimates

$$m_0 = \frac{1}{n} \sum z, \tag{28}$$

$$s_0^2 = \frac{1}{n-1} \sum (z - m_0)^2. \tag{29}$$

These estimates do not, however, satisfy the condition of depending only on the numbers x_i. With the aid of m_0 and s_0 we then compute

$$p_{i0} = \Phi\left(\frac{t_i - m_0}{s_0}\right) - \Phi\left(\frac{t_{i-1} - m_0}{s_0}\right) \tag{30}$$

and determine estimates m and s by the method of least squares so as to minimize the expression

$$\chi_0^2 = \sum \frac{(x_i - n p_i)^2}{n p_{i0}}. \tag{31}$$

We replace the p_i in (31) by linear functions of m and s:

$$p_i = p_{i0} + (m - m_0)\, q_i + (s - s_0)\, r_i, \tag{32}$$

where the q_i and r_i are to be determined from (27) by differentiation:

$$q_i = \frac{\partial p_i}{\partial m}(m_0, s_0); \qquad r_i = \frac{\partial p_i}{\partial s}(m_0, s_0). \tag{33}$$

The method of least squares then gives rise in the usual manner to two linear equations in $m - m_0$ and $s - s_0$ and these in turn yield m and s.

This method of calculation is indeed complicated. Hence there arises the question of the existence of a simpler method.

Cramér recommended computing m and s^2 from the grouped z-values using Sheppard's correction for s^2. That is, all z between t_{i-1} and t_i are thought of as being equal to the midpoint $\frac{1}{2}(t_{i-1} + t_i)$ of the interval, and the mean m and standard deviation s are calculated from these modified z-values. Use of Sheppard's correction, however, requires that the intervals all have equal length h. The m and s determined in this manner depend only on the numbers x_i.

A recent investigation by D. Koller[2] shows that this method of estimating μ and σ^2 and then applying the χ^2-test is justified.

The problem of choosing r, the number of classes, and the endpoints $t_1, \ldots, t_{r-1}$ in the best manner possible was the object of some noteworthy research by Mann and Wald[3]. To be sure, they do not give a complete answer, but they do provide us with useful guidelines. For $n = 200$, 400, and 1,000, their research indicates that the classes should be chosen so that the numbers of observations falling into each class are, respectively, about 12, 20, and 30. Following their recommendation, one would choose classes considerably smaller than is the usual practice and would encounter computational work which is correspondingly more laborious.

J. Testing the Normality of a Response Curve

Let $n_1, \ldots, n_r$ animals receive doses with logarithms $l_1, \ldots, l_r$, and let the numbers of animals reacting be $x_1, \ldots, x_r$ respectively. Then the methods of § 54 yield a normal response curve fitted to the observed frequencies $h_i = x_i/n_i$. If it is then desired to test the goodness of fit of

---

[2] D. Koller, Prüfung der Normalität einer Verteilung. Z. Wahrscheinlichkeitstheorie verw. Geb. 2 (1963) pp. 147–166.

[3] For a very good comprehensive review, see W. G. Cochran, The χ^2-test of goodness of fit. Ann. Math. Stat. 23 (1952) p. 315.

this response curve with the observations, the procedure is to compute from it the probabilities $p_1, \ldots, p_r$ and the complementary probabilities $q_i = 1 - p_i$ and form

$$\chi^2 = \sum \frac{(x_i - n_i p_i)^2}{n_i p_i} + \sum \frac{(y_i - n_i q_i)^2}{n_i q_i}. \tag{34}$$

Here the $y_i = n_i - x_i$ are the numbers of animals which have not reacted. Once more we have

$$(x_i - n_i p_i) + (y_i - n_i q_i) = 0,$$

so that χ^2 may be written more briefly as

$$\chi^2 = \sum \frac{(x_i - n_i p_i)^2}{n_i p_i q_i}. \tag{35}$$

The constants L and s, which determine the intercept and slope of the response curve, must be determined by an *efficient* method, such as the probit method or the minimum χ^2-method (§ 51). A graphical determination of the response line does not suffice in this case, because the resulting χ^2 can then be larger.

To determine the degrees of freedom, we note that there are $2r$ observed numbers $x_1, \ldots, x_r, y_1, \ldots, y_r$ among which r linear equations

$$x_i + y_i = n_i$$

hold. Also two parameters L and s have been fitted to the observations. Therefore the degrees of freedom are

$$f = 2r - r - 2 = r - 2.$$

If we have several empirical response curves for the same stimulus, then we can calculate a χ^2 for each and add these χ_k^2. The sum of a χ_1^2 with f_1 degrees of freedom and a χ_2^2 with f_2 degrees of freedom has, according to § 23, a χ^2-distribution with $f_1 + f_2$ degrees of freedom.

The larger the number of summands comprising the composite χ^2 the more we can rely on the asymptotic χ^2-distribution, a fact which follows from the Central Limit Theorem (§ 24 D).

Even when the resulting χ^2's, as well as their sum, lie in their respective acceptance regions, there is still cause for skepticism with respect to the hypothesis of normality. Only in the event that with extensive data it is found that the χ^2 fluctuate about their expected values f (=degrees of freedom) and therefore that the sum of all χ^2 remains close to the sum of all f can we attach somewhat more confidence to the normal response curve.

K. How large must the expected values np be to allow application of the χ^2-distribution?

Frequently one sees in the literature remarks to the effect that the expected values np should be at least 5 or 10 to justify application of the χ^2-distribution. Such remarks seem to be dictated only by too much concern on the part of the authors. Cochran and others who have examined the problem more carefully come to much more optimistic conclusions[4].

Cochran denotes by X^2 a discrete variable of the type which is in effect employed in the χ^2-test:

$$X^2 = \sum \frac{(x - np)^2}{np},$$

and by χ^2 a continuous variable which has a χ^2-distribution with the same number degrees of freedom f. He compares, as we have done in § 56 B, the X^2-distribution with the χ^2-distribution, particularly for those values for which the probability P of observing an even larger value lies between 0.05 and 0.01. His results show that the agreement is quite good, especially if the number of degrees of freedom is not too small. If it is larger than six, then one of the expected values may be as low as $\frac{1}{2}$, or two may both be as small as 1, and the agreement still remains quite good. With more than 60 degrees of freedom and small expected values, the exact tail probability is even considerably smaller than that given by the χ^2-distribution, because χ^2 has a larger variance than X^2. Hence in using the χ^2-distribution one remains on the conservative side. The approximation can be improved by using a normal distribution with variance equal to the exact variance as calculated by Haldane[5] instead of the χ^2-distribution.

The example considered in § 56 B shows that with two degrees of freedom the expected values may be as low as 2.

Only in the case of a single degree of freedom must one be more careful and again demand that the expected values be at least four, or still better, multiply χ^2 by $(N-1)/N$, where N is the total number of observations (see § 9).

[4] W. G. Cochran, The χ^2-test of goodness of fit. Ann. Math. Stat. 23 (1952) p. 328.

[5] J. B. S. Haldane, Biometrika 29 (1937) p. 133 and Biometrika 31 (1939) p. 346.

L. Examples Using the χ^2-Test

Example 37. Three-factor heterozygous *Primula* were back-crossed with the triple recessive[6]. The pairs of hereditary factors were

Ch – ch: *Sinensis* flower – *Stellata* flower
G – g : Green stigma – Red stigma
W – w : White eye – Yellow eye.

Twelve families[7] yielded the following numbers for the eight phenotypes:

Type	Family Number												Total
	107	110	119	121	122	127	129	131	132	133	135	178	
ChGW	12	17	9	10	24	9	3	16	20	9	11	10	150
ChGw	20	16	10	7	23	3	6	24	18	2	13	12	154
ChgW	14	10	6	8	19	5	5	23	18	10	7	12	137
Chgw	13	13	9	8	9	6	3	12	18	1	9	12	113
chGW	5	5	16	2	30	3	8	21	19	4	9	12	134
chGw	12	6	14	3	16	5	7	13	14	4	13	10	117
chgW	7	3	18	2	11	5	4	14	23	4	6	13	110
chgw	10	8	10	4	23	5	4	22	23	7	8	16	140
Total	93	78	92	44	155	41	40	145	153	41	76	97	1,055
$\chi^2=$	12.6	19.2	10.1	12.4	18.1	4.9	4.8	9.2	3.2	14.2	5.0	2.0	115.7

If the three hereditary factors are not linked and if neither lethal factors nor incompatibility factors play a role, then in every class the frequency should be $\frac{1}{8}$, and therefore the expected values for family No. 107, for example, should all be $\frac{93}{8}$. The sum of squared deviations from the expected values for all families together yields $\chi^2=115.7$. In each family there are seven degrees of freedom and hence a total of 84. The 5%-cut-off for 84 degrees of freedom is 106.4, which is, therefore, exceeded. The 5%-cut-off for seven degrees of freedom is 14.1, which is exceeded by three families. Family No. 110 also exceeds the 1%-cut-off 18.5. The observed frequencies therefore deviate considerably from those expected under Mendelian law.

In order to investigate which of the hereditary factors are irregular and whether linkage is exhibited, we shall, following Fisher, break the total χ^2 down into components which correspond to the individual factors and to the factor pairs. It will then be apparent which components are especially large.

Let the numbers within a family be $x_1, \ldots, x_8$ with $\sum x=n$. Then the χ^2 for this family is

$$\chi^2=\sum \frac{(x-\frac{1}{8}n)^2}{\frac{1}{8}n}=\frac{8}{n}\left(\sum x^2-\frac{n^2}{8}\right).$$

We now make an orthogonal transformation from $x_1, \ldots, x_8$ to new variables $y_1, \ldots, y_8$, where y_1 is to correspond to the factor Ch in such a way that $y_1\sqrt{8}$ is the excess of Ch

[6] Gregory, de Winton, and Bateson, Genetics of *Primula Sinensis*. J. of Genetics 13 (1923) p. 236. The statistical analysis follows R. A. Fisher, Statist. Methods for Research Workers, 11th edition, Ex. 15, p. 101.

[7] Families 54, 55, 58, and 59 were omitted because the data quoted by Fisher did not agree with those from the J. of Genetics 13.

over ch:

$$y_1 \sqrt{8} = z_1 = x_1 + x_2 + x_3 + x_4 - x_5 - x_6 - x_7 - x_8. \qquad \text{(Ch)}$$

y_2 and y_3 correspond in like manner to the factors G and W:

$$y_2 \sqrt{8} = z_2 = x_1 + x_2 - x_3 - x_4 + x_5 + x_6 - x_7 - x_8, \qquad \text{(G)}$$

$$y_3 \sqrt{8} = z_3 = x_1 - x_2 + x_3 - x_4 + x_5 - x_6 + x_7 - x_8. \qquad \text{(W)}$$

The next variable y_4 corresponds to the coupling GW:

$$y_4 \sqrt{8} = z_4 = x_1 - x_2 - x_3 + x_4 + x_5 - x_6 - x_7 + x_8. \qquad \text{(GW)}$$

If the factors G and W are not linked, the expected value of z_4 is zero. The definitions of z_5 and of z_6 are analogous:

$$y_5 \sqrt{8} = z_5 = x_1 - x_2 + x_3 - x_4 - x_5 + x_6 - x_7 + x_8, \qquad \text{(ChW)}$$

$$y_6 \sqrt{8} = z_6 = x_1 + x_2 - x_3 - x_4 - x_5 - x_6 + x_7 + x_8. \qquad \text{(ChG)}$$

In order to complete the orthogonal transformation we need yet two more variables:

$$y_7 \sqrt{8} = z_7 = x_1 - x_2 - x_3 + x_4 - x_5 + x_6 + x_7 - x_8,$$

$$y_8 \sqrt{8} = z_8 = x_1 + x_2 + x_3 + x_4 + x_5 + x_6 + x_7 + x_8.$$

z_7 has no simple biological significance. $z_8 = n$ is simply the number of plants in the family.

In practice one naturally avoids division by $\sqrt{8}$ by using the z's instead of the y's in the calculations. Our χ^2 expressed in terms of the z's becomes

$$\chi^2 = \frac{8}{n}\left(\sum y^2 - \frac{n^2}{8}\right) = \frac{1}{n}\left(\sum z^2 - z_8^2\right)$$
$$= \frac{1}{n} z_1^2 + \cdots + \frac{1}{n} z_7^2.$$

In this manner we have achieved our intended goal of breaking χ^2 into components. Every z_k is approximately normally distributed with mean zero and variance n. Every term $\frac{1}{n} z_k^2$ therefore has approximately a χ^2-distribution with a single degree of freedom. Computation yields the following values for these terms:

Family	(Ch)	(G)	(W)	(GW)	(ChW)	(ChG)	(z_7)	Total
107	**6.72**	0.27	3.11	1.82	0.10	0.27	0.27	12.56
110	**14.82**	1.28	0.82	0.82	0.20	1.28	0	19.22
119	**6.26**	0.39	0.39	0.17	2.13	0.04	0.70	10.08
121	**11.00**	0	0	0.36	0.82	0.09	0.09	12.36
122	0.16	**6.20**	1.09	1.86	0.52	0.32	**7.90**	18.05
127	0.61	0.02	0.22	0.61	1.20	0.22	1.98	4.86
129	0.90	1.60	0	0.40	0.10	0.90	0.90	4.80
131	0.17	0.06	0.06	0.06	0.06	0.34	**8.45**	9.20
132	0.16	0.79	0.32	0.32	0.06	1.47	0.06	3.18
133	0.22	0.22	**4.12**	0.02	**8.80**	0.22	0.61	14.21
135	0.21	3.37	1.32	0.05	0.05	0	0.05	5.05
178	0.26	0.84	0.09	0.09	0.01	0.26	0.50	2.05
Total	**41.49**	15.04	11.54	6.58	14.05	5.41	**21.51**	**115.62**

The 1%-cut-offs are 6.6 for the individual values (one degree of freedom) and 26.2 for the column sums (12 degrees of freedom). The 5%-cut-offs are 3.8 and 21.0. Those observations which exceed the 5%-cut-offs are printed in boldface type. The boldface numbers 6.26 in the column (Ch), 6.20 in the column (G), and 4.12 in the column (W) are meaningless, since given 84 numbers, on the average four must exceed the 5%-cut-off even if everything is in order. The remaining significant observations all take place in the columns (Ch), (ChW), and (z_7). In these columns, the 1%-cut-off is exceeded six times and the sum for column (Ch) exceeds even the 0.1%-cut-off 32.9. The factor Ch therefore surely does not behave in the expected way and the greatest part of the discrepancy is to be ascribed to this factor. Possibly the factor Ch is linked to a recessive lethal factor or to an incompatibility factor.

Pairwise linkage among the genes Ch, G, and W is apparently not present, since the sums for the columns (GW), (ChW), and (ChG) are not especially large.

Example 38 (from Cramér, Mathematical Methods of Statistics, p. 440). Johannsen's data on the breadths of $n=12{,}000$ beans were divided into 16 cells, the first cell consisting of all breadths under 7.00 mm, the second those from 7.00 to 7.25 mm, and so on, each increment being 0.25 mm. The numbers $x_1, \ldots, x_{16}$ observed in those cells are given in column 2 of the table below. In order to test the hypothesis that the breadths are normally distributed, m and s were calculated from the grouped data, using Sheppard's correction. In this computation the semi-infinite cells (up to 7.00 and exceeding 10.50) were first of all divided into subintervals of length 0.25. The resulting estimated values were

$$m=8.512, \qquad s=0.6163.$$

The normal distribution with mean m and standard deviation s was used to calculate the expected numbers $n\,p_i$ (column 3). The differences $x_i-n\,p_i$ are given in column 4, and χ^2 was found to be 196.5. The 0.1%-cut-off for 13 degrees of freedom is 34.5. The distribution is therefore most certainly not normal. A glance at the differences $x-n\,p$ shows that the distribution is considerably skew: there are more very large and fewer very small beans than can be expected with a normal distribution.

Class	Number x	Normal $n\,p$	$x-n\,p$
less than 7.00	32	68	− 36
7.00 – 7.25	103	132	− 29
7.25 – 7.50	239	310	− 71
7.50 – 7.75	624	617	+ 7
7.75 – 8.00	1,187	1,046	+141
8.00 – 8.25	1,650	1,506	+144
8.25 – 8.50	1,883	1,842	+ 41
8.50 – 8.75	1,930	1,920	+ 10
8.75 – 9.00	1,638	1,698	− 60
9.00 – 9.25	1,130	1,277	−147
9.25 – 9.50	737	817	− 80
9.50 – 9.75	427	444	− 17
9.75 – 10.00	221	205	+ 16
10.00 – 10.25	110	81	+ 29
10.25 – 10.50	57	27	+ 30
more than 10.50	32	10	+ 22
Total	12,000	12,000	0

§ 57. The Variance-Ratio Test (F-Test)

Let s_1^2 and s_2^2 be independent estimates of two variances σ_1^2 and σ_2^2. How do we test the hypothesis $\sigma_1=\sigma_2$?

If s_1^2 is obtained from n_1 observations according to the usual formula and if the observations themselves come from a normal distribution, then the quantity

$$\chi_1^2=\frac{(n_1-1)\,s_1^2}{\sigma_1^2} \tag{1}$$

has a χ^2-distribution with $f_1=n_1-1$ degrees of freedom, and likewise for

$$\chi_2^2=\frac{(n_2-1)\,s_2^2}{\sigma_2^2}, \tag{2}$$

but with $f_2=n_2-1$ degrees of freedom. If it is now assumed that $\sigma_1=\sigma_2$, then

$$\frac{\chi_1^2}{\chi_2^2}=\frac{f_1\,s_1^2}{f_2\,s_2^2}. \tag{3}$$

To test the hypothesis $\sigma_1=\sigma_2$ the variance ratio

$$F=\frac{s_1^2}{s_2^2} \tag{4}$$

is taken to be the test statistic. Once the ratio F exceeds a bound F_α the hypothesis $\sigma_1=\sigma_2$ is rejected. This is the F-test[8]. The bound is chosen so that the probability of the event $F>F_\alpha$ under the hypothesis $\sigma_1=\sigma_2$ is exactly α, where α is the preassigned error probability.

In order to determine F_α, we must examine the distribution function of F under the hypothesis $\sigma_1=\sigma_2$. This is known, once the distribution function $H(w)$ of the ratio

$$\frac{\chi_1^2}{\chi_2^2}=\frac{f_1\,s_1^2}{f_2\,s_2^2}=\frac{f_1}{f_2}\,F \tag{5}$$

is known.

The probability density of χ_1^2 is

$$g_1(t)=\alpha_1\,t^{\frac{1}{2}f_1-1}\,e^{-\frac{1}{2}t}, \quad \text{where} \quad \alpha_1=\Gamma(\tfrac{1}{2}f_1)^{-1}\,2^{-\frac{1}{2}f_1},$$

and likewise for χ_2^2. The probability that the ratio (5) is smaller than w is therefore

$$H(w)=\alpha_1\,\alpha_2\iint t^{\frac{1}{2}f_1-1}\,e^{-\frac{1}{2}t}\,u^{\frac{1}{2}f_2-1}\,e^{-\frac{1}{2}u}\,dt\,du, \tag{6}$$

integrated over the region

$$t>0, \qquad u>0, \qquad \frac{t}{u}<w.$$

---

[8] R. A. Fisher used $z=\frac{1}{2}\ln F$ as test statistic.

The double integral may be expressed as an iterated integral:

$$H(w)=\alpha_1\,\alpha_2 \int_0^\infty du \int_0^{uw} t^{\frac{1}{2}f_1-1}\,u^{\frac{1}{2}f_2-1}\,e^{-\frac{1}{2}t-\frac{1}{2}u}\,dt. \tag{7}$$

If a new variable of integration y defined by $t=u\,y$ is introduced and if f is defined by $f=f_1+f_2$, the iterated integral becomes

$$H(w)=\alpha_1\,\alpha_2 \int_0^\infty du \int_0^{w} u^{\frac{1}{2}f-1}\,y^{\frac{1}{2}f_1-1}\,e^{-\frac{1}{2}uy-\frac{1}{2}u}\,dy. \tag{8}$$

Interchanging the order of integration yields

$$H(w)=\alpha_1\,\alpha_2 \int_0^w y^{\frac{1}{2}f_1-1}\,dy \int_0^\infty u^{\frac{1}{2}f-1}\,e^{-\frac{y+1}{2}u}\,du. \tag{9}$$

The integral with respect to u is a gamma function:

$$\int_0^\infty u^{\frac{1}{2}f-1}\,e^{-\frac{y+1}{2}u}\,du=\left(\frac{2}{y+1}\right)^{\frac{1}{2}f}\Gamma(\tfrac{1}{2}f). \tag{10}$$

Hence the integral becomes

$$H(w)=C\int_0^w y^{\frac{1}{2}f_1-1}(y+1)^{-\frac{1}{2}f}\,dy, \tag{11}$$

where

$$C=\Gamma(\tfrac{1}{2}f_1)^{-1}\,\Gamma(\tfrac{1}{2}f_2)^{-1}\,\Gamma(\tfrac{1}{2}f). \tag{12}$$

The integral (11) is an incomplete beta function. Obviously the integral may be calculated by elementary methods when f_1 and f_2 are integers. The distribution function $H(w)$ is therefore known.

By making the substitution

$$w=\frac{f_1}{f_2}\,w' \tag{13}$$

we obtain the distribution function $G(w')$ of the ratio F from $H(w)$. The desired bound $w'=F_\alpha$ is then the solution of the equation

$$G(w')=1-\alpha. \tag{14}$$

Hence F_α depends not only on α but also on f_1 and f_2. The bounds F_α for $\alpha=0.05$, 0.01, and 0.001 are tabulated in Tables 8A, 8B, and 8C.

Example 39. Analyses of natural gas were made in 30 laboratories in the USA. Each laboratory repeated the analysis several (usually 10) times. The detailed results of these analyses have been published by M. Shepherd[9].

[9] M. Shepherd, Analysis of a standard sample of natural gas. Journal of Research, National Bureau of Standards 38 (1947) p. 19.

In attempting to determine the standard deviation of such measurements within laboratories, we encounter the difficulty that the variances s^2 differ greatly among laboratories, for some laboratories are better than others. If we wish to calculate an average variance s^2 representative of the good and average laboratories, which can then serve as a standard for all laboratories, we must omit the very bad ones from the averaging. Those s^2, however, which are by pure chance somewhat larger than the others must be included, since omitting them would yield an average which would be systematically too small.

As a criterion for the rejection of large values of s^2, an F-test may be used. To illustrate the method, we have selected the determination of methane according to "Method A" (combustion), which was used by most of the laboratories.

In some laboratories the analyses were made by more than one analyst. There is evidence that the standard deviation of measurements by two analysts is usually somewhat larger than the standard deviation of those by an analyst working alone. In order to obtain the standard deviations for the individual analysts, the results of the various analysts had to be separated out from one another, and for each analyst a variance had to be calculated according to the formula

$$s^2 = Q/(n-1) \quad \text{where} \quad Q = \sum (x-\bar{x})^2$$

(x = percentage methane). The results, ordered according to increasing size of the variance s^2, were:

No.	Q	$n-1$	s^2	No.	Q	$n-1$	s^2
1	0.0 :	1 =	0.00	16	3.6 :	5 =	0.72
2	0.6 :	9 =	0.07	17	6.7 :	9 =	0.74
3	0.8 :	9 =	0.09	18	7.3 :	9 =	0.81
4	0.4 :	4 =	0.10	19	4.0 :	4 =	1.00
5	1.3 :	9 =	0.14	20	8.3 :	8 =	1.04
6	0.6 :	4 =	0.15	21	9.1 :	8 =	1.14
7	0.6 :	4 =	0.15	22	18.1 :	15 =	1.20
8	1.5 :	9 =	0.17	23	5.9 :	4 =	1.47
9	1.8 :	9 =	0.20	24	13.8 :	9 =	1.53
10	2.2 :	10 =	0.22	25	10.2 :	6 =	1.70
11	0.9 :	4 =	0.23	26	21.1 :	9 =	2.34
12	3.5 :	9 =	0.39	27	28.5 :	9 =	3.17
13	3.8 :	9 =	0.42	28	29.9 :	9 =	3.32
14		=	0.60	29	16.9 :	4 =	4.23
15		=	0.65	30	43.4 :	9 =	4.82
				31	15.2 :	2 =	7.60

As we can see, the variance for analyst 31 is much larger than the others. In order to decide whether it can to attributed to chance, we divide the variance $s_1^2 = 7.60$ by the average variance for the remaining analysts, which is calculated according to the formula

$$s_2^2 = \frac{\sum Q}{\sum (n-1)}.$$

Hence we obtain $s_2^2 = 239.9/215 = 1.116$ and

$$F_{31} = \frac{s_1^2}{s_2^2} = \frac{7.60}{1.116} = 6.81.$$

In like manner we could calculate $F_1, F_2, \ldots, F_{30}$: each of the 31 analysts has, compared with the others, his own F_j. The largest of these F_j is our F_{31}.

That this F_{31} exceeds the 5%-bound does not say much. The probability that a particular F_j exceeds the bound $F_{0.05}$ is, to be sure, only 5%. We have, however, taken for F_{31} the largest F_j and it is not at all improbable that one among 31 F-ratios exceeds the limit $F_{0.05}$. If the σ^2 are in fact all equal, then the probability that all F_j are less than $F_{0.05}$ is $0.95^{31}=0.20$; therefore the probability that at least one is greater than $F_{0.05}$ is

$$1-0.20=0.80.$$

That F_{31} also exceeds the 1%-bound likewise does not say much, since the probability that such an event occurs by chance is still

$$1-0.99^{31}=0.27.$$

For F_{31} to exceed the 0.1%-bound would be significant, but that bound, 7.15, is not exceeded.

The situation with F_{30} is more favorable, because the number of degrees of freedom is now larger. We find

$$F_{30}=\frac{4.82}{1.066}=4.53.$$

The 0.1%-bound 3.26 is well exceeded. The probability that this would happen by chance is only $1-0.999^{31}=0.03$. Hence the hypothesis that the variance σ^2 for analyst 30 is the same must be rejected.

After analyst 30 has been dropped from consideration, we can compare 31 once again with the others (1 through 29). We now find

$$F'_{31}=\frac{7.60}{1.00}=7.60.$$

The 0.1%-bound 7.15 is now exceeded and therefore 31 is likewise to be omitted.

In the same way as 30 and 31 have been dropped from consideration, we can likewise drop analysts 28, 27, 29, and 26. In each case the error probability is less than 3%, and hence in total (since six conclusions were drawn one after another) less than 18%.

A conclusion with an error probability of 18% could at first appear dangerous. Yet closer consideration reveals that most of the conclusions would still be possible at the 0.05%-level, thus immediately reducing the error probability by half. Moreover, the mean methane determinations obtained by analysts 29 and 30 are much too large and that by 31 is much too small. The rejection of analysts 29, 30, and 31 was therefore not in error. As for the rejection of the remaining three, each error probability is less than 1.4%, and therefore the total is less than 4.2%. If a composite error rate of 5% is allowed, then it appears that the rejection of these six analysts is probably correct.

For those remaining (1 through 25) the average variance of the analysts is

$$s^2=\frac{\sum Q}{\sum(n-1)}=\frac{110.1}{175}=0.63.$$

This s^2, which refers to the individual analyst and not to differences among analysts, is called the *repeatability variance*. Hence s is 0.8 (% methane).

It should be pointed out that the deviations among laboratories are almost twice as large as s, even if the analysts with the widest deviations are omitted. The standard deviation among laboratories is $S=1.4$. By comparison with more precise measurements the systematic error of the combustion method was found to be 2.6. The combustion Method A is, therefore, not very reliable.

§ 58. The Analysis of Variance

A. Variation Within and Between Groups

Let $x_1, \ldots, x_n$; $y_1, \ldots, y_n$; and $z_1, \ldots, z_n$ be independent, normally distributed random variables. Let the x_i all be observed under the same experimental conditions, so that they may be assumed to have the same mean and the same standard deviation; let the same assumption hold for the y_j and for the z_k. Furthermore, assume that the x_i, y_j, and z_k all have the same standard deviation σ, a hypothesis which can be tested by means of the F-test. We want to test the hypothesis that they all have the same expected value as well.

The statistical technique known as the analysis of variance serves this purpose. The main idea is that the total sum of squares, or total variation,

$$Q = \sum (x-M)^2 + \sum (y-M)^2 + \sum (z-M)^2, \tag{1}$$

where M is the grand mean of all x, y, and z, may be partitioned into two components, the first of which corresponds to variation *within the three groups* and the second to variation *between groups*. These two components are then compared with one another by means of an F-test.

The mathematical technique giving rise to this partition is the orthogonal transformation. Such a transformation of $x_1, \ldots, x_n$ yields new variables $u_1, \ldots, u_n$, of which the first

$$u_1 = \frac{x_1 + \cdots + x_n}{\sqrt{n}} = \bar{x}\sqrt{n} \tag{2}$$

is proportional to the arithmetic mean. Because of the orthogonality condition, the equations

$$\sum x^2 = \sum u^2 = u_1^2 + \cdots + u_n^2 \tag{3}$$

and, therefore,

$$u_2^2 + \cdots + u_n^2 = \sum x^2 - u_1^2 = \sum x^2 - n\bar{x}^2 = \sum (x-\bar{x})^2 \tag{4}$$

hold. This component of the sum of squares therefore corresponds to the variation among the x.

Analogously defined variables $v_1, \ldots, v_n$ replace $y_1, \ldots, y_n$, and $w_1, \ldots, w_n$ replace $z_1, \ldots, z_n$. Hence,

$$v_2^2 + \cdots + v_n^2 = \sum (y-\bar{y})^2, \tag{5}$$

$$w_2^2 + \cdots + w_n^2 = \sum (z-\bar{z})^2. \tag{6}$$

Addition of (4), (5), and (6) yields

$$Q_2 = (u_2^2 + \cdots + u_n^2) + (v_2^2 + \cdots + v_n^2) + (w_2^2 + \cdots + w_n^2)$$
$$= \sum (x - \bar{x})^2 + \sum (y - \bar{y})^2 + \sum (z - \bar{z})^2. \tag{7}$$

This sum of squares suffices for the estimation of the *variation within classes*. An estimate of this variance is, as always,

$$s_2^2 = \frac{Q_2}{3(n-1)}. \tag{8}$$

If the three samples consist of unequal numbers n_1, n_2, n_3 of observations, then instead of (8), we have to take

$$s_2^2 = \frac{Q_2}{(n_1 - 1) + (n_2 - 1) + (n_3 - 1)},$$

or, if N is defined by $N = n_1 + n_2 + n_3$,

$$s_2^2 = \frac{Q_2}{N-3}. \tag{9}$$

In order to determine the *variation between classes*, u_1, v_1, w_1 are subjected to a second orthogonal transformation taking u_1, v_1, w_1 into u', v', w', where u' is proportional to the grand mean M of all x_i, y_j, z_k:

$$u' = M\sqrt{N} = \frac{\sum x + \sum y + \sum z}{\sqrt{N}} = \frac{u_1\sqrt{n_1} + v_1\sqrt{n_2} + w_1\sqrt{n_3}}{\sqrt{N}}. \tag{10}$$

Such a transformation is possible, according to § 13, because the sum of the squares of the coefficients on the right-hand side of (10) is one:

$$\frac{n_1}{N} + \frac{n_2}{N} + \frac{n_3}{N} = 1.$$

From the orthogonality of the transformation it follows that

$$u_1^2 + v_1^2 + w_1^2 = u'^2 + v'^2 + w'^2, \tag{11}$$

and therefore that, if Q_1 is defined by $Q_1 = v'^2 + w'^2$,

$$Q_1 = u_1^2 + v_1^2 + w_1^2 - u'^2 = u_1^2 + v_1^2 + w_1^2 - NM^2. \tag{12}$$

If this equation is added to the previously obtained

$$Q_2 = (u_2^2 + \cdots + u_n^2) + (v_2^2 + \cdots + v_n^2) + (w_2^2 + \cdots + w_n^2),$$

the result is

$$\begin{aligned} Q_1+Q_2&=\sum u_i^2+\sum v_j^2+\sum w_k^2-NM^2 \\ &=\sum x_i^2+\sum y_j^2+\sum z_k^2-NM^2 \\ &=\sum (x-M)^2+\sum (y-M)^2+\sum (z-M)^2. \end{aligned} \tag{13}$$

The right-hand side is our previously defined Q. Thus we have the required partition:

$$Q=Q_1+Q_2. \tag{14}$$

The second component Q_2 determines, according to (9), the variation within groups. The first component

$$Q_1=v'^2+w'^2 \tag{15}$$

depends only on the differences between the group means $\bar{x}$, $\bar{y}$, and $\bar{z}$; i.e.,

$$\begin{aligned} Q_1&=u_1^2+v_1^2+w_1^2-NM^2 \\ &=n_1\,\bar{x}^2+n_2\,\bar{y}^2+n_3\,\bar{z}^2-NM^2, \end{aligned} \tag{16}$$

where M is the weighted mean of $\bar{x}$, $\bar{y}$, and $\bar{z}$ with weights n_1, n_2, and n_3:

$$M=\frac{n_1\,\bar{x}+n_2\,\bar{y}+n_3\,\bar{z}}{N}. \tag{17}$$

Hence in place of (16), Q_1 may also be written as

$$Q_1=n_1(\bar{x}-M)^2+n_2(\bar{y}-M)^2+n_3(\bar{z}-M)^2. \tag{18}$$

The same expression (18) is also the result if $\bar{x}$, $\bar{y}$, and $\bar{z}$ are considered as estimates with varying precision of an unknown true value ϑ and the method of least squares is applied. Since $\bar{x}$, $\bar{y}$, and $\bar{z}$ are means of n_1, n_2, and n_3 observations respectively, they must be given weights proportional to n_1, n_2, and n_3 to calculate the grand mean from (17). From the method of least squares, then, the "variance of an observation of unit weight" is estimated by

$$s_1^2=\frac{n_1(\bar{x}-M)^2+n_2(\bar{y}-M)^2+n_3(\bar{z}-M)^2}{3-1}=\frac{Q_1}{r-1}.$$

The denominator is the number of groups less one: if there are r groups, it is $r-1$.

If it is assumed that not only the standard deviations but also the expected values of all the x, y, and z in the three groups are equal, then it follows from the theory of least squares that

$$s_1^2=\frac{Q_1}{r-1} \tag{19}$$

is an unbiased estimate of the common variance σ^2.

We now give a second proof, independent of the theory of least squares, that the estimate (19) is unbiased, that is, that the mean value of Q_1 is equal to $(r-1)\,\sigma^2$, or, in the case at hand, equal to $2\sigma^2$.

Let ϑ be the common expected value of the x, y, and z. In as much as we can make the transformation to the new variables $x-\vartheta$, $y-\vartheta$, and $z-\vartheta$, there is no loss of generality in taking this common expected value to be zero. The expected values of x_i^2, y_j^2, and z_k^2 are then equal to σ^2 and those of all other products $x_i\,x_j$, $x_i\,y_j$, *etc.*, are zero. These properties of the expected values of the squares and of the products are preserved by orthogonal transformation; therefore the expected values of u_i^2, v_j^2, w_k^2, and of u'^2, v'^2, w'^2 are likewise equal to σ^2. Consequently the mean value of (15) is equal to $2\sigma^2$. Q.E.D.

Analogously, from (7) and (8) the expected value of s_2^2 is equal to σ^2. This fact obviously remains true even if the expected values of the x, y, and z are different: a parallel shift

$$x_i'=x_i-a, \qquad y_j'=y_j-b, \qquad z_k'=z_k-c$$

does not affect s_2^2.

The simplest computational forms of Q_1 and Q_2 are (18) and (7). As a check, either (13) or

$$Q_1+Q_2=Q=\sum(x-a)^2+\sum(y-a)^2+\sum(z-a)^2-N(M-a)^2 \tag{20}$$

for arbitrary a suffices.

Once Q_1 and Q_2 are computed, each needs only to be divided by its respective "number of degrees of freedom" to yield s_1^2 and s_2^2:

$$s_1^2=\frac{Q_1}{r-1}, \qquad s_2^2=\frac{Q_2}{N-r}. \tag{21}$$

B. The *F*-Test

If s_1^2 is smaller, or only slightly larger, than s_2^2, there is no justification for assuming that a real difference exists among the expected values of the groups. If, however, s_1^2 is considerably larger than s_2^2, then there arises the suspicion that the expected values in the three groups are in fact different. To investigate whether this suspicion is well-founded, the F-test is applied to the ratio

$$F=\frac{s_1^2}{s_2^2}=\frac{Q_1/(r-1)}{Q_2/(N-r)}. \tag{22}$$

To determine the exact distribution function of the quotient F, the further assumption must be made that the x_i, y_j, and z_k are all normally distributed. If this is so, then u', v', w' and $u_2, v_2, w_2, \ldots, u_n, v_n, w_n$ are likewise independent and normally distributed with common variance σ^2, since an orthogonal transformation takes the probability density

$$C \cdot \exp\left[-\frac{\sum(x-\vartheta)^2+\sum(y-\vartheta)^2+\sum(z-\vartheta)^2}{2\sigma^2}\right]$$

into a probability density of the same form:

$$C \cdot \exp\left[-\frac{(u'-\vartheta\sqrt{N})^2+v'^2+w'^2+u_2^2+v_2^2+\cdots+w_n^2}{2\sigma^2}\right]. \tag{23}$$

Hence

$$\frac{Q_1}{\sigma^2}=\frac{v'^2+w'^2}{\sigma^2} \tag{24}$$

has a χ^2-distribution with

$$f_1=2 \qquad \text{(or, in general, } f_1=r-1) \tag{25}$$

degrees of freedom, and likewise

$$\frac{Q_2}{\sigma^2}=\frac{u_2^2+v_2^2+w_2^2+\cdots+u_n^2+v_n^2+w_n^2}{\sigma^2} \tag{26}$$

has a χ^2-distribution with

$$f_2=N-3 \qquad \text{(or, in general, } f_2=N-r)$$

degrees of freedom. The two ratios (24) and (26) are independent, since the probability density (23) is a product of two factors, the first of which depends only on u', v', w' and the second only on $u_2, v_2, w_2, \ldots, u_n, v_n, w_n$. Hence:

The ratio (22) *has, under the assumptions that all x_i, y_j, z_k are independent and normally distributed with common expected value ϑ and common standard deviation σ, an F-distribution. Therefore the F-test may be applied. The degrees of freedom for numerator and denominator are $r-1$ and $N-r$ respectively.*

For applications it is expedient to make an "analysis of variance table" of the following sort:

	Sum of squares	Degrees of freedom	Variance
Variation between groups	Q_1	$f_1=r-1$	$Q_1/f_1=s_1^2$
Variation within groups	Q_2	$f_2=N-r$	$Q_2/f_2=s_2^2$
Total variation	Q	$N-1$	$Q/(N-1)=s^2$

C. Non-Normal Distributions

The question now arises whether the F-test can also be applied if the x, y, and z are not normally distributed. In this discussion of the question we shall assume that the number n of observations in each group is not too small, say $n \geqq 4$. Under this assumption, $f_1 = r-1$ is considerably smaller than $f_2 = (n-1)r$. Hence it follows that s_1^2 exhibits (percentagewise) considerably larger chance deviations than s_2^2. The distribution function of the quotient depends, therefore, chiefly on that of the numerator. The numerator was defined by (19) and (18) in terms of $\bar{x}$, $\bar{y}$, and $\bar{z}$, each of which is a mean of at least four quantities x_i, y_j, or z_k. According to the Central Limit Theorem, such means have approximately a normal distribution, even if the individual distributions deviate strongly from normality. Forming two linear combinations v' and w' and adding their squares further compensates for any deviation from a normal distribution. Formation of the ratio (22) does not make the approximation appreciably worse.

For small n the situation is somewhat less favorable. Even so, in the case $n=2$, the formation of the sums and differences,

$$u_1 = \frac{x_1 + x_2}{\sqrt{2}}, \qquad u_2 = \frac{x_1 - x_2}{\sqrt{2}}, \ldots,$$

and the subsequent addition of the squares (formation of $u_1^2 + v_1^2 + w_1^2$ and of $u_2^2 + v_2^2 + w_2^2$) compensate to a great extent for the non-normality. This general conclusion has been confirmed numerically in specific examples.

In conclusion, *the F-test can be applied without great error even if nothing is known about the normality of the observed quantities.*

Example 40 (from R. A. Fisher, Statistical Methods, 11th ed. Ex. 38). In an experiment involving the accuracy of soil bacteria counts, a piece of land was divided into four sections and seven plots in each section were inoculated. The numbers in the colonies on the 28 plots are given in the following table:

Plot	Soil sample			
	I	II	III	IV
1	72	74	78	69
2	69	72	74	67
3	63	70	70	66
4	59	69	58	64
5	59	66	58	62
6	53	58	56	58
7	51	52	56	54
Total	426	461	450	440
Mean	60.9	65.9	64.3	62.9

The analysis of variance technique was used to determine whether or not, within the limits of chance, there was agreement among the outcomes of the four soil samples. The analysis of variance table shows

	Sum of squares	Degrees of freedom	Variance
Between groups	$Q_1 = 95$	$f_1 = 3$	$s_1^2 = 32$
Within groups	$Q_2 = 1{,}446$	$f_2 = 24$	$s_2^2 = 60$
Total	$Q = 1{,}541$	$N-1=27$	$s^2 = 57$

The variance within these groups is even larger than that between groups. Hence it follows immediately that the differences between the plot-means are not significant: there is no need at all to calculate $F = 32/60$. The 28 plots can be considered as a homogeneous sample.

The grand mean is 63.5 and the best estimate of the variance σ^2 is

$$s^2 = \frac{1{,}541}{27} = 57.1.$$

If the distribution of the individual observations were a Poisson distribution with expected value 63.5, then the true value of the variance would be $\sigma^2 = 63.5$ (§ 10 A). In order to test this hypothesis, we calculate the quotient

$$\chi^2 = \frac{27\, s^2}{\sigma^2} = \frac{1{,}541}{63.5} = 24.3.$$

Now if the observed values were normally distributed, this χ^2 would have a χ^2-distribution with 27 degrees of freedom. The Poisson distribution with as large an expected value as 63.5 deviates only slightly from a normal distribution. Therefore, according to formulas (9) and (10) § 23, the expected value of χ^2 would be 27 and the standard deviation $\sqrt{54} = 7.4$, and a value in the interval

$$\chi^2 = 27 \pm 7.4$$

would not be unreasonable. The observed value 24.3 thus does not deviate from the expected value by more than we should expect by pure chance. The assumption – in itself plausible – of a Poisson distribution is therefore not refuted by experiment. The best estimate of the mean value ϑ of the Poisson distribution is $\tilde{\vartheta} = 63.5$. The variance of the Poisson distribution is equal to the expected value.

D. Relation to the *t*-Test

If there are only two sequences of observed random variables, then

$$\begin{aligned} s_1^2 = Q_1 &= n_1(\bar{x} - M)^2 + n_2(\bar{y} - M)^2 \\ &= n_1\left[\bar{x} - \frac{n_1\bar{x} + n_2\bar{y}}{n_1 + n_2}\right]^2 + n_2\left[\bar{y} - \frac{n_1\bar{x} + n_2\bar{y}}{n_1 + n_2}\right]^2 \\ &= \frac{n_1 n_2}{n_1 + n_2}(\bar{x} - \bar{y})^2, \end{aligned}$$

and

$$s_2^2 = \frac{Q_2}{N-2} = \frac{\sum(x-\bar{x})^2 + \sum(y-\bar{y})^2}{n_1 + n_2 - 2},$$

and therefore

$$F=\frac{s_1^2}{s_2^2}=\frac{(\bar{x}-\bar{y})^2}{\left(\frac{1}{n_1}+\frac{1}{n_2}\right)s_2^2}. \tag{27}$$

The right-hand expression is the square of the test statistic t used in Student's test. Hence: *With two groups the F-test is equivalent to the two-sided t-test.*

Consequently, the two-sided t-test can also be used if the distributions of the x and the y are not normal, provided n_1 and n_2 are not too small (say, both $\geqq 4$).

E. Intraclass Correlation

If every group contains only two observations, then we have in fact a sequence of r pairs (x, x'). For the calculation of Q_1 and Q_2, we form for each pair the average and difference

$$\bar{x}=\frac{x+x'}{2} \quad \text{and} \quad d=x-x'.$$

If M is the average of the averages $\bar{x}$, then from (18)

$$Q_1=2\sum(\bar{x}-M)^2 \tag{28}$$

and from (7)

$$Q_2=\tfrac{1}{2}\sum d^2. \tag{29}$$

The formula

$$Q_1+Q_2=Q=\sum\{(x-M)^2+(x'-M)^2\} \tag{30}$$

provides a check.

Furthermore, from (21)

$$s_1^2=\frac{Q_1}{r-1}, \qquad s_2^2=\frac{Q_2}{r}, \tag{31}$$

and finally, as always,

$$s^2=\frac{Q}{N-1}=\frac{Q_1+Q_2}{2r-1}. \tag{32}$$

The *intraclass correlation* r^* is defined as

$$r^*=\frac{2\sum(x-M)(x'-M)}{\sum\{(x-M)^2+(x'-M)^2\}}. \tag{33}$$

Expression (33) is similar to a correlation coefficient (§ 66 B). A simple calculation yields

$$r^*=\frac{Q_1-Q_2}{Q_1+Q_2}. \tag{34}$$

The intraclass correlation would be $+1$ if the variance within groups were zero; it would be -1 if the variance between groups were zero.

In practice, these extremes almost never occur. If the two variances are approximately equal, then r^* is nearly zero.

Example 41. Hadorn, Bertani, and Gallera[10] divided the male genital-imaginal disk of *Drosophila melanogaster* sagittally and implanted the pieces in a host of the same type. An ejaculatory duct of approximately normal size developed from each of the two pieces of embryonic tissue. The size, however, varied greatly from case to case. We shall now

x	x'	$\bar{x}$	$(\bar{x}-M)^2$	d	d^2
394	328	361	2,256	66	4,356
382	344	363	2,450	38	1,444
375	328	351.5	1,444	47	2,209
369	319	344	930	50	2,500
369	319	344	930	50	2,500
369	293	331	306	76	5,776
363	350	356.5	1,849	13	169
357	325	341	756	32	1,024
357	300	328.5	225	57	3,249
356	331	343.5	900	25	625
353	347	350	1,332	6	36
350	297	323.5	100	53	2,809
347	325	336	506	22	484
344	313	328.5	225	31	961
335	319	327	182	16	256
331	319	325	132	12	144
328	269	298.5	225	59	3,481
325	300	312.5	1	25	625
322	300	311	6	22	484
319	313	316	6	6	36
319	306	312.5	1	13	169
319	303	311	6	16	256
316	303	309.5	16	13	169
313	272	292.5	441	41	1,681
313	234	273.5	1,600	79	6,241
309	300	304.5	81	9	81
309	294	301.5	144	15	225
306	300	303	110	6	36
297	287	292	462	10	100
297	253	275	1,482	44	1,936
294	281	287.5	676	13	169
287	253	270	1,892	34	156
281	272	276.5	1,369	9	81
281	269	275	1,482	12	144
275	263	269	1,980	12	144
250	234	242	5,112	16	256
11,811	10,763	11,287	31,611	1,048	46,012

[10] E. Hadorn, G. Bertani, and J. Gallera, Regulationsfähigkeit und Feldorganisation der männlichen Genital-Imaginalscheibe von *Drosophila*. Wilhelm Roux' Archiv für Entwicklungsmechanik der Organismen 114 (1949) p. 31.

determine whether this variation can be attributed to the fact that the two implanted halves were frequently unequally large. In such a case, one would expect that in the implantation pairs each especially large duct would be paired with an extremely small one; i.e., that the variation within pairs would be larger than the variation between pairs. Calculation, however, shows that, on the contrary, the variation within pairs is smaller than that between pairs. The lengths x, x' of the ejaculatory ducts are given in the preceding table, in which the larger of the two lengths is always given first.

Computation yields

$$M=\tfrac{1}{36}\sum \bar{x}=313.5,$$

$$Q_1=2\sum(\bar{x}-M)^2=63{,}222,$$

$$Q_2=\tfrac{1}{2}\sum d^2=23{,}006.$$

The intraclass correlation is

$$r^*=\frac{Q_1-Q_2}{Q_1+Q_2}=\frac{40{,}216}{86{,}228}=0.47.$$

The analysis of variance table shows

	Sum of squares	Degrees of freedom	Variance
Between pairs	$Q_1=63{,}222$	$f_1=35$	$s_1^2=1{,}806$
Within pairs	$Q_2=23{,}006$	$f_2=36$	$s_2^2=\ \ 639$
Total	$Q\ =86{,}228$	$f\ =71$	$s^2=1{,}214$

Finally we have

$$F=\frac{1{,}806}{639}=2.83.$$

The differences between pairs are therefore considerably larger than those within pairs. The 1%-cut-off for F is 2.21. The differences in the lengths therefore do not depend on the different sizes of the two pieces.

F. Further Applications

The method of partitioning the total variation is also applicable to more complicated situations. For example, it can happen that the $n\cdot r$ observed quantities x_{ik} are divided into groups not only row-wise but also column-wise and that we wish to know whether there are actual differences at hand not only among the rows but also among the columns. In such cases, however, additional assumptions must be made about the distributions of the x_{ik} to apply the F-test. In our case we have to assume that the random variables x_{ik} may be represented as a sum

$$x_{ik}=a_i+b_k+z_{ik},$$

where the z_{ik} are independent, normally distributed random variables with zero mean and common standard deviation.

For further development of the ideas only mentioned here, the reader is referred to the literature on the subject: for example, R. A. Fisher, The Design of Experiments. London: Oliver & Boyd 1935, or O. Kempthorne, The Design and Analysis of Experiments. New York: John Wiley & Sons 1952.

§ 59. General Principles. Most Powerful Tests

A. Basic Concepts

J. Neyman and E. S. Pearson[11] considered the problem of hypothesis testing from a generalized point of view. We shall discuss the fundamental results of their research.

Let the possible outcomes of an experiment be represented by points X in a space E. It does not matter whether only discrete points (e.g., finitely many) or the points of a continuum represent these possible outcomes. Let the experiment be relevant to testing a hypothesis H.

Let every measurable set B in the space have a probability $\mathscr{P}(B)=\mathscr{P}(B|H)$ under the hypothesis. This probability can be obtained, for example, by summation of the probabilities of the points themselves or by integration of a probability density over B.

All tests take the form that the hypothesis H is rejected whenever the observed point belongs to a certain *critical region*, or *rejection region*, C. The set C is determined so that, if the hypothesis H is true, it has only a small probability. Therefore C should satisfy

$$\mathscr{P}(C|H) \leqq \alpha,$$

where α is a preassigned allowable error probability, or *significance level* (*e.g.*, 5 or 1%).

The question of the rationale behind the choice of the set C now arises, for indeed, there can exist very different sets with probability $\leqq \alpha$. The concept of two types of error will be introduced in order to answer this question.

A *Type* I *error* is the error of *rejecting a true hypothesis*. Hence the probability of a Type I error is $\leqq \alpha$. If only the Type I error needed to be taken into consideration, then the choice of C would be completely arbitrary. One could even conclude that the set C should be the empty set, so that the resulting significance level would be zero. This is not a solution to the problem, because consideration has yet to be given to the Type II error.

---

[11] See, in particular, J. Neyman and E. S. Pearson, Phil. Trans. Roy. Soc., London A 231 (1932) p. 332.

A *Type* II *error* is the error of *acceptance of a false hypothesis*. If the hypothesis H is false, it can very easily happen that the observed point does not fall into the set C and the hypothesis is therefore not rejected. One would like a high probability of avoiding this error. The purpose of the experiment is, after all, to decide whether the hypothesis H is true or false with a high probability of a correct decision, whether or not it is true. Therefore, one endeavors *to make the probability of a Type* II *error as small as possible.*

At this point a difficulty arises: the probability of a Type II error depends on what hypothesis H' is correct instead of H and can not be computed without this additional information.

We shall first assume that there is only *one* alternative hypothesis H' to be taken into consideration. The *power* P' of a given test against the alternative H' is defined as the probability that the test rejects the hypothesis H computed under the assumption that H' is correct:

$$P' = \mathscr{P}'(C) = \mathscr{P}(C|H').$$

The probability that H is accepted when H' is correct, that is, the probability of a Type II error, is equal to $1-P'$. This probability is to be minimized, and therefore the power P' is to be maximized. If P' is maximized subject to the condition $\mathscr{P}(C) \leqq \alpha$, the test is said to be *most powerful* against the alternative H'.

Hence we may formulate our problem as follows. Let two measures $\mathscr{P}(B) = \mathscr{P}(B|H)$ and $\mathscr{P}'(B) = \mathscr{P}(B|H')$ be defined on a space E and satisfy the axioms of probability theory. A set C is to be determined which maximizes $\mathscr{P}'(C)$ subject to the condition

$$\mathscr{P}(C) \leqq \alpha. \tag{1}$$

The continuous and discrete cases require separate consideration here.

B. The Case of Continuous Variables

Let E be the space of continuous variables $x_1, \ldots, x_n$. We shall assume that the two measures $\mathscr{P}$ and $\mathscr{P}'$ are defined by continuous probability density functions

$$f(X) = f(x_1, \ldots, x_n) \quad \text{and} \quad g(X) = g(x_1, \ldots, x_n).$$

If f is zero on a set T of the space E, then we can include this set in C without affecting the condition (1). This enlargement of the set C can at most increase $\mathscr{P}'(C)$. Therefore we need be concerned only with the complementary set $E-T$ in the space E.

On the set $E-T$ we have $f\neq 0$, so that

$$U=U(X)=\frac{g}{f} \tag{2}$$

is a continuous function of X. For every positive t, the event $U<t$ has a probability defined by the measure $\mathscr{P}$:

$$G(t)=\mathscr{P}(U<t)=\mathscr{P}(g<tf). \tag{3}$$

We shall now assume that the distribution function $G(t)$ takes on the value $1-\alpha$ exactly, so that there exists c such that

$$\mathscr{P}(g<cf)=1-\alpha,$$

and that

$$\mathscr{P}(g\geqq cf)=\alpha. \tag{4}$$

The assertion then is that the set C defined by

$$g\geqq cf, \quad \text{or} \quad g/f\geqq c,$$

possesses the desired maximal property.

By definition, condition (1) is satisfied, as is expressed exactly by equality (4). If W is any other set which also satisfies condition (1), then we must show that $\mathscr{P}'(W)\leqq\mathscr{P}'(C)$.

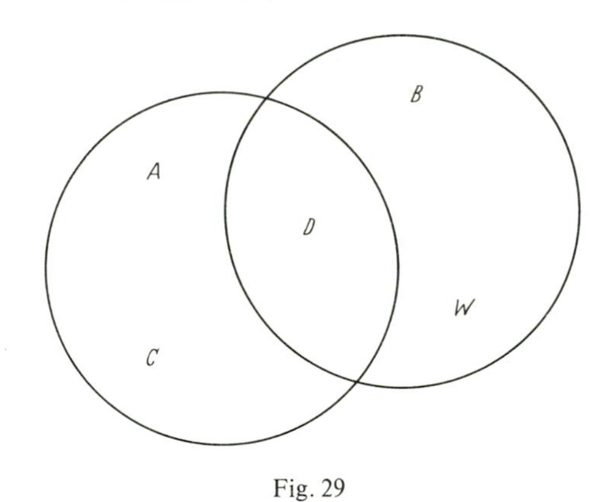

Fig. 29

Let the intersection of C and W be D, and let

$$C=D+A$$
$$W=D+B.$$

That is, A is that part of C which does not belong to W, and B is that part of W which does not belong to C. Hence we have

$$\mathscr{P}(C)=\mathscr{P}(D)+\mathscr{P}(A)=\alpha,$$
$$\mathscr{P}(W)=\mathscr{P}(D)+\mathscr{P}(B)\leqq\alpha,$$

and therefore

$$\mathscr{P}(A) \geqq \mathscr{P}(B),$$

or, what is the same thing,

$$\int_A f\, dX \geqq \int_B f\, dX. \tag{5}$$

On A we have $g \geqq cf$, since A is a subset of C. Therefore we have

$$\begin{aligned}\mathscr{P}'(C) &= \mathscr{P}'(D) + \mathscr{P}'(A) = \mathscr{P}'(D) + \int_A g\, dX \\ &\geqq \mathscr{P}'(D) + \int_A cf\, dX \\ &\geqq \mathscr{P}'(D) + \int_B cf\, dX, \qquad \text{from (5)}, \\ &\geqq \mathscr{P}'(D) + \int_B g\, dX = \mathscr{P}'(D) + \mathscr{P}'(B) = \mathscr{P}'(W).\end{aligned}$$

Hence the maximal property of the set C is proved.

If the function $G(t)$ does not take on the value $1-\alpha$, but instead has a discontinuity and takes on only values $<1-\alpha$ and $>1-\alpha$, there exists an alternative method for forming the set C. First of all, the entire set on which $g > cf$ is included in C, and then a sufficiently large subset of the set on which $g = cf$ is also included to make the total probability $\mathscr{P}(C) = \alpha$. This subset can, moreover, be chosen arbitrarily. The proof remains the same. In the future we shall disregard this case, which seldom occurs in applications.

The probability density $f(X)$ is also called the *likelihood* of the hypothesis H, and likewise $g(X)$ is called the likelihood of the hypothesis H'. The ratio (2) is, for this reason, known as the *likelihood ratio.*

The most powerful test of the hypothesis H against the alternative H' can thus be formulated as:

The hypothesis H is rejected whenever the likelihood ratio (2) *is at least equal to c. This critical value c is determined by the requirement that the probability of a Type* I *error – that is, $\mathscr{P}(U \geqq c)$ – be equal to α.*

This test is called the *likelihood ratio test.* It is the most powerful test of the hypothesis H against the alternative H', and it can always be used when one adheres strictly to the assumption that H' is the only possible alternative to H.

C. The Case of Discrete Variables

The case that the space E consists of a countable number of discrete points X and the probability of a point set is equal to the sum of the probabilities of the individual points can be treated in exactly the same manner. The probabilities of the individual points, $\mathscr{P}(X)$ and $\mathscr{P}'(X)$, occur in place of the probability densities $f(X)$ and $g(X)$. We shall again use the notation $f(X)$ and $g(X)$, but to refer, now, to these probabilities.

If there are points of E with probability zero under the hypothesis H, these points are always included in the rejection region C. For all remaining points, $f(X) \neq 0$, so that we may define the random variable

$$U = \frac{g(X)}{f(X)} = \frac{\mathscr{P}'(X)}{\mathscr{P}(X)}.$$

Let us again assume at first that the distribution function $G(u)$ of U takes on the value $1-\alpha$, so that for a well-defined c the equation

$$\mathscr{P}(g \geqq cf) = \alpha$$

holds. Then we may take the set C to be the set $g \geqq cf$ and proceed as above, with the exception of replacing integration by summation. Again, if W is any other set with $\mathscr{P}(W) \leqq \alpha$, we set

$$C = D + A$$
$$W = D + B$$

and have as above

$$\mathscr{P}(A) \geqq \mathscr{P}(B),$$

or

$$\sum_A f(X) \geqq \sum_B f(X).$$

On A we have $g \geqq cf$ and on B we have $g < cf$. Hence we have

$$\begin{aligned}
\mathscr{P}'(C) = \mathscr{P}'(D) + \mathscr{P}'(A) &= \mathscr{P}'(D) + \sum_A g(X) \\
&\geqq \mathscr{P}'(D) + \sum_A c f(X) \\
&\geqq \mathscr{P}'(D) + \sum_B c f(X) \\
&\geqq \mathscr{P}'(D) + \sum_B g(X) = \mathscr{P}'(D) + \mathscr{P}'(B) = \mathscr{P}'(W).
\end{aligned}$$

Hence we have established the maximal property of the set C.

If $G(u)$ does not assume the value $1-\alpha$, but jumps from a value $< 1-\alpha$ to one $> 1-\alpha$, then the entire set defined by $g > cf$ is included in C and, if possible, enough points X for which $g = cf$ to make the total probability $\mathscr{P}(C) = \alpha$ exactly. The proof is as above.

If it is not possible to find such points X to yield $\mathscr{P}(C)$ exactly equal to α, as a first step enough points are included in C to make $\mathscr{P}(C)$ the fraction nearest α but not exceeding it. Let $\mathscr{P}(C)$ be, say, equal to $\alpha - \varepsilon$.

If yet another point X of those points for which $g(X)=cf(X)$ were added to C, then $\mathscr{P}(C+X)$ would be equal to $\alpha+\delta$, or $C+X$ would be too large. Therefore the point X must be split into two "points" X_1 and X_2 with probabilities $\mathscr{P}(X_1)=\varepsilon$ and $\mathscr{P}(X_2)=\delta$ and the "point" X_1 included in the set C.

In order to achieve this splitting, we employ the following trick. We play a game of chance in which the probability of winning is exactly

$$p=\frac{\varepsilon}{\varepsilon+\delta}.$$

If the experiment yields the point X and if we win the game of chance, then the hypothesis H is rejected, but if we lose, then we accept the hypothesis.

We conclude by noting that $C+X_1$ is in fact the desired maximal set. The probability of the point X was $\varepsilon+\delta$. The event X_1 is the intersection of the independent events observing X and winning the game of chance, and therefore its probability is

$$(\varepsilon+\delta)\,p=\varepsilon.$$

The probability of $C+X_1$ is thus

$$(\alpha-\varepsilon)+\varepsilon=\alpha,$$

as was to be shown.

In practice one seldom plays such a game of chance, which gives no information at all whether the hypothesis is true or false, but instead takes C, without X_1, as the rejection region. To be sure, this results in a somewhat larger probability of a Type II error, but as compensation, the corresponding probability of a Type I error is smaller, namely $\alpha-\varepsilon$ instead of α. If $\alpha-\varepsilon$ is taken to be the preassigned significance level instead of α, then C is a maximal region; i.e., the test corresponding to C is most powerful at level $\alpha-\varepsilon$.

D. Examples

Example 42. Let E be the space of the variables $x_1, \ldots, x_n$. Under the hypothesis H let the variables $x_1, \ldots, x_n$ each be independently normally distributed with zero mean and unit standard deviation. The probability density is then

$$f(X)=(2\pi)^{-\frac{n}{2}}e^{-\frac{1}{2}(x_1^2+\cdots+x_n^2)}.$$

Under the alternative hypothesis H' let the $x_1, \ldots, x_n$ again be independent and each normally distributed with unit standard deviation, but with a larger mean a:

$$g(X)=(2\pi)^{-\frac{n}{2}}e^{-\frac{1}{2}[(x_1-a)^2+\cdots+(x_n-a)^2]}, \qquad a>0.$$

The likelihood ratio is

$$U = \frac{g}{f} = e^{a\sum x - \frac{1}{2} n a^2},$$

which is an increasing function of

$$\bar{x} = \frac{1}{n} \sum x.$$

The hypothesis H is therefore to be rejected whenever the sample mean $\bar{x}$ exceeds a critical value c determined so that the probability of exceeding it will be equal to α under the hypothesis. The distribution of $\bar{x}$ is normal with standard deviation $1/\sqrt{n}$ and expectation zero under the hypothesis H. Therefore c must satisfy

$$c = \frac{\sigma}{\sqrt{n}} \Psi(1-\alpha), \tag{6}$$

where Ψ is the inverse function of the normal distribution function Φ. In our case σ was taken to be 1, but formula (6) is valid in general.

It is worth noting in this example that the resulting test depends not on the value of a but only on a being positive. The one-sided test which rejects H whenever $\bar{x} > c$ is therefore *uniformly most powerful* against all alternatives H' with $a > 0$. If instead we were working with the hypothesis of a normal distribution with a negative, we would then have to reject the hypothesis H that the expectation is zero whenever $\bar{x} < -c$.

Example 43. Let an event have probability p under the hypothesis H and a larger probability p' under the alternative hypothesis H'. Let the event occur x times in n independent trials. When is the hypothesis to be rejected?

Under the hypothesis H the probability of x occurrences is

$$f(x) = \binom{n}{x} p^x (1-p)^{n-x}.$$

Under the hypothesis H' this probability is

$$g(x) = \binom{n}{x} p'^x (1-p')^{n-x}.$$

The likelihood ratio is

$$u = \frac{g}{f} = \left(\frac{p'}{p}\right)^x \left(\frac{1-p'}{1-p}\right)^{n-x}.$$

Since u is an increasing function of x, we have to reject H whenever $x > c$. Here the bound c is to be determined so that the sum of the probabilities of the x-values in the rejection region remains $\leqq \alpha$:

$$\binom{n}{c+1} p^{c+1} (1-p)^{n-c-1} + \binom{n}{c+2} p^{c+2} (1-p)^{n-c-2} + \cdots + p^n \leqq \alpha. \tag{7}$$

The left side of (7) is an increasing function of p, since the derivative

$$(c+1)\binom{n}{c+1} p^c (1-p)^{n-c-1}$$

is always positive. Since the left side of (7) is zero for $p=0$ and one for $p=1$, there exists exactly one value p_α for which the left side is exactly equal to α. For $p \leqq p_\alpha$ inequality (7) is satisfied, but not for any larger p. *Therefore on the basis of the test which rejects the hypothesis whenever x exceeds c satisfying* (7), *those hypotheses H with $p \leqq p_\alpha$ are to be rejected but not those with $p > p_\alpha$.*

The bound p_α is just the one-sided (lower) confidence bound for p found by Clopper and Pearson (cf. § 7). Hence the theory of confidence bounds as developed earlier is a special case of the general theory of Neyman and Pearson.

§ 60. Composite Hypotheses

A *simple hypothesis* is a hypothesis which assigns to every subset of the event set E a completely specified probability. If, however, the hypothesis assigns probabilities which depend on parameters, the hypothesis is said to be *composite*. We can think of such a hypothesis as being composed of the simple hypotheses obtained by assigning specific values to the parameters. Hence, an equivalent definition is that *a composite hypothesis is a collection of simple hypotheses.*

If we wish to test a simple hypothesis H against a simple alternative H', then, as we have seen in § 59, there always exists a most powerful test of H against the alternative H'. If, however, H' is composite, then either there exists a *uniformly most powerful test* against all the simple hypotheses composing H' or there exists no such test.

Example 42 (§ 59) illustrates both cases. The hypothesis H in this example is simple and states that all x_i are normally distributed with mean 0 and standard deviation 1. The alternative hypothesis H' depends on a parameter a and is therefore composite: it states that the x_i are normally distributed with unspecified mean value a and standard deviation 1. If only positive values of a are allowed, then there exists a uniformly most powerful test, which rejects the hypothesis H whenever $\bar{x}$ is observed to be larger than $c n^{-\frac{1}{2}}$. If, however, negative values of a are also allowed, then there exists no such uniformly most powerful test. A test which rejects the hypothesis H for large $\bar{x}$-values loses power against negative values of a, and likewise for a test which rejects H for small $\bar{x}$-values when a is in fact positive.

As an optimality criterion for choosing a test in such cases, the concept of *unbiasedness* has been introduced. A test of a simple hypothesis H is said to be *unbiased* if the probability of rejecting H when H is true is at most equal to the probability of rejecting H when one of the alternative hypotheses composing H' is true:

$$\mathscr{P}(C|H) \leqq \mathscr{P}(C|H') \qquad \text{for all } H'. \tag{1}$$

Or, equivalently, the probability of rejecting the hypothesis H when it is true should not be larger than the probability of rejecting it when it is false. This is certainly a reasonable requirement.

If in Example 42 one allows all strictly positive and all strictly negative mean values a to compose the hypothesis H', then neither one-sided test – rejecting the hypothesis H when $\bar{x}$ is larger than $c\,n^{-\frac{1}{2}}$, or rejecting when $\bar{x}$ is smaller than $-c\,n^{-\frac{1}{2}}$ – is unbiased. One has an unbiased test if the hypothesis H is rejected whenever the absolute value $|\bar{x}|$ exceeds $c'n^{-\frac{1}{2}}$. If c' is determined so that the probability of rejecting the hypothesis H when it is true is exactly α, then this test is a *most powerful unbiased test* of H at level α against each alternative in H', and hence is a *uniformly most powerful unbiased test* of H against H'. For a proof, see J. Neyman and E. S. Pearson, On the problem of the most efficient tests of statistical hypotheses, Philosophical Transactions of the Royal Society, London A 231 (1933) pp. 289–337.

The problem becomes still more complicated if H is also taken to be a composite hypothesis. For example, let H be the hypothesis that $x_1, \ldots, x_n$ are independent normal variables with mean zero and arbitrary (unspecified) standard deviation σ. Under the hypothesis H, the probability density of the variables $x_1, \ldots, x_n$ is

$$f(x|\sigma)=(2\pi\,\sigma)^{-\frac{n}{2}}\exp\left(-\frac{x_1^2+\cdots+x_n^2}{2\sigma^2}\right). \tag{2}$$

If the critical region is taken to be C – that is, if the hypothesis is to be rejected whenever the observed point X lies in C – then the probability of a Type I error,

$$\mathscr{P}(C|\sigma)=\int_C f(x|\sigma)\,dX, \tag{3}$$

is, in general, dependent on σ. If $\mathscr{P}(C|\sigma)\leqq\alpha$ for all σ, then we say that the test, or the region C, has *level of significance* α. If in fact

$$\mathscr{P}(C|\sigma)=\alpha$$

for all σ, then we say that the test, or the region C, is *similar to the sample space*, in the terminology of Neyman and Pearson.

Neyman, Lehmann, and Scheffé[12] have developed general methods which yield such similar regions C. We shall discuss the methods with reference to the above-mentioned example, but refer the reader to the literature for the proofs.

From the form of the probability density (2) we see immediately that

$$Q=x_1^2+\cdots+x_n^2 \tag{4}$$

[12] See in particular, E. L. Lehmann and H. Scheffé, Completeness, similar regions, and unbiased estimation, Sankhyā 10 (1950) p. 305 and Sankhyā 15 (1956) p. 219, and the additional literature cited there.

is a sufficient estimate of $n\sigma^2$. The probability density of Q is

$$f(u|\sigma)=C\sigma^{-n}u^{\frac{n}{2}-1}\exp(-\tfrac{1}{2}u\sigma^{-2}) \quad \text{with} \quad C^{-1}=\Gamma\left(\frac{n}{2}\right)2^{\frac{n}{2}}. \tag{5}$$

The probability densities $f(u|\sigma)$ form a *boundedly complete family* in the sense of Lehmann and Scheffé. In their terminology, the family of densities $f(u|\sigma)$ is said to be boundedly complete if whenever a bounded integrable function $\varphi(t)$ satisfies the integral equation

$$\int_0^\infty \varphi(u)\,f(u|\sigma)\,du=0 \qquad \text{for all } \sigma>0, \tag{6}$$

then $\varphi(t)=0$. Completeness in the present situation is immediately clear once the integral condition (6) is written without the factor $C\sigma^{-n}$ as

$$\int_0^\infty u^{\frac{n}{2}-1}\,\varphi(u)\,e^{-\lambda u}\,du=0 \qquad \text{for all } \lambda>0. \tag{7}$$

Lehmann and Scheffé have proved that if the probability densities form a boundedly complete family, then all similar regions C can be found by a method due to Neyman. This method involves finding a region C_u corresponding to each individual value u of the sufficient statistic Q and having conditional probability given $Q=u$ equal to α. Then if the union C of all such C_u is measurable, it is similar to the sample space.

In our case C_u is a set on the sphere

$$x_1^2+\cdots+x_n^2=u. \tag{8}$$

The conditional density of $x_1, \ldots, x_n$ on this sphere is

$$\frac{f(x|\sigma)}{\int f(x|\sigma)\,d\omega_{n-1}}, \tag{9}$$

where the integration in the denominator is over the sphere (8). The region C_u on the sphere is to be chosen so that the integral of (9) over C_u is exactly equal to α. But $f(x|\sigma)$ is constant over the entire sphere. Therefore the factors $f(x|\sigma)$ appearing in the numerator and the denominator cancel and the integral becomes simply the area of C_u divided by the area of the entire sphere. Hence the area of C_u must be set equal to α times the area of the sphere. Otherwise the choice of C_u is arbitrary (within the limits imposed by the measurability restriction on the union C).

Which region is chosen depends heavily on what alternative hypothesis is under consideration. We shall assume that the alternative H' is the composite hypothesis that the x_i are independent, normally distributed variables with an unspecified standard deviation σ and a *positive* mean value a. The probability density is then

$$f_a(u|\sigma)=(2\pi\,\sigma)^{-\frac{n}{2}}\exp\left(-\frac{(x_1-a)^2+\cdots+(x_n-a)^2}{2\sigma^2}\right). \tag{10}$$

The determination of a most powerful similar test against this alternative hypothesis H' is now easy. We choose a value u of σ and determine a region C_u which is most powerful against the simple alternative H'_u. Since the integral of (9) is independent of σ, we can take this same value of σ in (9) as well. Straightforward application of the methods of § 59 yields the likelihood ratio test

$$\frac{f_a(u|\sigma)}{f(u|\sigma)}\geqq c, \tag{11}$$

and hence yields in our example

$$\exp\left(\frac{(x_1+\cdots+x_n)\,a}{\sigma^2}-\frac{n\,a^2}{2\sigma^2}\right)\geqq c.$$

Therefore we must reject the hypothesis H whenever the sample mean

$$\bar{x}=\frac{1}{n}(x_1+\cdots+x_n)$$

exceeds a critical value w to be determined in the following manner. The plane $\bar{x}=w$ divides the sphere (8) into two segments, and w is to be chosen so that the area of the segment $\bar{x}>w$ is exactly α times the area of the entire sphere. This process leads just to the one-sided t-test.

Therefore, among all tests similar to the sample space, the one-sided t-test is the most powerful against the alternative H' with $a>0$.

With these same methods one can also prove that the one-sided t-test for comparing the theoretical means μ and ν of two independent normal samples $x_1, \ldots, x_m$ and $y_1, \ldots, y_n$ is, among all tests of the hypothesis $\mu=\nu$ which are similar to the sample space, most powerful against all alternatives with $\mu>\nu$.

One might ask whether Student's test is also uniformly most powerful among all tests having level of significance α. Unfortunately the answer is no[13].

[13] E. L. Lehmann and C. Stein, Most powerful tests of composite hypotheses I. Ann. of Math. Stat. 19 (1948) p. 495.

Chapter Twelve

Order Tests

Order tests are tests which use not the exact values of the observations, but only their order relations – i.e., only the relations $x<y$ and $x>y$ between the observed x and y. Since such tests are not based upon definite distribution functions for the quantities x and y, they are also called *distribution-free*, or *nonparametric*, tests.

The theory of nonparametric tests does not require much background knowledge, and, in fact, we have had to assume only that of Chapters 1 and 2.

§ 61. The Sign Test

A. The Principle

If we observe an increase in blood pressure in all ten out of ten experimental animals subjected to a certain kind of treatment, then purely on intuitive grounds we conclude, "This can not be by chance!" To justify this spontaneous response, we note that if the observed changes in blood pressure were purely chance fluctuations, then in probability approximately half of the differences would be positive and half negative. The probability of a positive difference would then be one-half for each animal and the probability that all differences would be positive would therefore be $(\frac{1}{2})^{10}=\frac{1}{1024}$. Since one need not take into account so improbable an event, it is to be assumed that the observed effect is real.

This completely simple decision procedure can be formulated as an exact nonparametric test with a tolerable error probability, or *level*, α, which can be chosen arbitrarily. Let n differences x_i-y_i be observed ($i=1,\ldots,n$), of which k are positive and $n-k$ are negative, and for the moment, let us ignore the possibility that $x_i=y_i$. The hypothesis H to be tested is the hypothesis that for every i the two observations x_i and y_i are independent random variables with the *same* distribution function. Under this hypothesis the probability of a positive difference x_i-y_i is equal to the probability of a negative difference. If the event $x_i=y_i$ has probability zero, then the probabilities of positive and negative differences must both be exactly equal to $\frac{1}{2}$. This is the conclusion which is to be verified by means of the sign test.

We can set $z_i = x_i - y_i$, in which case the differences $z_1, \ldots, z_n$ are then independent variables. The hypothesis H to be tested implies then that *for every i the positive and negative z_i are equally probable:*

$$\mathscr{P}(z_i > 0) = \mathscr{P}(z_i < 0). \tag{1}$$

Even if the z are not differences, the sign test can still be used to test the hypothesis (1). In the case that the event $z_i = 0$ has probability zero, then from (1) it follows that

$$\mathscr{P}(z_i > 0) = \tfrac{1}{2}. \tag{2}$$

Under this assumption, the probability that more than m of the z_i are observed to be positive is

$$\left[\binom{n}{m+1} + \binom{n}{m+2} + \cdots + \binom{n}{n}\right]\left(\frac{1}{2}\right)^n. \tag{3}$$

If m is the smallest integer for which expression (3) remains $\leqq \alpha$, the sign test can then be formulated as:

Whenever k, the number of positive z_i, exceeds m, the hypothesis H is to be rejected. The error probability of this test – that is, the probability of rejecting the hypothesis H even though it is true – is obviously $\leqq \alpha$. Hence the test is an exact test. This is the *one-sided sign test*. The *two-sided* sign test rejects the hypothesis H not only whenever the number k of positive z exceeds the cut-off value m but also whenever the number $n-k$ of negative z exceeds m. If the cut-off m is not changed, then the error probability of the two-sided test is twice as large as that of the one-sided, thus it is $\leqq 2\alpha$.

Table 9 gives the cut-off points m for $n \leqq 50$ and for the most common levels,

two-sided $2\alpha = 5\%$, 2%, 1%,
one-sided $\alpha = 2\frac{1}{2}\%$, 1%, $\frac{1}{2}\%$.

B. Ties

There now arises the question of what procedure to use in the presence of ties, that is, in the event that some differences $x_i - y_i = z_i$ are zero. Half the differences could perhaps be treated as positive and the other half as negative. Also a coin could be tossed whenever a tie occurs and the difference treated as positive if the coin falls heads. *The best solution, however, is simply to discard the tied observations*[1].

[1] J. Hemelrijk, A theorem on the sign test when ties are present. Proc. Kon. Ned. Akad. Section on Sciences A 55 (1952) p. 322.

Let the number of positive differences z_i be k, the number of negative be l, and their sum be $k+l=n$. Then if the sign test is applied to these n, the error probability is guaranteed to be $\leqq\alpha$ (or, $\leqq 2\alpha$ in the two-sided application). To prove the above assertion, assume that the hypothesis (1) is true. Let the total number of observations or observed differences be N and let the probability that n of these observations are different from zero be p_n. The sum of all p_n is, of course, one:

$$\sum_0^N p_n = 1. \tag{4}$$

If the number of non-zero differences is n, then the conditional probability that k exceeds m, where in each case m is the critical value corresponding to a sample size n, is at most equal to α. If we call this conditional probability P_n, then

$$P_n \leqq \alpha. \tag{5}$$

The unconditional probability that the test leads to the rejection of H is, according to the total probability law,

$$P = \sum p_n P_n \leqq \sum p_n \alpha = \alpha. \tag{6}$$

Hence the assertion is proved.

Example 44. In the research of H. Fritz-Niggli[2], *Drosophila* eggs were exposed to soft and hard radiation ($18 \cdot 10^4$ and $31 \cdot 10^6$ eV). The mean frequency of lethality was determined for each of the different groups of animals which had received the same radiation dose. Then in each case the difference d of means for soft and hard radiation was formed and subjected to Student's test. The signs of the differences d for the various ages of the eggs and for the various radiation doses were

Age	1 hour	+ + + + − + + −	(8 cases)
	$1\frac{3}{4}$ hours	+ + + − − − +	(7 cases)
	3 hours	+ + + + +	(5 cases)
	4 hours	+ + + + +	(5 cases)
	$5\frac{1}{2}$ hours	+ +	(2 cases)
	7 hours	+ + + +	(4 cases).

In the age groups from 1 through 3 hours, Student's test detected a difference only in a single case, and then in this case ($1\frac{3}{4}$-hour-eggs) a difference only at the 5%-level, but in no case did it detect a difference at the 1%-level. In the age groups from 4 through 7 hours, on the other hand, Student's test led to the detection of a difference at the 5%-level in 7 out of 11 cases and in 5 of these 7 cases a difference at the 1%-level as well. It is thus practically certain (at least in the higher age brackets) that the soft rays are more strongly lethal than the hard ones at the same dose.

---

[2] H. Fritz-Niggli, Vergleichende Analyse der Strahlenschädigung von *Drosophila*-Eiern. Fortschr. auf dem Geb. d. Röntgenstrahlen 83 (1955) p. 178.

Student's test requires a great deal of calculation and, moreover, assumes the normality of the distribution. Therefore the question arises whether a conclusion could not be drawn simply by mere consideration of the signs + and −.

If we combine the age groups from 4 through 7 hours, we then find the + sign in 11 out of 11 cases. Table 9 gives 1 and 10 as the two-sided cut-off points at the 1%-level. Since 11 lies outside these bounds, the effect is "strongly assured", that is, it is assured at the 1%-level.

In the age brackets from 1 through 3 hours we find the + sign in 15 out of 20 cases. The two-sided bounds at the 5%-level are 6 and 14. Since 15 lies outside these bounds, the effect is "weakly assured", that is, it is assured with an error probability of 5%.

Therefore, almost without calculation the sign test leads to the decision that certainly in the higher age groups and probably in the lower groups the softer radiation has the stronger effect.

C. Symmetry of a Distribution

The distribution function of a variable z is said to be *symmetric about zero* if

$$\mathscr{P}(z>u)=\mathscr{P}(z<-u) \tag{7}$$

for all u. If the distribution has a probability density $g(u)$, then (7) implies that $g(u)$ is an even function:

$$g(u)=g(-u). \tag{8}$$

Since (7) implies (1), the sign test may be used as a test for symmetry of a distribution. Hemelrijk[3] has investigated other tests of symmetry.

D. Confidence Limits for the Median

If we restrict attention to *continuous* distribution functions $F(u)$, then (1) is equivalent to

$$\mathscr{P}(z<0)=\mathscr{P}(z>0)=\tfrac{1}{2}. \tag{9}$$

But, condition (9) states that zero is the (true) *median* of the distribution (cf. § 17). Consequently, we may use the sign test to test the hypothesis that the median of a distribution is zero.

If we wish to test the hypothesis that the median is ζ, we can make the transformation to new variables $z-\zeta$ and use the sign test. Employing the one-sided sign test, we reject the value ζ as the median whenever there are more the m positive differences $z_i-\zeta$ in a sample $z_1, \ldots, z_n$.

We can formulate the above problem in a somewhat different way. Let $z^{(1)}, \ldots, z^{(n)}$ denote the sample $z_1, \ldots, z_n$ ordered according to increasing size. These $z^{(i)}$ are the *order statistics* mentioned earlier. Then consider the order statistic with index $(n-m)$, i.e., $z^{(n-m)}$. If $\zeta<z^{(n-m)}$,

[3] J. Hemelrijk, A family of parameter-free tests for symmetry. Proc. Kon. Ned. Akad. Section on Sciences 53 (1950) pp. 945−955 and 1186−1198.

then more than m differences $z^{(i)}-\zeta$ are positive, and all $\zeta<z^{(n-m)}$ are to be rejected as the median. With the two-sided sign test, all $\zeta>z^{(m+1)}$ are also to be rejected as the median. *Hence we get* $z^{(n-m)}$ *and* $z^{(m+1)}$ *as two-sided confidence bounds for the median* ζ.

As a confidence interval we get

$$z^{(n-m)} \leqq \zeta \leqq z^{(m+1)}. \tag{10}$$

Assertion (10) has probability $\geqq 1-2\alpha$. The result so formulated is also valid for discontinuous distributions, as we can easily verify.

§ 62. The Two-sample Problem

A. Statement of the Problem

Let $n=g+h$ independent random variables

$$x_1, \ldots, x_g; \qquad y_1, \ldots, y_h$$

be observed. Let the x all be observed under the same experimental conditions, so that we can assume that they all have the same distribution function, and let the same be true of the y. Suppose that we have observed certain differences between the empirical distribution of x and that of y, e.g., that the x are in general larger than the y, or that they are more widely dispersed. We should like to know whether these differences are real or attributable to chance.

The null hypothesis H_0 to be tested states that the x and the y all have the same distribution function and that therefore the observed differences are purely due to chance. No special assumptions about this distribution function of x and y are, however, to be made.

Two tests which we have mentioned previously – namely, Student's test and the variance-ratio test – both depend on the normality assumption. For this reason we shall drop these tests from consideration immediately. To be sure, both tests are applicable, at least as a certain approximation, to non-normal distributions as well, but here we are concerned with finding an exact test which uses only the order relations $x<y$ and $x>y$. It will be seen that these rank-order tests are even more powerful against certain alternatives than Student's test. That is, they correctly detect a difference between two distribution functions in cases in which Student's test does not detect the difference by correctly rejecting the hypothesis H_0.

We assume that the distribution function $F(t)$ common to the x_i and y_k under the hypothesis H_0 is continuous. Hence it follows that events such as $x_i=x_j$ or $x_i=y_k$ have probability zero. Strictly speaking, however, this continuity assumption is never satisfied in practice, since any result of measurement is rounded to a certain number of decimals. Thus it occurs quite frequently in applications that, for example, an x_i is equal to a y_k. In the use of rank-order tests, such "ties" lead to minor difficulties. How these are to be overcome we shall investigate later.

The transformation

$$t' = F(t)$$

takes the variables x_i and y_k into new variables x'_i and y'_k whose distribution function is "rectangular":

$$F'(t') = t', \quad 0 \leqq t' \leqq 1.$$

Since the rank-ordering of the x' and y' is exactly the same as that of the x and y, for rank-order tests it does not matter whether we work with the x and y or with the x' and y'. Thus, whenever it simplifies the calculation of the probabilities, we can always assume that the distribution of the x and y is rectangular. Also, should we wish to do so, we can make the assumption that we have at hand any other type of continuous distribution, such as a normal distribution with zero mean and unit standard deviation.

Under the null hypothesis H_0 all the permutations of the $n = g + h$ observations $x_1, \ldots, x_g; y_1, \ldots, y_h$ are equally probable. There exist $n!$ such permutations, each of which has, therefore, probability $1/n!$.

To prescribe a test of the hypothesis H_0 is to give a *critical region* C consisting of part of the $n!$ permutations. If the observed permutation belongs to the region C, the hypothesis H_0 is to be rejected. If the error probability of the test is to be $\leqq \alpha$, the region C may include at most $\alpha n!$ permutations.

B. Smirnov's Test

The test of Smirnov is completely analogous to that of Kolmogorov (§16). Kolmogorov's test compares an empirical distribution function with one based on theoretical considerations. Smirnov's test compares two empirical distribution functions with one another.

Let $F_g(t)$ be the empirical distribution function of $x_1, \ldots, x_g$: If $k(t)$ is the number of the $x_i < t$, then

$$F_g(t) = \frac{k(t)}{g}.$$

Likewise let $G_h(t)$ be the empirical distribution function of $y_1, \ldots, y_h$. Let the supremum of the differences $|F_g - G_h|$ be D. Then Smirnov's test rejects the hypothesis H_0 whenever $D > D_\alpha$, where D_α is determined so that the probability of the event $D > D_\alpha$ is at most α.

Smirnov[4] proved that the probability of the event

$$D > \lambda (g^{-1} + h^{-1})^{\frac{1}{2}}$$

is, for large n, asymptotically equal to the sum of the infinite series

$$2e^{-2\lambda^2} - 2e^{-2^2 \cdot 2\lambda^2} + 2e^{-3^2 \cdot 2\lambda^2} - \cdots. \quad (1)$$

Therefore if we determine λ_α so that the sum of this series is α, then for large n we can set

$$D_\alpha = \lambda_\alpha (g^{-1} + h^{-1})^{\frac{1}{2}}. \quad (2)$$

Since the series (1) converges very quickly, for practical purposes it can be replaced by its first term. Furthermore, in doing so we remain on the safe side. Hence we obtain the quite useable approximation

$$\lambda_\alpha \sim \sqrt{-\tfrac{1}{2} \ln \alpha/2}, \quad (3)$$

which we need only to insert into (2) to obtain quite a good approximation to D_α.

The beauty of Smirnov's test is the fact that it detects any discrepancy between the distribution functions of the x and y with arbitrarily high probability, provided n is sufficiently large. We may thus apply the test whenever we have available a very large collection of data with which to test for complete agreement between the distribution functions $F(t)$ and $G(t)$ of the x and y throughout their domains.

If, on the other hand, it is a matter of establishing whether or not the x are on the average larger than the y, we shall apply more powerful tests which can detect a real difference with a smaller value of n. Two such tests are Wilcoxon's test and the X-test which are to be discussed now.

§ 63. Wilcoxon's Test

A. The Test Procedure

Suppose that the observations x_i and y_k are ordered according to increasing size. If the indices are dropped, the result is a sequence of letters x and y such as

$$y\,y\,x\,y\,x\,y\,y\,x\,x. \quad (1)$$

In this sequence the *rank number* of the first x is $r_1 = 3$. The rank number of the second x is $r_2 = 5$, and so on. The sum of the rank numbers,

$$\sum r_i = W,$$

[4] N. Smirnov, Estimation of the discrepancy between empirical distributions for two samples. Bull. Math. Univ. Moscow 2 (1939) p. 1.

is in our case

$$3+5+8+9=25.$$

Wilcoxon's test rejects the null hypothesis whenever the sum W *exceeds a bound* W_α. The bound W_α is determined in such a way that the probability under the null hypothesis for W to exceed W_α is at most α.

Mann and Whitney proposed an equivalent test based upon the *number of inversions* in the sequence. Whenever an x follows a y in such a sequence as this we have an inversion. In our example the first x forms $s_1=2$ inversions with the two preceding y, the second x forms $s_2=3$ inversions with the three preceding y, and so on. The total number of inversions,

$$\sum s_i = U,$$

is in our case

$$2+3+5+5=15.$$

In general,

$$s_1=r_1-1,$$

$$s_2=r_2-2,$$

$$s_3=r_3-3, \quad \text{etc.},$$

for if the smallest x has rank r_1, then it forms exactly r_1-1 inversions with the r_1-1 preceding y, the next larger x, which has rank r_2, forms r_2-2 inversions with the r_2-2 preceding y, and so on. Hence

$$\sum s_i = \sum r_i - (1+2+\cdots+g),$$

or

$$U = W - \tfrac{1}{2} g(g+1). \tag{2}$$

Therefore the Wilcoxon test can be expressed in the following equivalent Mann-Whitney form: *The null hypothesis is to be rejected whenever the number of inversions* U *exceeds a bound* U_α. The bound U_α is to be chosen so that under the null hypothesis the number of permutations with $U>U_\alpha$ is at most $\alpha n!$. For the moment we formulate the test as a one-sided test.

For small numbers g and h we can determine the cut-off point U_α by explicit enumeration of the sequences with the largest numbers of inversions. To simplify the enumeration we drop all indices from the x and y, as in (1), so that the number of possible sequences is then not $n!$ but rather only

$$\binom{n}{g} = \frac{n!}{g!\,h!}.$$

We begin enumeration with the sequence

$$y\,y \ldots y\,y\,x\,x \ldots x\,x, \tag{3}$$

which has $g\,h$ inversions, then we write down the sequence

$$y\,y \ldots y\,x\,y\,x \ldots x\,x, \tag{4}$$

with $g\,h-1$ inversions, and so on, until we have more than $\alpha\binom{n}{g}$ sequences. The number of inversions in the last sequence we write down is then U_α.

The following example illustrates the procedure:

$$g=h=5, \qquad \alpha=2\tfrac{1}{2}\%$$

$$\alpha\binom{n}{g}=\frac{1}{40}\cdot 252=6.3.$$

1. $y\,y\,y\,y\,y\,x\,x\,x\,x\,x$
2. $y\,y\,y\,y\,x\,y\,x\,x\,x\,x$
3. $y\,y\,y\,x\,y\,y\,x\,x\,x\,x$
4. $y\,y\,y\,y\,x\,x\,y\,x\,x\,x$
5. $y\,y\,x\,y\,y\,y\,x\,x\,x\,x$
6. $y\,y\,y\,x\,y\,x\,y\,x\,x\,x$
7. $y\,y\,y\,y\,x\,x\,x\,y\,x\,x$

Since the last sequence in the enumeration has 22 inversions, $U_\alpha=22$.

In this example there are only four sequences (1 through 4) with more than 22 inversions. The error probability of the corresponding test is $\frac{4}{252}=1.6\%$ and lies, therefore, considerably below the allowable level. This same problem arises in other examples. Usually there are several rank-orderings with the same number of inversions. In our example, sequences 5, 6, and 7 each have 22 inversions. If we were to include all three in the rejection region, then the error probability α would be exceeded. If, however, we include none of the three in the rejection region, then the power of the test is unnecessarily low.

For large g and h the calculation of the exact cut-off point U_α becomes very cumbersome. We shall see, however, that for large g and h the distribution of U can be approximated by a normal distribution.

In the two-sided application of the Wilcoxon test, the null hypothesis is rejected not only whenever the number of inversions exceeds the bound U_α but also whenever the number $g\,h-U$ of non-inversions exceeds this same bound. The error probability is then doubled.

We shall now examine the distribution of U somewhat more closely. In doing so we shall again assume that the null hypothesis, which states that the x and y all have the same (continuous) distribution and are independent, is true.

B. Expectation and Variance of U

For every pair of observations x_i, y_k we define a function z_{ik} which takes on only the values 0 and 1 and satisfies

$$z_{ik}=\begin{cases}1 & \text{if} \quad x_i>y_k,\\ 0 & \text{otherwise.}\end{cases}$$

Then obviously

$$U=\sum z_{ik}. \tag{5}$$

Under the null hypothesis, for every z_{ik} the values 0 and 1 are equally probable, so that the expectation of z_{ik} is $\frac{1}{2}$. From (5) we thus get immediately the expectation of U:

$$\hat{U}=\tfrac{1}{2}gh. \tag{6}$$

In place of (5) we can now write

$$U-\hat{U}=\sum(z_{ik}-\tfrac{1}{2}). \tag{7}$$

In order to find the variance σ^2 of U, we take the expectation of the square of (7):

$$\sigma^2=\mathscr{E}(U-\hat{U})^2=\sum\mathscr{E}(z_{ik}-\tfrac{1}{2})(z_{jh}-\tfrac{1}{2}). \tag{8}$$

The terms with $i \neq j$ and $k \neq h$ are zero, because z_{ik} and z_{jh} are independent and have expectation $\frac{1}{2}$. The terms with $i=j$ and $k=h$ are all equal to $\frac{1}{4}$. The products $(z_{ik}-\frac{1}{2})(z_{jh}-\frac{1}{2})$ with $i=j$ and $k \neq h$ take on the value $-\frac{1}{4}$ if x_i lies between y_k and y_h and $+\frac{1}{4}$ otherwise. The expectation of such a product is therefore

$$-\tfrac{1}{4}\cdot\tfrac{1}{3}+\tfrac{1}{4}\cdot\tfrac{2}{3}=\tfrac{1}{12}.$$

The same is true for $k=h$ and $i \neq j$. Hence from (8) we get

$$\sigma^2=\tfrac{1}{4}gh+\tfrac{1}{12}gh(h-1)+\tfrac{1}{12}hg(g-1)=\tfrac{1}{12}gh(g+h+1). \tag{9}$$

C. Asymptotic Distribution of U for $g\to\infty$ and $h\to\infty$

Mann and Whitney[5] have determined not only the expectation and variance but also the higher central moments of U asymptotically for large g and h. The moments of odd order are zero, since the distribution

[5] H. B. Mann and D. R. Whitney, On a test whether one of two random variables is stochastically larger than the other. Ann. of Math. Stat. 18 (1947) p. 50.

of U is symmetric with respect to the expectation $\frac{1}{2}gh$. The moments of even order are, if $u=U-\frac{1}{2}gh$,

$$\mathscr{E}u^{2r}=1\cdot 3\cdot 5\cdot \cdots \cdot(2r-1)\cdot(gh)^r(g+h+1)^r\,12^{-r}+\cdots, \tag{10}$$

where the terms $+\cdots$ are of smaller order of magnitude than the initial term. If we divide (10) by

$$\sigma^{2r}=(\mathscr{E}u^2)^r=(gh)^r(g+h+1)^r\,12^{-r}$$

and let g and h go to infinity, we have

$$\lim \mathscr{E}(\sigma^{-1}u)^{2r}=1\cdot 3\cdot 5\cdot \cdots \cdot(2r-1). \tag{11}$$

From this it follows, according to the "second limit theorem" (§ 24F), that $\sigma^{-1}u$ is, for $g\to\infty$ and $h\to\infty$, asymptotically normally distributed with zero expectation and unit standard deviation, or:

If g and h both tend to ∞, then U is asymptotically normally distributed with expectation $\frac{1}{2}gh$ and standard deviation σ.

The *method of moments* used here to prove asymptotic normality can also be applied in many other cases for proofs of asymptotic normality of test statistics – e.g., for the proof of that of the variable U even if the x and y have different distributions[6].

D. Asymptotic Distribution of U as $h\to\infty$

If h alone becomes infinite while g remains fixed, then we must use another method to determine the asymptotic distribution of U. The line of thought becomes clear most easily if, to begin with, we take $g=2$.

Let x_1, x_2 and $y_1, \ldots, y_h$ be independent random variables. Let their common distribution function be $F(t)=t$; i.e., let all variables x_i and y_k be uniformly distributed between zero and one. Let the number of inversions which x_1 and x_2 form with the preceding y_k be u_1 and u_2 respectively. The total number of inversions is then $U=u_1+u_2$.

We shall first of all keep x_1 and x_2 fixed. For fixed x_1, the probability that a y is smaller than x_1 is $F(x_1)=x_1$. The frequency of this event is

$$v_1=\frac{u_1}{h}, \tag{12}$$

since of the h quantities $y_1, \ldots, y_h$, u_1 are smaller than x_1. For large h it is highly probable that the frequency lies close to the probability, and

[6] See E. L. Lehmann, Consistency and unbiasedness of nonparametric tests. Ann. of Math. Stat. 22 (1951) p. 167, Theorem 3.2, and the literature cited there, as well as W. Hoeffding, A combinatorial central limit theorem. Ann. of Math. Stat. 22 (1951) p. 558.

therefore that v_1 lies close to x_1. Likewise, v_2 lies close to x_2. Therefore

$$v_1+v_2=\frac{u_1+u_2}{h}=\frac{U}{h} \tag{13}$$

is close to x_1+x_2 in probability.

The problem is to calculate the distribution function of U, that is, to calculate the probability of the event $U<u$. Equivalently, we can consider

$$v_1+v_2=\frac{U}{h}<\frac{u}{h}=t \tag{14}$$

instead of $U<u$, so that our problem becomes that of calculating the probability of the event $v_1+v_2<t$.

Since v_1+v_2 is in probability close to x_1+x_2, we shall begin by calculating the probability of the event $x_1+x_2<t$. The probability density of the pair (x_1, x_2) is 1, since x_1 and x_2 are uniformly distributed between zero and one. The desired probability is therefore equal to the area of the region G_t:

$$0<x_1<1, \quad 0<x_2<1, \quad x_1+x_2<t.$$

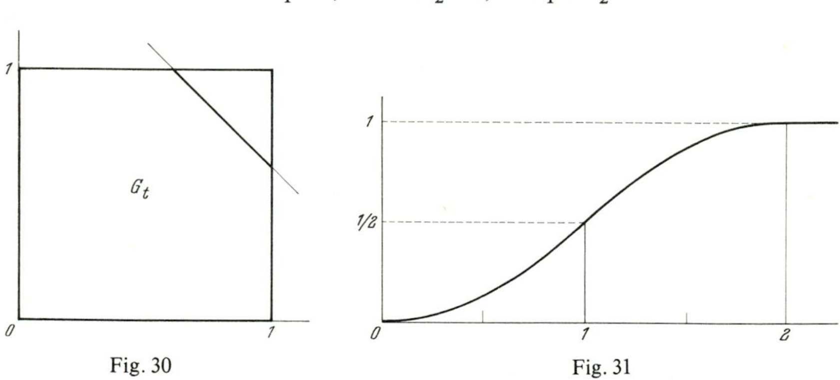

Fig. 30 Fig. 31

As is shown in Fig. 30, the region G_t is the portion of the unit square below the line $x_1+x_2=t$. Its area is

$$H(t)=\begin{cases}\frac{1}{2}t^2 & (t\leqq 1)\\ 1-\frac{1}{2}(2-t)^2 & (t\geqq 1).\end{cases} \tag{15}$$

As is to be seen in Fig. 31, the graph of the function is composed of two parabolic arcs. For the corresponding probability density see § 25, Fig. 16.

For $g=3$ the results are analogous: The probability is equal to the volume of that portion of the unit cube cut off by the plane $x_1+x_2+x_3=t$ (Fig. 32).

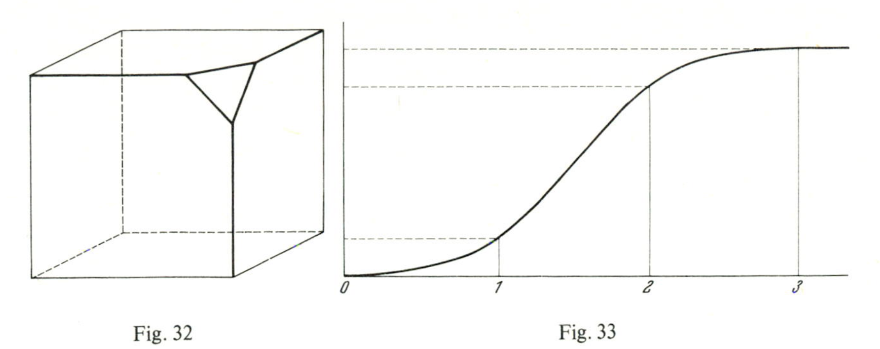

Fig. 32

Fig. 33

Calculation yields

$$H(t)=\begin{cases}\frac{1}{6}t^3 & (t\leqq 1)\\ \frac{1}{6}t^3-\frac{1}{2}(t-1)^3 & (1\leqq t\leqq 2)\\ 1-\frac{1}{6}(3-t)^3 & (2\leqq t\leqq 3).\end{cases} \tag{16}$$

The curve in Fig. 33 represents the function. For the corresponding probability density function see § 25, Fig. 17.

Even for $g=2$ and $g=3$ the curves resemble a normal distribution function. For $g=4$ the curve almost coincides with a normal curve; for larger g the agreement is even better.

To make the transition from x_1+x_2 to v_1+v_2 we need the following lemma:

For every g the function H(t) has the following property:

$$H(t+\varepsilon)-H(t)\leqq\varepsilon. \tag{17}$$

Proof. The left side of (17) is the g-fold integral

$$I=\int\cdots\int dx_1\ldots dx_g \tag{18}$$

integrated over the region

$$t\leqq x_1+x_2+\cdots+x_g<t+\varepsilon, \tag{19}$$

$$0<x_i<1 \qquad (i=1,2,\ldots,g). \tag{20}$$

If we carry out the integration first with respect to x_g for constant $x_1,\ldots,x_{g-1}$, then we must integrate over an interval of length at most ε, since inequality (19) defines an interval of length ε and the effect of the inequalities (20) can only be to shorten it. If we now integrate with respect to $x_1, \ldots, x_{g-1}$ over the unit cube in $(g-1)$-dimensional space, we still get at most ε. Hence (17) is proved.

Since the distribution function $H(t)$ of x_1+x_2 is known, we can now use this lemma to place upper and lower bounds on the distribution function of v_1+v_2. For given ε we wish to show that, for sufficiently large h, the probability of the event $v_1+v_2<t$ differs from $H(t)$ by an amount not exceeding 2ε.

We already know that v_1+v_2 differs in probability from x_1+x_2 only by an arbitrarily small amount. From this it follows that the probability that x_1+x_2 differs from v_1+v_2 by more than ε is smaller than ε for sufficiently large h. Now if $v_1+v_2<t$, then either $x_1+x_2<t+\varepsilon$ or x_1+x_2 differs from v_1+v_2 by more than ε. The event $v_1+v_2<t$ is therefore contained in the union of the events $x_1+x_2<t+\varepsilon$ and $(x_1+x_2)-(v_1+v_2)>\varepsilon$. From this it follows that

$$\mathscr{P}(v_1+v_2<t) \leqq \mathscr{P}(x_1+x_2<t+\varepsilon)+\varepsilon = H(t+\varepsilon)+\varepsilon \leqq H(t)+2\varepsilon.$$

Likewise, we find

$$\mathscr{P}(v_1+v_2<t) \geqq \mathscr{P}(x_1+x_2<t-\varepsilon)-\varepsilon = H(t-\varepsilon)-\varepsilon \geqq H(t)-2\varepsilon.$$

Therefore $\mathscr{P}(v_1+v_2<t)$ differs from $H(t)$ by at most 2ε, as was to be proved.

We can make this same transition for arbitrary g, as for $g=2$, to get the following result:

For fixed g and for $h\to\infty$, the asymptotic distribution function of the variable U is that of a sum of g independent variables each uniformly distributed between 0 *and h.*

The expectation of this sum is $\frac{1}{2}gh$, which thus agrees with the exact expectation of U. Its standard deviation is

$$\sqrt{\tfrac{1}{12}gh^2},$$

whereas the exact standard deviation of U is

$$\sigma_U = \sqrt{\tfrac{1}{12}gh(g+h+1)}.$$

For $g \geqq 4$ the distribution of the sum can be approximated quite exactly by a normal distribution. Therefore, the normal approximation found by Mann and Whitney for large g and h is useable not only for large g *and* h but also for moderate g and large h, so long, only, as $g>3$.

Numerical examples show that h need not be very large. Fig. 34 shows a portion of the exact distribution of U (step function) for $g=h=10$ and the corresponding portion of the normal distribution. The agreement is very satisfactory in the region between 95% and 99%, which is most important for practical applications. Beyond 99% the step function even lies above the normal distribution, which is to say that in using the normal distribution we remain on the safe side.

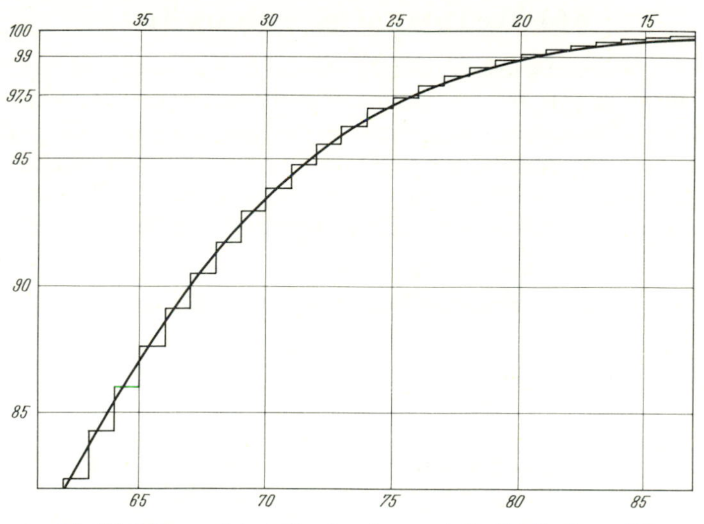

Fig. 34. Wilcoxon's test: Exact and asymptotic distributions of U

The result of this investigation can be formulated as follows: *For* $g>3$ *and* $g+h\geqq 20$ *the normal distribution is sufficiently exact for the Wilcoxon test statistic. For smaller* g *and* h *the exact distribution must be used.*

E. Table for Small g and h

Up to now our convention has been to reject the null hypothesis whenever U, the number of inversions, is too large – i.e., larger than the bound U_α. We can also reject the null hypothesis whenever U is too small, say smaller than a bound u. The two procedures are equivalent if the roles of the x and the y are interchanged. Table 10 at the back of the book was calculated for this latter procedure and gives the *significance probabilities* $p(u)$ for the Wilcoxon test for all g and h satisfying

$$g\leqq h\leqq 10.$$

The significance probability $p(u)$ is the probability of the event $U\leqq u$ under the null hypothesis. If the experimental results show the number of inversions to be u and if $p(u)\leqq\alpha$, then the null hypothesis is rejected at the level α. In a two-sided application the above procedure is repeated with u replaced by $gh-u$. In the one-sided application the error probability is $\leqq\alpha$, in the two-sided application it is $\leqq 2\alpha$.

The table was computed from a table by H. R. Van der Vaart, published by the Mathematical Center Amsterdam (Report S 32).

§ 64. The Power of the Wilcoxon Test

By the power of a test of the hypothesis H_0 against the alternative H' we mean, as in § 60, the probability that the test rejects H_0 when H' is true. The null hypothesis in our case is the hypothesis that the x_i and y_k all have the same distribution function. We now take as our alternative the hypothesis H' that the x_i have a distribution function $F(t)$ and the y_k have a specific distribution function $G(t)$ which is different from $F(t)$.

A. Expectation and Standard Deviation of U under the Hypothesis H'

We again set

$$U=\sum z_{ik}, \tag{1}$$

with $z_{ik}=1$ if $x_i>y_k$ and 0 otherwise, and determine the expectation and variance of U by the same method as in § 63 B. The result is [7]

$$\mathscr{E}U=g\,h\,p, \tag{2}$$

$$\sigma_U^2=g\,h[(g-1)\,r^2+(h-1)\,s^2+p\,q], \tag{3}$$

where

$$p=\int G(t)\,dF(t), \tag{4}$$

$$q=1-p=\int F(t)\,dG(t), \tag{5}$$

$$r^2=\int [F(t)-q]^2\,dG(t), \tag{6}$$

$$s^2=\int [G(t)-p]^2\,dF(t), \tag{7}$$

and all integrals are to be taken from $-\infty$ to $+\infty$.

Let us assume, say, that the x_i are normally distributed with expectation $\mu>0$ and standard deviation one and that the y_k are likewise normal random variables, but with expectation zero and standard deviation one.

---

[7] See D. Van Dantzig, Consistency and power of Wilcoxon's test. Proc. Kon. Ned. Akad. Amsterdam (Section on Sciences) A 54, p. 1.

Then

$$F(t)=\Phi(t-\mu),$$
$$G(t)=\Phi(t),$$
$$p=\int_{-\infty}^{\infty}\Phi(t)\,d\Phi(t-\mu)=\int_{-\infty}^{\infty}\Phi(x+\mu)\,d\Phi(x)$$
$$=\frac{1}{2\pi}\int_{-\infty}^{\infty}e^{-\frac{1}{2}x^2}\,dx\int_{-\infty}^{x+\mu}e^{-\frac{1}{2}y^2}\,dy$$
$$=\frac{1}{2\pi}\iint e^{-\frac{1}{2}(x^2+y^2)}\,dx\,dy$$

integrated over $y<x+\mu$. To calculate the integral we use an orthogonal transformation to introduce new variables t and u:

$$x+y=t\sqrt{2},$$
$$-x+y=u\sqrt{2},$$

in which case

$$p=\frac{1}{2\pi}\iint e^{-\frac{1}{2}(t^2+u^2)}\,dt\,du$$

integrated over $u\sqrt{2}<\mu$. If we integrate first with respect to t and then with respect to u, we get

$$p=\Phi\left(\frac{\mu}{\sqrt{2}}\right). \tag{8}$$

We now have

$$r^2=\int[\Phi(t-\mu)-q]^2\,d\Phi(t), \tag{9}$$
$$s^2=\int[\Phi(t+\mu)-p]^2\,d\Phi(t). \tag{10}$$

Since the substitution $t=-t'$ takes (9) into (10), we have $r^2=s^2$. Furthermore, the substitution $\mu=-\mu$ takes p into q and s^2 into r^2, and hence into itself, so that s^2 is an even function of μ.

We could transform the integral s^2 further, but there is little point in doing so, as we still could not achieve an explicit expression in terms of known functions. We shall therefore be satisfied with the remark that p, r^2, and s^2 are bounded functions of μ (r^2 and s^2 even tend to zero as μ tends to $+\infty$ or $-\infty$) and that for small μ they can be expanded into power series with respect to μ:

$$p=\frac{1}{2}+\frac{\mu}{2\sqrt{\pi}}+\cdots, \tag{11}$$
$$r^2=s^2=\tfrac{1}{12}+\cdots, \tag{12}$$

where only the constant term and first power of μ have been written out.

Our aim is to calculate the power of the one-sided Wilcoxon test, that is, the probability of the event $U > U_\alpha$, as a function of μ. Let us name this function $P(\mu)$. In particular, we wish to compare the power of the Wilcoxon test with that of Student's test. From § 60 we know that under the assumption that the x and y are normally distributed with equal standard deviations, Student's test is most powerful among all one-sided level-α similar tests. Since we are concerned with one-sided tests, we shall assume in our discussion that $\mu > 0$.

We can assume that, say, $g \leqq h$. We now consider two cases:

First case: g and h large and of the same order of magnitude.

Second case: g large and h large relative to g.

In both cases the only values of μ which are relevant to our discussion are those which are so small as to be of the order of magnitude $g^{-\frac{1}{2}}$. If in fact μ is large relative to $g^{-\frac{1}{2}}$, then the desired power function of Wilcoxon's test, as well as that of Student's test, is close to 1, whereby the whole comparison becomes uninteresting. This we can prove as follows:

If we set $U - \frac{1}{2} g h = V$ and $U_\alpha - \frac{1}{2} g h = V_\alpha$, then the expectation of V is $(p - \frac{1}{2}) g h$, and therefore it is of the order of magnitude $\mu g h$. If μ is large relative to $g^{-\frac{1}{2}}$, then the expectation of V is large relative to $h\sqrt{g}$. The standard deviation of V is, according to (3), of the order of magnitude $h\sqrt{g}$, and the bound V_α has likewise the order of magnitude $h\sqrt{g}$. The probability that V exceeds the bound V_α therefore is close to 1. Student's test is at least as powerful as Wilcoxon's test, so that its power is likewise close to 1.

We shall assume from now on that μ is at most of the order of magnitude $g^{-\frac{1}{2}}$. Then in (11) and (12) we can neglect the terms of order of magnitude $\mu^2 = g^{-1}$ and limit ourselves to the terms already written out. If we substitute these into (2) and (3), we get

$$\mathscr{E} U \sim \frac{1}{2} g h \left(1 + \frac{\mu}{\sqrt{\pi}}\right), \tag{13}$$

$$\sigma_U^2 \sim \tfrac{1}{12} g h (g + h). \tag{14}$$

B. First Case: *g* and *h* of the Same Order of Magnitude

To make the discussion precise, let us assume that g and h tend to infinity in such a way that the quotient g/h remains constant. Lehmann[8] proved that in this case U is asymptotically normally distributed with expectation $g h p$ and standard deviation σ_U. The probability $P(\mu)$ that

[8] See second footnote in § 63 C.

$U > U_\alpha$ is therefore asymptotically equal to the probability that a normally distributed variable with expectation $g\,h\,p$ and standard deviation σ_U exceeds the bound U_α:

$$P(\mu) \sim \Phi\left(\frac{g\,h\,p - U_\alpha}{\sigma_U}\right). \tag{15}$$

If we substitute (11) and (14) on the right-hand side, we get

$$P(\mu) \sim \Phi(b\,\mu - c), \tag{16}$$

where

$$b = \left(\frac{3}{\pi}\,\frac{g\,h}{g+h}\right)^{\frac{1}{2}}, \tag{17}$$

while the constant c depends on U_α. Now U_α was determined so that when $\mu = 0$ the probability that $U > U_\alpha$ is at most equal to α and exactly equal to α asymptotically. Therefore when $\mu = 0$ (16) must take on the value α:

$$\Phi(-c) = \alpha. \tag{18}$$

Hence (16), (17), and (18) define the asymptotic power function of the test.

C. Comparison with Student's Test

Student's test for the comparison of two means (§ 29) uses the test statistic

$$t = \frac{D}{S} = \frac{\bar{x} - \bar{y}}{S} \tag{19}$$

where

$$\bar{x} = \frac{1}{g}(x_1 + \cdots + x_g), \tag{20}$$

$$\bar{y} = \frac{1}{h}(y_1 + \cdots + y_h), \tag{21}$$

$$S^2 = \left(\frac{1}{g} + \frac{1}{h}\right)\left(\frac{\sum (x-\bar{x})^2 + \sum (y-\bar{y})^2}{g+h-2}\right). \tag{22}$$

In (19) the numerator D has expectation μ, while S^2 has expectation

$$\sigma_D^2 = \left(\frac{1}{g} + \frac{1}{h}\right)\sigma^2, \tag{23}$$

where, as before, we can assume that $\sigma = 1$. The numerator D is normally distributed and its standard deviation is of the same order of magnitude as the expectation μ. The standard deviation of the denominator S is of smaller order of magnitude than its expectation. From here it follows easily that asymptotically we get the same distribution function if we replace the denominator simply by σ_D. That is to say, the distribution function of t is asymptotically the same as that of

$$t' = \frac{\bar{x} - \bar{y}}{\sigma_D}.$$

This quantity is, however, asymptotically normally distributed with expectation μ/σ_D and unit standard deviation, and hence the probability of the event $t' > c$ is

$$\Phi\left(\frac{\mu}{\sigma_D} - c\right). \tag{24}$$

For $\mu = 0$, $\Phi(-c)$ must again be equal to α, so that the constant c is exactly the same as before. If we substitute the value for σ_D from (23) into (24), we find the power function of Student's test to be

$$P'(\mu) \sim \Phi(b'\mu - c), \tag{25}$$

where

$$b' = \left(\frac{gh}{g+h}\right)^{\frac{1}{2}}. \tag{26}$$

Comparison of (16) and (25) shows that the power functions $P(\mu)$ and $P'(\mu)$ differ asymptotically for large g and h only by a factor $\sqrt{3/\pi}$ in the coefficients of μ. This may be stated in another fashion: If we compare Wilcoxon's test using numbers g and h with Student's test using numbers g' and h' and set

$$g' \sim \frac{3}{\pi} g \quad \text{and} \quad h' \sim \frac{3}{\pi} h, \tag{27}$$

then the power functions $P(\mu)$ and $P'(\mu)$ are asymptotically equal. We shall also use the expression *the asymptotic efficiency of the Wilcoxon test is* $3/\pi$. By this we mean that with Student's test, which has the greatest power possible, we can use numbers g and h which are smaller in the proportion $3/\pi$ and get asymptotically the same power function as with the Wilcoxon test using the original numbers.

Since $3/\pi$ is approximately equal to $\frac{21}{22}$, we can also say that the Wilcoxon test applied with 22 observations is approximately equally as powerful as Student's test with 21 observations. The loss in power in using the Wilcoxon test is hence very small.

The great advantage of the Wilcoxon test is, of course, that it can also be applied to non-normal distributions. In addition, it requires much less calculation than Student's test. These two advantages are gained for large g and h with a small loss in power.

D. Second Case: *h* Large Relative to *g*

Let h now be large relative to g. If we keep g fixed, then we can apply the methods of § 63 D. We shall again assume at first that $g=2$ and keep x_1 and x_2 fixed. The probability of observing a y which is smaller than x_1 is $G(x_1)$. The frequency with which a y is smaller than x_1 is, as in (12) § 63,

$$v_1 = \frac{u_1}{h}.$$

Since it is highly probable that the frequency lies close to the probability, v_1 lies near $G(x_1)$, and likewise v_2 lies near $G(x_2)$; hence

$$v_1 + v_2 = \frac{u_1 + u_2}{h} = \frac{U}{h}$$

lies near

$$G(x_1) + G(x_2).$$

The probability of the event $U < u$, or

$$\frac{U}{h} < \frac{u}{h} = t, \tag{28}$$

is thus asymptotically equal to the probability of the event

$$G(x_1) + G(x_2) < t. \tag{29}$$

The distribution function of U/h is therefore asymptotically that of a sum of two, or more generally of g, independent variables, each of which is distributed as $G(x_1)$, where x_1 has the distribution function $F(t)$. The distribution function of $G(x_1)$ is the probability of the event $G(x_1) < t$, or, if G^{-1} is the inverse function of G, of the event $x_1 < G^{-1}(t)$; i.e., it is equal to

$$K(t) = F\big(G^{-1}(t)\big). \tag{30}$$

If we assume in addition that the x_i and y_k are normally distributed with standard deviation 1 and expectation μ for the x and 0 for the y, then

$$F(t) = \Phi(t-\mu), \qquad G(t) = \Phi(t),$$

and if Ψ again denotes the inverse function of the normal distribution function Φ,

$$K(t) = \Phi[\Psi(t) - \mu].$$

Under the null hypothesis, $F = G$ (or, $\mu = 0$) and $K(t) = t$; i.e., $G(x_1)$ is uniformly distributed between zero and one. For small μ, or in general if F does not deviate strongly from G, $K(t)$ does not deviate strongly from a uniform distribution. In every case $G(x_1)$ lies between zero and one; its distribution is therefore that of a bounded random variable and all its moments are bounded.

Now we consider the distribution of the sum $G(x_1)+G(x_2)$, or in the general case, that of the sum

$$G(x_1)+\cdots+G(x_g).$$

According to the Central Limit Theorem (§ 24 D) the sum is, for large g, asymptotically normally distributed. Here g need not be so very large: even for moderately large g the approximation is very good. In the case of a uniform distribution the approximation is excellent even for g as small as 4. If $K(t)$ deviates somewhat from the uniform distribution, then the approximation is not much worse. Therefore the distribution of

$$v_1+\cdots+v_g=\frac{u_1+\cdots+u_g}{h}=\frac{U}{h}$$

does not deviate much from a normal distribution, so long as $g \geqq 4$.

In order to obtain a theoretically rigorous asymptotic result, we would have to let g tend to infinity. In doing so, it would not matter whether we let h/g remain bounded, or let it go to infinity, for in both cases we would get asymptotically a normal distribution for U. In practice it is enough to assume that both g and h are at least 4 and that $g+h$ is at least 20, provided μ is not too large.

E. Further Cases

For small g (say, $g=2$) and large h, we can apply the same methods, except that we can not now replace the distribution of $G(x_1)+G(x_2)$ by a normal distribution, but rather we must actually calculate it as in § 4 B Theorem III. For $g=2$ and $h\to\infty$ the following values have been found[9] for the power functions of Wilcoxon's test and of Student's test in the case $\alpha=0.05$ and under the assumption that the x and the y are normally distributed with expectations $\mu=1.5$ and 0 respectively:

$$P(\mu)=0.64, \qquad P'(\mu)=0.68. \tag{31}$$

For large g and h we can calculate $P(\mu)$ and $P'(\mu)$ according to formulas (16) and (25), whereby we determine b and b' according to (17) and (26) and c according to (18). If we again set $\alpha=0.05$ and choose $b\,\mu=2.03$, we get $b'\,\mu=2.08$ and

$$P(\mu)=0.64, \qquad P'(\mu)=0.67, \tag{32}$$

which is almost the same result as (31).

[9] B. L. van der Waerden, Proc. Kon. Ned. Akad. Amsterdam, A 55 (1952) p. 456.

A further case which may be easily evaluated numerically is $g=h=3$ and $\alpha=0.05$[10]. The Wilcoxon test in this case simply rejects the hypothesis H_0 whenever x_1, x_2, x_3 and y_1, y_2, y_3 occur in the sequence

$$y\ y\ y\ x\ x\ x.$$

The error probability is exactly $\alpha=\frac{1}{20}$. For $\mu=2$ we find the power functions of Wilcoxon's and Student's tests to be

$$P(\mu)=0.62, \qquad P'(\mu)=0.65, \tag{33}$$

and therefore the difference to be small, similar to the previous case.

For small g and h the Wilcoxon test has, however, a drawback which reduces its power function considerably in some circumstances. This is the fact that there are sometimes many permutations with the same number of inversions. In § 63 A we have already referred to this problem and given an example. Such examples abound.

Let us take, say, $g=4$ and $h=6$, and $\alpha=0.05$. The Wilcoxon test rejects the null hypothesis in the following cases with 21 or more inversions:

1. $y\ y\ y\ y\ y\ y\ x\ x\ x\ x$
2. $y\ y\ y\ y\ y\ x\ y\ x\ x\ x$
3. $y\ y\ y\ y\ x\ y\ y\ x\ x\ x$
4. $y\ y\ y\ y\ y\ x\ x\ y\ x\ x$
5. $y\ y\ y\ x\ y\ y\ y\ x\ x\ x$
6. $y\ y\ y\ y\ x\ y\ x\ y\ x\ x$
7. $y\ y\ y\ y\ y\ x\ x\ x\ y\ x$

At the 5%-level we could, in all, reject the hypothesis for 5% of 210, or 10, sequences. If, however, we were to include those sequences with 20 inversions, there would be 12 in all. which is too many. Therefore, the Wilcoxon test has no greater power at the 5%-level than at the $3\frac{1}{3}$%-level, whereas Student's test is naturally considerably more powerful at the 5%-level than at the $3\frac{1}{3}$%-level.

Likewise we find in the same case, $g=4$ and $h=6$, that the Wilcoxon test is no more powerful at the $2\frac{1}{2}$%-level than at the 2%-level, or in the two-sided application no more powerful at the 5%-level than at the 4%-level, etc.

We could increase the power of the test by drawing a card from a suitably constituted deck whenever the borderline sequences occur and rejecting the hypothesis H_0 whenever, say, a black card appears. It is preferable, however, to use a more powerful test, namely the X-test, which we shall now consider.

[10] See p. 452 of the above note.

§ 65. The *X*-Test

A. An Heuristic Derivation

Let us again restrict attention to the case $g=2$ and $h\to\infty$, and thus let $x_1, x_2, y_1, \ldots, y_h$ denote the observations. Let the number of y-observations smaller than x_1 and x_2 respectively again be u_1 and u_2. The frequencies v_1 and v_2 are therefore

$$v_1=\frac{u_1}{h}, \qquad v_2=\frac{u_2}{h}, \tag{1}$$

and the number of inversions is

$$U=u_1+u_2. \tag{2}$$

The Wilcoxon test rejects the null hypothesis whenever $u_1+u_2>U_\alpha$, or

$$v_1+v_2>b, \quad \text{where} \quad b=\frac{U_\alpha}{h}. \tag{3}$$

The probability of this event is asymptotically equal to that of the event

$$G(x_1)+G(x_2)>b \tag{4}$$

(cf. § 64D).

Let us now assume that the y are normally distributed with zero expectation and unit standard deviation and therefore set

$$G(t)=\Phi(t). \tag{5}$$

We can now write

$$\Phi(x_1)+\Phi(x_2)>b \tag{6}$$

in place of (4).

If x_1 and x_2 are likewise normally distributed with unit standard deviation but with non-negative expectation μ, then the most powerful test of the null hypothesis $\mu=0$ rejects this hypothesis whenever

$$x_1+x_2>c, \tag{7}$$

where the constant c is chosen so that the probability of the event (7) under the hypothesis $\mu=0$ is exactly α. This gives rise to the condition

$$\Phi\left(\frac{c}{\sqrt{2}}\right)=1-\alpha, \tag{8}$$

or

$$c=\sqrt{2}\cdot\Psi(1-\alpha). \tag{9}$$

Asymptotically, the test given by (7) and Student's test have the same power. This we can see immediately, once we calculate the power of the test given by (7) and compare it with the asymptotic power of Student's test, as evaluated in § 64C.

The difference between the asymptotic powers of Wilcoxon's test and Student's test thus arises from the fact that the left-hand side of (6) is $\Phi(x_1)+\Phi(x_2)$, whereas that of (7) is x_1+x_2. The latter test is somewhat better.

It is, however, very easy to modify test (3) so that the substitution $v_i=G(x_i)=\Phi(x_i)$, which led from (3) to (6), leads instead to (7). We need only replace v_i by $\Psi(v_i)$, where Ψ is the inverse function of the function Φ. Thus we have a modified test which rejects the null hypothesis whenever the sum

$$S=\Psi(v_1)+\Psi(v_2)=\Psi\left(\frac{u_1}{h}\right)+\Psi\left(\frac{u_2}{h}\right) \tag{10}$$

exceeds a suitably chosen constant c. If we substitute

$$v_i=\Phi(x_i) \tag{11}$$

into (10), the sum S becomes x_1+x_2, or we have the test given by (7).

For arbitrary g we should have to consider the sum

$$S=\Psi\left(\frac{u_1}{h}\right)+\Psi\left(\frac{u_2}{h}\right)+\cdots+\Psi\left(\frac{u_g}{h}\right) \tag{12}$$

instead of (10).

The test derived in this way still has, however, one drawback: The terms of the sum (12) can be $-\infty$ (for $u_i=0$) or $+\infty$ (for $u_i=h$), which makes the calculation of the sum impossible. In order to remedy this situation, we can arrange the x_i according to increasing size, and thus simultaneously order the u_i likewise, and then replace the u_i by

$$r_1=u_1+1,\ r_2=u_2+2,\ \ldots,\ r_g=u_g+g \tag{13}$$

and the denominator h by

$$n+1=g+h+1.$$

Hence we get the final expression

$$X=\Psi\left(\frac{r_1}{n+1}\right)+\Psi\left(\frac{r_2}{n+1}\right)+\cdots+\Psi\left(\frac{r_g}{n+1}\right). \tag{14}$$

These $r_1, \dots, r_g$ defined by (13) are simply the rank numbers of $x_1, \dots, x_g$ in the combined sample of x_i and y_k ordered according to increasing size. Since the rank numbers range from 1 to $n=g+h$, it is impossible for the terms of the sum (14) to become $\pm\infty$.

If in (12) we do not take into account the fact that some u_i may lie near 0 or h (which is highly improbable for very large h), then the sum (14) behaves asymptotically, as $h\to\infty$, exactly as the sum (12). Therefore the above heuristic derivation, which led first to the sum S and then to the test which rejects the null hypothesis whenever $S>c$, leads also to the more useable sum X and to the test to be discussed in the next section.

B. The X-Test

Let the $n=g+h$ quantities $x_1, \dots, x_g$ and $y_1, \dots, y_h$ be ordered according to increasing magnitude. Let the ranks of the x_i be denoted by r_i, or simply r, and those of the y_k by s_k, or s. Let us then consider the sums

$$X=\sum_r \Psi\left(\frac{r}{n+1}\right), \tag{15}$$

$$Y=\sum_s \Psi\left(\frac{s}{n+1}\right). \tag{16}$$

The function $\Psi(t)$ is finite for $0<t<1$ and has the property

$$\Psi(1-t)=-\Psi(t). \tag{17}$$

Therefore the sum $X+Y$ is always zero:

$$X+Y=\Psi\left(\frac{1}{n+1}\right)+\Psi\left(\frac{2}{n+1}\right)+\cdots+\Psi\left(\frac{n}{n+1}\right)=0. \tag{18}$$

Interchanging the roles of x and y takes X into $Y=-X$. Reversing the rank-ordering as well (i.e., arranging according to decreasing instead of increasing size) takes $-X$ back again to X.

Let the cut-off point X_α be determined so that the event

$$X>X_\alpha \tag{19}$$

has probability $\leqq\alpha$ under the assumption that all $n!$ rank-orderings of the x_i and y_k are equally probable. This assumption is a consequence of the *null hypothesis* H_0, which states that the x and y are independent and all have the same distribution function. For the time being, we shall assume that this distribution function is continuous, and, as a consequence, we need not be concerned with such events as $x_i=y_k$.

The *one-sided X-test* rejects the null hypothesis whenever the sum X exceeds the cut-off point X_α. It is applicable whenever the alternative to the null hypothesis states that the x are in general larger than the y. The error probability of this test is at most α.

The *two-sided X-test* rejects the null hypothesis whenever either X or Y exceeds the cut-off point X_α. If $X > X_\alpha$, then the conclusion is that the x are in general larger than the y. If $Y > X_\alpha$, then the conclusion is that, on the contrary, the y are in general larger than the x. The error probability of the two-sided test is at most 2α.

C. The Calculation of X_α

For small g and h the cut-off point X_α can be calculated exactly by enumerating the equally probable cases. As an example we shall choose the case $g=4$, $h=6$ and take

$$\alpha = \tfrac{1}{40} = 2.5\%, \quad \text{and hence} \quad 2\alpha = \tfrac{1}{20} = 5\%.$$

The number of rank-orderings $x\, y\, y \ldots x$ is 210, one-fortieth of which is five. Hence we must determine the five rank-orderings with the largest X-values.

We first make a table of Ψ-values, rounded-off to two decimal places:

$$\begin{aligned} \Psi(\tfrac{1}{11}) &= -1.34, & \Psi(\tfrac{6}{11}) &= 0.11, \\ \Psi(\tfrac{2}{11}) &= -0.91, & \Psi(\tfrac{7}{11}) &= 0.35, \\ \Psi(\tfrac{3}{11}) &= -0.60, & \Psi(\tfrac{8}{11}) &= 0.60, \\ \Psi(\tfrac{4}{11}) &= -0.35, & \Psi(\tfrac{9}{11}) &= 0.91, \\ \Psi(\tfrac{5}{11}) &= -0.11, & \Psi(\tfrac{10}{11}) &= 1.34. \end{aligned}$$

Then we use (15) to calculate the X-value corresponding to each rank-ordering $y\,y\,y\,y\,x\,y\,y\,x\,x\,x$. The six arrangements with the largest X-values are

1. $y\,y\,y\,y\,y\,y\,x\,x\,x\,x$ $\quad X=3.31$
2. $y\,y\,y\,y\,y\,x\,y\,x\,x\,x$ $\quad X=2.96$
3. $y\,y\,y\,y\,x\,y\,y\,x\,x\,x$ $\quad X=2.74$
4. $y\,y\,y\,y\,y\,x\,x\,y\,x\,x$ $\quad X=2.71$
5. $y\,y\,y\,x\,y\,y\,y\,x\,x\,x$ $\quad X=2.50$
6. $y\,y\,y\,y\,x\,y\,x\,y\,x\,x$ $\quad X=2.49$.

If we choose $X_\alpha=2.49$, then only five arrangements have a larger X-value. *In the practical application of the X-test it is always assumed that only two decimal places are retained and that the null hypothesis is* *rejected only in the event that the X calculated from the observations is strictly larger than X_α.*

In order to have a check, it is always advisable to calculate Y as well as X. The sum $X+Y$ (even with the rounded-off Ψ-values) must be exactly zero.

Such enumeration as this of all possible cases is feasible in practice only for $g+h\leqq 20$. For larger g and h there exist asymptotic approximations.

D. Expectation and Standard Deviation of X

Let us denote the Ψ-values by $a_1, \ldots, a_n$:

$$a_i = \Psi\left(\frac{i}{n+1}\right). \tag{20}$$

From (15) we see that the test statistic X is a sum

$$X = \sum a_r = a_{r_1} + a_{r_2} + \cdots + a_{r_g}, \tag{21}$$

where $r_1, \ldots, r_g$ are a combination of g out of the n possible indices $i = 1, 2, \ldots, n$. Under the null hypothesis all such combinations are equally probable.

For each individual summand a_r in (21), the values $a_1, \ldots, a_n$ are equally probable. Therefore the expectation of each individual term of (21) is zero and likewise also

$$\mathscr{E}X = 0. \tag{22}$$

To calculate the variance of X, we first determine the expectation of a_r^2. Since for a_r^2 the values $a_1^2, \ldots, a_n^2$ are equally probable, we have

$$\mathscr{E}a_r^2 = \frac{1}{n}(a_1^2 + \cdots + a_n^2) = Q. \tag{23}$$

We then determine the expectation of a product $a_{r_1} a_{r_2}$. This product takes on all values $a_i a_k$ for $i \neq k$ with equal probabilities. Therefore we have

$$\begin{aligned} \mathscr{E}(a_{r_1} a_{r_2}) &= \frac{1}{n(n-1)} \sum_{i \neq k} a_i a_k \\ &= \frac{1}{n(n-1)} \left[\left(\sum a_i\right)\left(\sum a_k\right) - \sum a_i^2\right] \\ &= -\frac{1}{n(n-1)} \sum a_i^2 = -\frac{Q}{n-1}. \end{aligned} \tag{24}$$

If we now square (21) and take its expectation, we get

$$\begin{aligned}\mathscr{E}X^2 &= g\,\mathscr{E}a_r^2 + g(g-1)\,\mathscr{E}(a_{r_1}a_{r_2})\\ &= g\,Q - \frac{g(g-1)}{n-1}\,Q\\ &= \frac{g(n-g)}{n-1}\,Q,\end{aligned}$$

or

$$\sigma_X^2 = \frac{g\,h}{n-1}\,Q, \tag{25}$$

where Q is defined as in (23) by

$$Q = \frac{1}{n}\sum_1^n \Psi^2\left(\frac{i}{n+1}\right).$$

The expectation and standard deviation of X are thus known from (22) and (25). The quantity Q is tabulated in Table 12[11].

E. The Asymptotic Distribution of X

Let $x_1, \dots, x_g$ and $y_1, \dots, y_h$ be independent random variables with the same distribution function, and for the moment let h be large relative to g. It is immaterial whether g is large as well or not. The following theorem is then true: *X is asymptotically normally distributed with expectation zero and standard deviation σ_X.* The proof, which is not at all difficult, is to be found in my paper on the X-test in Mathematische Annalen 126 (1953) p. 94.

According to the above-quoted theorem, for every ε there exists M such that for $h/g > M$ the distribution function of X differs from a normal distribution function by an amount less than ε. The same is true, of course, for $g/h > M$. Hence we have only to consider the case in which both g/h and h/g remain $\leqq M$, but $n = g + h$ nevertheless increases without bound. In this case the asymptotic normality is still valid, but it is much more difficult to prove. In his review of my Annalen paper quoted above, G. E. Noether (Mathematical Reviews 15 (1954) p. 46) remarks that the proof follows from a theorem of Wald and Wolfowitz, which in turn can be proved by the method of moments indicated in § 63 C. D. J. Stoker carried out the proof completely in his dissertation "Oor 'n klas van toetsingsgroothede vir die probleem van twee steekproewe" (Amsterdam, 1955).

[11] Table 12 agrees with Table 5 of van der Waerden and Nievergelt, Tafeln zum Vergleich zweier Stichproben. There the calculation of the table is also explained.

Consequently, as $n \to \infty$, X is asymptotically normally distributed regardless of whether g or h alone tends to ∞ or both become infinite simultaneously. In this respect X thus behaves differently from the number of inversions U.

It is on the basis of these theorems that tables for the X-test have been calculated[12]. For small n (i.e., for small g and h) the cut-off points X_α were determined exactly by enumeration of the possible cases. For large n the asymptotic normal distribution was used. It is possible to improve the approximation considerably by paying particular attention to the terms a_1 and a_n, which can occur in the sum (21).

F. Treatment of Equal *x* and *y*

Up to now we have assumed that the x and y have continuous distribution functions and consequently we have had no need to be concerned with the case $x_i = y_k$. In practice, however, the x_i and y_k are always subject to round-off error, which introduces discrete distributions so that the case $x_i = y_k$ can very well occur. The problem of determining the rank numbers r_i and s_k which we need for calculation of X and Y from (15) and (16) thus arises here as it also does with the Wilcoxon test.

Various procedures have been suggested. Whether x_i or y_k is to be treated as the larger can be determined by tossing a coin. The tied observations $x_i = y_k$ which ought to receive the ranks r and $r+1$ can also be assigned the average rank $r+\frac{1}{2}$.

The following procedure, however, appears to be the best, and we shall consider at once its most general case, namely, that $c = a + b$ equal observations $x_1, \ldots, x_a$ and $y_1, \ldots, y_b$ occupy the rank positions $r, r+1, \ldots, r+c-1$. The procedure, then, is to assign the c available ranks in all $c!$ possible ways to the observations $x_1, \ldots, x_a, y_1, \ldots, y_b$, calculate X corresponding to each permutation, and take the arithmetic mean of all these X-values.

For practical computation the procedure can be simplified, for we need not add $c!$ terms. We instead form the sum of c terms

$$S_c = \Psi\left(\frac{r}{n+1}\right) + \Psi\left(\frac{r+1}{n+1}\right) + \cdots + \Psi\left(\frac{r+c-1}{n+1}\right) \tag{26}$$

with the c available rank numbers $r, r+1, \ldots, r+c-1$ and take as contributions to the sums X and Y the proportional parts

$$\frac{a}{a+b} S_c \quad \text{and} \quad \frac{b}{a+b} S_c. \tag{27}$$

[12] B. L. van der Waerden and E. Nievergelt, Tafeln zum Vergleich zweier Stichproben mittels X-test und Zeichentest. Berlin, Göttingen, Heidelberg: Springer-Verlag 1956.

Should the problem of tied observations reoccur elsewhere in the sample with $a'+b'=c'$ equal x_i and y_k, we repeat the above procedure there as well. The sums of all such contributions calculated in this way yield X and Y respectively.

The distribution function of X is affected only slightly by these modifications. The standard deviation of X is now somewhat smaller, and likewise, presumably, also the error probability of the test. Therefore we remain on the safe side if we leave the cut-off point X_α unchanged.

G. Comparison with Student's Test

We now assume that the x_i and y_k are normally distributed with standard deviations one and expectations $\mu \geqq 0$ for the x and 0 for the y. Furthermore, we assume that g remains fixed whereas h tends to infinity. Under these assumptions we wish to evaluate the power of the X-test asymptotically and to compare it with that of Student's test.

The power function $P(\mu)$ of the X-test is the probability of the event

$$\Psi\left(\frac{r_1}{n+1}\right)+\Psi\left(\frac{r_2}{n+1}\right)+\cdots+\Psi\left(\frac{r_g}{n+1}\right)>X_\alpha. \tag{28}$$

Since all rank-orderings of the observations $x_1, \ldots, x_g$ are equally probable, we can assume that $x_1<x_2<\cdots<x_g$. Then (13) holds once again and instead of (28) we can write

$$\Psi\left(\frac{u_1+1}{n+1}\right)+\Psi\left(\frac{u_2+2}{n+1}\right)+\cdots+\Psi\left(\frac{u_g+g}{n+1}\right)>X_\alpha. \tag{29}$$

If from (1) we substitute in

$$u_i=h\,v_i$$

and $n=g+h$ here, we get

$$\Psi\left(\frac{h\,v_1+1}{h+g+1}\right)+\Psi\left(\frac{h\,v_2+2}{h+g+1}\right)+\cdots+\Psi\left(\frac{h\,v_g+g}{h+g+1}\right)>X_\alpha. \tag{30}$$

Each frequency v_i lies in a neighborhood of its corresponding probability. As in § 64 D we now replace these frequencies by the corresponding probabilities, in which case the expression

$$\Psi\left(\frac{h\,v_i+i}{h+g+1}\right)$$

becomes

$$\Psi\left(\frac{h\,\Phi(x_i)+i}{h+g+1}\right).$$

If we drop the terms in the numerator and denominator which are small relative to h, we get

$$\Psi[\Phi(x_i)] = x_i,$$

and (30) becomes

$$x_1 + x_2 + \cdots + x_g > X_\alpha. \tag{31}$$

Finally we replace X_α by its asymptotic expression for $h \to \infty$ to get

$$x_1 + x_2 + \cdots + x_g > \sqrt{g} \cdot \Psi(1-\alpha). \tag{32}$$

The result is the generalization to arbitrary g of the test (7). That we come back to this test is not surprising, for in § 65 A we proceeded directly from (7) to the X-test. We have simply traversed the same path in the reverse direction.

From the considerations indicated here it is clear that the X-test has asymptotically for $h \to \infty$ the same power function as the test (32). This test is, as we have already seen, the most powerful among all level α tests of the null hypothesis. The power of this test (32) is easily calculated to be

$$P'(\mu) = \Phi(b'\mu - c), \tag{33}$$

where

$$b' = \sqrt{g} \quad \text{and} \quad c = \Psi(1-\alpha). \tag{34}$$

Therefore (33) is also the asymptotic power function of the X-test. In § 64 C we have seen that (33) is also the asymptotic power function of Student's test. *Therefore, for fixed g and $h \to \infty$ the X-test is asymptotically equally as powerful as Student's test.*

Here the intention has been only to indicate the line of reasoning of the proof. An exact derivation is to be found in my Annalen paper already quoted (Mathematische Annalen 126 (1953) p. 103 § 5).

I suspect that the same result is also true if g and h both tend to infinity.

H. Non-Normal Distributions

The great advantage of rank tests is that they are completely independent of the hypothesis of normality, an assumption which must always be made with Student's test. With rank tests the error probability is always at most α, independent of the choice of continuous distribution, whereas with Student's test the error probability can very well exceed α if the distribution of either the x or the y is non-normal.

Now it is quite true that under suitable regularity conditions this increase in the error probability with Student's test is not very significant for large g and h. That is, the average $\bar{x}$ of the independent variables

$x_1, \dots, x_g$ is approximately normally distributed for sufficiently large g and likewise for the average $\bar{y}$ for sufficiently large h. Consequently the same is also true for the difference $D = \bar{x} - \bar{y}$. The denominator S in Student's test statistic can be replaced by the true standard deviation σ_D of D as an approximation when $n = g + h$ is large. The quotient D/S has therefore an approximately normal distribution with a standard deviation which approaches unity as $g + h \to \infty$. Thus, so long as the x and y have the same, but not too irregular distribution function, and g and h are large, the error probability of the test is approximately equal to α, as under normality.

For non-normal distributions Student's test has, however, another drawback in the face of rank tests, namely its smaller power. In § 6 of my Annalen paper already mentioned (Mathematische Annalen 126 (1953) p. 106) I have constructed an example in which a one-to-one transformation

$$x' = \tau(x), \qquad y' = \tau(y) \tag{35}$$

takes the distributions F and G of the x and y into normal distributions with equal standard deviations, but different expectations. The power of the X-test, as well as that of every other rank test, remains unchanged by this transformation. The power of Student's test can, however, be reduced quite considerably by the transformation (35). This reduction occurs especially if the transformation (35) effects a greater increase in the standard deviations σ_x and σ_y than in the difference of expectations $\hat{x} - \hat{y}$.

In a further note (Proc. Kon. Ned. Akad. A 56 (1953) p. 311), I considered the case that $x_1, \dots, x_4$ are uniformly distributed between 0 and 1 and $y_1, \dots, y_6$ are uniformly distributed between 0 and $1 + \mu$. The result there was that the power functions of the rank tests tend to unity as $\mu \to \infty$, whereas that of Student's test does not.

I am familiar with applied problems in which the hypothesis of normal distributions with equal standard deviations for the x and the y was obviously not fulfilled, and in which the X-test led to the rejection of the null hypothesis, whereas Student's test at the same level was not sufficient to make the differentiation.

Example 45. In an industrial concern, measurement of waiting times produced very widely dispersed results. The numerical values, however, I have unfortunately forgotten; therefore let us assume that they are the same as those in the previously mentioned tables by van der Waerden and Nievergelt:

$$x_1 = 11, \quad x_2 = 34, \quad x_3 = 13, \quad x_4 = 18.$$

After a reorganization, the times were shorter and also much less dispersed, say:

$$y_1 = 8, \quad y_2 = 10, \quad y_3 = 7, \quad y_4 = 6.$$

The applicability of Student's test here is very questionable, for it appears that the distributions are not normal and that the standard deviations are quite unequal; also g and h are not very large. If we nevertheless use Student's test (two-sided, at the 5%-level), the null hypothesis is not rejected: The quotient t is only 2.1 and the cut-off point (from Table 7) is 2.4.

Wilcoxon's test leads immediately to the rejection of the null hypothesis: The number of inversions is 16, since all x are larger than all y; or, if we interchange the designations x and y to use Table 10, it is then zero. Under 4;4 with $u=0$ we find the significance probability 1.43%. Since it is smaller than 2.5%, the null hypothesis is to be rejected at the 2.5%-level (one-sided), or at the 5%-level (two-sided).

To apply the X-test we must order the x and y according to increasing magnitude (whereby the x receive the rank numbers 5, 8, 6, and 7) and compute

$$\begin{aligned} X &= \Psi(\tfrac{5}{9}) + \Psi(\tfrac{8}{9}) + \Psi(\tfrac{6}{9}) + \Psi(\tfrac{7}{9}) \\ &= 0.14 + 1.22 + 0.43 + 0.76 = 2.55. \end{aligned}$$

For this computation we can use either Table 2 of this book or, more conveniently, Table 2 of van der Waerden-Nievergelt. The 5%-cut-off point (two-sided) with $n=8$ and $g-h=0$ is 2.40 (Table 11). The X-test therefore also rejects the null hypothesis.

Chapter Thirteen

Correlation

This chapter requires knowledge of the material which appears in Chapters 1 through 6 only.

§ 66. Covariance and the Correlation Coefficient

A. The True Correlation Coefficient ρ

If x and y are two dependent random variables, then the variance of the sum $\lambda x+y$ is no longer simply the sum of the variances of λx and y, but involves a linear term as well:

$$\mathscr{E}(\lambda x+y-\lambda\hat{x}-\hat{y})^2=\lambda^2\,\mathscr{E}(x-\hat{x})^2+2\lambda\,\mathscr{E}(x-\hat{x})(y-\hat{y})+\mathscr{E}(y-\hat{y})^2. \quad (1)$$

The coefficient of 2λ in (1) is called the *covariance* of x and y. If we divide the covariance by the product $\sigma_x\,\sigma_y$ of the standard deviations, which are assumed to be different from zero, we get what is known as the *true correlation coefficient* ρ:

$$\rho=\frac{\mathscr{E}(x-\hat{x})(y-\hat{y})}{\sigma_x\,\sigma_y}. \quad (2)$$

Use of (2) enables us to write (1) as

$$\sigma^2_{\lambda x+y}=\lambda^2\,\sigma_x^2+2\lambda\,\rho\,\sigma_x\,\sigma_y+\sigma_y^2. \quad (3)$$

The correlation coefficient is closely related to the *regression coefficient* γ, which is defined by setting

$$y=\gamma\,x+z \quad (4)$$

and determining γ so as to minimize the variance of z. Expression (3) with $\lambda=-\gamma$ yields the variance of $z=y-\gamma\,x$:

$$\sigma_z^2=\gamma^2\,\sigma_x^2-2\gamma\,\rho\,\sigma_x\,\sigma_y+\sigma_y^2. \quad (5)$$

The polynomial (5) takes on its minimum value when

$$\gamma = \rho \frac{\sigma_y}{\sigma_x}, \tag{6}$$

which shows the relationship between the regression coefficient γ and the correlation coefficient ρ.

The minimum value of the variance is

$$\sigma_z^2 = \rho^2 \sigma_y^2 - 2\rho^2 \sigma_y^2 + \sigma_y^2 = (1-\rho^2)\sigma_y^2. \tag{7}$$

As a consequence of (7) we see that

$$1 - \rho^2 \geqq 0.$$

Therefore the correlation coefficient ρ always lies between -1 *and* $+1$.

If ρ assumes one of its extreme values, $\rho = \pm 1$, then it follows from (7) that $\sigma_z = 0$. According to the last theorem of § 3, this is possible only if z is constant in probability, i.e., if with probability one y is equal to a linear function of x:

$$y = \gamma x + \beta. \tag{8}$$

The correlation coefficient ρ is a measure of the (linear) dependence between x and y. In the case of independence, $\rho = 0$. In the case of exactly linear dependence (8), we have $\rho = \pm 1$. Eq. (6) implies that the sign of ρ is equal to the sign of the regression coefficient γ.

A decomposition of the variance of y yields an interpretation of the correlation coefficient. From formula (4) y is represented as the sum of two random variables γx and z, of which the one (γx) is proportional to x, whereas the other (z) can be shown to be uncorrelated with x. That is, the covariance of x and z is zero. The variance of y is therefore the sum of the variances of γx and z:

$$\sigma_y^2 = \gamma^2 \sigma_x^2 + \sigma_z^2. \tag{9}$$

If we substitute into the right-hand side of (9) the values of γ and σ_z^2 given by (6) and (7), we find the first term to be $\rho^2 \sigma_y^2$ and the second to be $(1-\rho^2)\sigma_y^2$, the sum of which is, of course, σ_y^2. *Therefore ρ^2 denotes the fraction of the variance of y which arises from the term γx in* (4).

B. The Empirical Correlation Coefficient r

If we observe n pairs of values $(x_1, y_1), \ldots, (x_n, y_n)$, and if we assume that these pairs of variables (x_i, y_i) are independent and all have the same two-dimensional distribution, then as an estimate of the variance $\sigma_{\lambda x + y}^2$

we have the *empirical variance*

$$\begin{aligned} s^2_{\lambda x+y} &= \frac{1}{n-1} \sum (\lambda x + y - \lambda \bar{x} - \bar{y})^2 \\ &= \lambda^2 \frac{\sum (x-\bar{x})^2}{n-1} + 2\lambda \frac{\sum (x-\bar{x})(y-\bar{y})}{n-1} + \frac{\sum (y-\bar{y})^2}{n-1}. \end{aligned} \tag{10}$$

Hence, as an estimate of the covariance $\mathscr{E}(x-\hat{x})(y-\hat{y})$ we have the *empirical covariance*

$$\frac{1}{n-1} \sum (x-\bar{x})(y-\bar{y}), \tag{11}$$

where $\bar{x}$ and $\bar{y}$ are, as always, the empirical mean values

$$\bar{x} = \frac{1}{n} \sum x, \quad \bar{y} = \frac{1}{n} \sum y. \tag{12}$$

Unbiasedness of the estimate (11) of the covariance follows from (3) and from the unbiasedness of the estimate (10) of $\sigma^2_{\lambda x+y}$.

In order to obtain an estimate of ρ, we divide (11) by $s_x s_y$. Thus we get the *empirical correlation coefficient*

$$r = \frac{\sum (x-\bar{x})(y-\bar{y})}{(n-1)\, s_x s_y} = \frac{\sum (x-\bar{x})(y-\bar{y})}{\sqrt{\sum (x-\bar{x})^2 \cdot \sum (y-\bar{y})^2}}. \tag{13}$$

We can subject r to considerations equivalent to those made above for the true correlation coefficient ρ. In particular, we can determine $\lambda = -c$ so that the polynomial (10) is minimized. For the minimum we obtain

$$c = r \frac{s_y}{s_x} = \frac{\sum (x-\bar{x})(y-\bar{y})}{\sum (x-\bar{x})^2} \tag{14}$$

and

$$s^2_{y-cx} = (1-r^2)\, s^2_y. \tag{15}$$

Since (15) is always positive or zero, *r always lies between -1 and $+1$*. If $r = \pm 1$, it follows from (15) that the $y_i - c x_i$ all have the same value b, i.e., that the observed points (x_i, y_i) all lie on a line

$$y = c x + b.$$

But even if the points do not lie on a line, we can draw a line through the center of gravity $(\bar{x}, \bar{y})$ with the slope (14). This line is the *empirical regression line*

$$y-\bar{y}=c(x-\bar{x}) \tag{16}$$

already introduced in § 33. The line was defined in § 33 by the requirement that the sum of squares of the distances, measured in the y-direction, of the points (x_i, y_i) from it be minimized. The slope c of this line is the *empirical regression coefficient*. The relationship between it and the empirical correlation coefficient is given by formula (14).

The numerator of (13) can be computed in a variety of ways, which provide a means of checking computations:

$$\begin{aligned}\sum(x-\bar{x})(y-\bar{y}) &= \sum(x-\bar{x})\,y = \sum x(y-\bar{y})\\ &= \sum xy - n\bar{x}\bar{y}\\ &= \sum(x-a)(y-b) - n(\bar{x}-a)(\bar{y}-b).\end{aligned}$$

The same is true of the denominator, as we remarked earlier.

Example 46. For various kinds of pollen, Tammes[1] observed a connection between the pollen-grain size and the number of existing places of exit on the pollen tube. Here Fuchsia Globosa will be taken as an example. The pollen-grains have from 0 to 4 places of exit arranged in an equatorial plane. Ten pollen-grains were measured for each case – 0, 1, 2, 3, or 4 places of exit – and the average of repeated readings was rounded-off to 5 μ. In the accompanying correlation table are given the numbers of pollen-grains.

Average	Number of places of exit				
	$x=0$	1	2	3	4
$y=10$	3				
15	7	3			
20		6			
25		1			
30			4		
35			5		
40			1	3	
45				4	
50				3	3
55					4
60					3

Rarely does one see a more beautiful case of linear regression! Calculations yield

$$\bar{x}=2, \quad \sum(x-\bar{x})^2=100,$$
$$\bar{y}=33.2, \quad \sum(x-\bar{x})(y-\bar{y})=1{,}090,$$
$$\sum(y-\bar{y})^2=12{,}588.$$

[1] P. M. L. Tammes, On the origin of number and arrangement of the places of exit on the surface of pollen-grains. Diss. Groningen 1930.

The empirical regression coefficient is

$$c=\frac{1{,}090}{100}=10.9$$

and the equation of the regression line reads

$$y-\bar{y}=c(x-\bar{x}),$$

or

$$y=10.9\,x+11.4.$$

In the strict sense, we can speak of a correlation coefficient only if the pairs (x, y) are determined purely by chance. In our case, the x-values were not used in the frequencies with which they occurred by chance, but instead 10 pollen-grains were used for every x-value. If we nevertheless calculate r according to (13), we find a very high empirical correlation:

$$r=\frac{1{,}090}{\sqrt{100\cdot 12{,}588}}=0.97.$$

§ 67. The Correlation Coefficient as a Characteristic of Dependence

Since r is an estimate of ρ and since $\rho=0$ in the case of independence, whenever r is considerably different from zero, we shall be able to draw the conclusion that $\rho \neq 0$ and the variables x and y are mutually dependent.

In order to decide how large an r is required to draw this conclusion confidently, we must answer the following question. If x and y are in fact independent and, therefore, $\rho=0$, by how much can the empirical correlation coefficient r differ from zero by chance?

We shall assume that x and y are independent and normally distributed. In as much as x can be replaced by $a(x-\hat{x})$ and y by $b(y-\hat{y})$, there is no loss of generality in assuming that both variables have expectation 0 and standard deviation 1. The probability density of the pair (x, y) is therefore

$$f(x, y)=f(x)\,f(y)=(2\pi)^{-1}\exp\left(-\tfrac{1}{2}x^2-\tfrac{1}{2}y^2\right). \tag{1}$$

Since the individual pairs $(x_1, y_1), \ldots, (x_n, y_n)$ have been assumed to be mutually independent, the joint probability density of all the variables $(x_1, y_1, \ldots, x_n, y_n)$ is the product

$$f(x_1, y_1)\ldots f(x_n, y_n)=(2\pi)^{-n}\exp\left(-\tfrac{1}{2}\sum x_i^2-\tfrac{1}{2}\sum y_i^2\right). \tag{2}$$

Our problem is: What is the distribution function of r?

We shall consider a somewhat more comprehensive question: What is the joint distribution of the five random variables $\bar{x}, \bar{y}, s_x^2, s_y^2$, and r? That is, what is the probability that simultaneously each lies within given bounds?

To begin with, use of an orthogonal transformation easily disposes of $\bar{x}$ and $\bar{y}$. The variables $x_1, \ldots, x_n$ may be transformed orthogonally into $u_1, \ldots, u_n$ in such a way that u_1 is proportional to the mean $\bar{x}$:

$$\begin{aligned} u_1 &= n^{-\frac{1}{2}} x_1 + n^{-\frac{1}{2}} x_2 + \cdots + n^{-\frac{1}{2}} x_n = \bar{x}\sqrt{n} \\ u_2 &= a_{21}\, x_1 + a_{22}\, x_2 + \cdots + a_{2n}\, x_n \\ &\cdots\cdots\cdots\cdots\cdots \end{aligned} \tag{3}$$

The same orthogonal transformation applied to $y_1, \ldots, y_n$ yields

$$\begin{aligned} v_1 &= n^{-\frac{1}{2}} y_1 + n^{-\frac{1}{2}} y_2 + \cdots + n^{-\frac{1}{2}} y_n = \bar{y}\sqrt{n} \\ v_2 &= a_{21}\, y_1 + a_{22}\, y_2 + \cdots + a_{2n}\, y_n \\ &\cdots\cdots\cdots\cdots\cdots \end{aligned} \tag{4}$$

Here, then, $\sum x_i^2 = \sum u_i^2$ and $\sum y_i^2 = \sum v_i^2$. But, at the same time the sums $x_i + y_i$ have been subjected to this same transformation: for example,

$$u_2 + v_2 = a_{21}(x_1 + y_1) + a_{22}(x_2 + y_2) + \cdots + a_{2n}(x_n + y_n).$$

Hence it follows that

$$\sum (x_i + y_i)^2 = \sum (u_i + v_i)^2.$$

If we subtract $\sum x^2 = \sum u^2$ and $\sum y^2 = \sum v^2$ from this and divide by 2, we have

$$\sum x\,y = \sum u\,v. \tag{5}$$

Hence we have

$$\begin{aligned} r^2 &= \frac{(\sum x\,y - n\,\bar{x}\,\bar{y})^2}{(\sum x^2 - n\,\bar{x}^2)(\sum y^2 - n\,\bar{y}^2)} = \frac{(\sum u\,v - u_1\,v_1)^2}{(\sum u^2 - u_1^2)(\sum v^2 - v_1^2)} \\ &= \frac{(u_2\,v_2 + \cdots + u_n\,v_n)^2}{(u_2^2 + \cdots + u_n^2)(v_2^2 + \cdots + v_n^2)} \end{aligned} \tag{6}$$

and

$$(n-1)\,s_x^2 = \sum x^2 - n\,\bar{x}^2 = \sum u^2 - u_1^2 = u_2^2 + \cdots + u_n^2, \tag{7}$$

$$(n-1)\,s_y^2 = \sum y^2 - n\,\bar{y}^2 = \sum v^2 - v_1^2 = v_2^2 + \cdots + v_n^2. \tag{8}$$

The variables u_1 and v_1 are mutually independent and independent of the remaining $u_2, v_2, \ldots, u_n, v_n$. *Therefore $\bar{x}$ and $\bar{y}$ are independent of one another and of s_x^2, s_y^2, and r.*

Obviously, $\bar{x}$ and $\bar{y}$ are normally distributed with standard deviation $n^{-\frac{1}{2}}$. There still remain to be considered, therefore, only s_x^2, s_y^2, and r, which are given by (6), (7), and (8) as functions of $u_2, v_2, \ldots, u_n, v_n$. We must calculate the probability of the event

$$s_x^2 < a, \qquad s_y^2 < b, \qquad r < c. \tag{9}$$

This probability is equal to the integral

$$P=(2\pi)^{-n+1}\int\cdots\int \exp\left(-\tfrac{1}{2}\sum_2^n u_i^2-\tfrac{1}{2}\sum_2^n v_i^2\right)du_2\,dv_2\ldots du_n\,dv_n, \quad (10)$$

integrated over the region G defined by (9).

In (10) we can integrate first with respect to $v_2, \ldots, v_n$ and then with respect to $u_2, \ldots, u_n$. For the inner integration with respect to $v_2, \ldots, v_n$, the u are to be considered constant. For constant u we now apply an orthogonal transformation to the v:

$$\begin{aligned} w_2 &= b_{22}v_2+\cdots+b_{2n}v_n \\ &\cdot\;\cdot\;\cdot\;\cdot\;\cdot\;\cdot\;\cdot\;\cdot\;\cdot\;\cdot \\ w_n &= b_{n2}v_2+\cdots+b_{nn}v_n, \end{aligned} \quad (11)$$

where we determine the coefficients in the first row so that

$$b_{2i}=(u_2^2+\cdots+u_n^2)^{-\frac{1}{2}}u_i, \qquad i=2,\ldots,n. \quad (12)$$

The sum of the squares of the coefficients b_{2i} is one. Therefore, according to § 13, it is possible to determine the coefficients in the remaining rows so that the entire transformation becomes orthogonal. Hence the inner integral becomes

$$\begin{aligned} I &= \int\cdots\int \exp\left(-\tfrac{1}{2}\sum_2^n v_i^2\right)dv_2\ldots dv_n \\ &= \int\cdots\int \exp\left(-\tfrac{1}{2}\sum_2^n w_i^2\right)dw_2\ldots dw_n. \end{aligned} \quad (13)$$

On account of (12),

$$w_2=\sum_2^n b_{2i}v_i=(u_2^2+\cdots+u_n^2)^{-\frac{1}{2}}(u_2v_2+\cdots+u_nv_n),$$

and therefore

$$r=(v_2^2+\cdots+v_n^2)^{-\frac{1}{2}}w_2=(w_2^2+\cdots+w_n^2)^{-\frac{1}{2}}w_2. \quad (14)$$

If we set

$$w_3^2+\cdots+w_n^2=\zeta^2, \quad (15)$$

we then see from (8) and (14) that r and s_y^2 depend only on w_2 and ζ^2:

$$(n-1)\,s_y^2=w_2^2+w_3^2+\cdots+w_n^2=w_2^2+\zeta^2, \quad (16)$$

$$r=(w_2^2+\zeta^2)^{-\frac{1}{2}}w_2. \quad (17)$$

To evaluate the integral in (13), we introduce polar coordinates $\zeta, \varphi_1, \ldots, \varphi_{n-3}$ in place of $w_3, \ldots, w_n$. Since the angles $\varphi_1, \ldots, \varphi_{n-3}$ do not occur at all in the limits of integration (9), we can carry out the integration with respect to these angular coordinates immediately. If we write simply w instead of w_2, the inner integral (13) is

$$I = C \iint e^{-\frac{1}{2}w^2 - \frac{1}{2}\zeta^2} \zeta^{n-3} \, dw \, d\zeta, \tag{18}$$

integrated over

$$\begin{aligned} &\zeta \geqq 0 \\ &w^2 + \zeta^2 = (n-1)\, s_y^2 < (n-1)\, b \\ &(w^2+\zeta^2)^{-\frac{1}{2}} w = r < c. \end{aligned} \tag{19}$$

The integral I does not depend on the u_i and, therefore, in (10) the factor I can be brought out in front of the integral. Hence we have

$$P = (2\pi)^{-n+1} I \int \cdots \int \exp\left(-\tfrac{1}{2} \sum_2^n u_i^2\right) du_2 \dots du_n, \tag{20}$$

where the integration is to be taken over the region

$$\sum_2^n u_i^2 = (n-1)\, s_x^2 < (n-1)\, a. \tag{21}$$

If we also introduce polar coordinates $\chi, \varphi_1', \dots, \varphi_{n-2}'$ here, we can again carry out the integration with respect to the angular coordinates and obtain finally

$$\begin{aligned} P &= C' I \int e^{-\frac{1}{2}\chi^2} \chi^{n-2} \, d\chi \\ &= \alpha \iiint e^{-\frac{1}{2}\chi^2} \chi^{n-2} \, d\chi \cdot e^{-\frac{1}{2}\zeta^2} \zeta^{n-3} \, d\zeta \cdot e^{-\frac{1}{2}w^2} \, dw, \end{aligned} \tag{22}$$

integrated over the region

$$\begin{aligned} &\chi^2 < (n-1)\, a, \qquad \chi \geqq 0, \\ &w^2 + \zeta^2 < (n-1)\, b, \qquad \zeta \geqq 0, \\ &(w^2+\zeta^2)^{-\frac{1}{2}} w < c. \end{aligned} \tag{23}$$

The result is obviously valid not only for a region of the special form (9), but for any region G in (s_x^2, s_y^2, r)-space. Let G' denote the transformed region in the space of the new variables χ, ζ, w, which are defined by

$$\chi^2 = (n-1)\, s_x^2/\sigma_x^2, \qquad \chi \geqq 0, \tag{24}$$

$$w^2 + \zeta^2 = (n-1)\, s_y^2/\sigma_y^2, \qquad \zeta \geqq 0, \tag{25}$$

$$w = (w^2+\zeta^2)^{\frac{1}{2}} r = (n-1)^{\frac{1}{2}} r\, s_y/\sigma_y. \tag{26}$$

Here the denominators σ_x^2, σ_y^2, and σ_y have been included so that the formulas are valid even if σ_x and σ_y have not been normed to 1. In this case, the probability of the region G is

$$\mathscr{P} G = P = \alpha \iiint e^{-\frac{1}{2}\chi^2} \chi^{n-2} \, d\chi \cdot e^{-\frac{1}{2}\zeta^2} \zeta^{n-3} \, d\zeta \cdot e^{-\frac{1}{2}w^2} \, dw, \tag{27}$$

where the integration extends over the transformed region G'. The constant α is, of course, determined so that the integral over the whole space $\chi \geqq 0$, $\zeta \geqq 0$ is equal to 1.

The result can also be formulated as:

$\chi^2 = u$, $\zeta^2 = v$, and w are independent variables with probability densities

$$f(u) = \alpha_1 \, e^{-\frac{1}{2}u} u^{\frac{n-3}{2}}, \qquad \alpha_1^{-1} = \Gamma\left(\frac{n-1}{2}\right) 2^{\frac{n-1}{2}}; \tag{28}$$

$$g(v) = \alpha_2 \, e^{-\frac{1}{2}v} v^{\frac{n-4}{2}}, \qquad \alpha_2^{-1} = \Gamma\left(\frac{n-2}{2}\right) 2^{\frac{n-2}{2}}; \tag{29}$$

$$h(w) = (2\pi)^{-\frac{1}{2}} e^{-\frac{1}{2}w^2}. \tag{30}$$

Hence the variables χ^2 and ζ^2 have χ^2-distributions with $n-1$ and $n-2$ degrees of freedom, respectively. The variable w is normally distributed with expectation 0 *and standard deviation* 1.

The distribution of the empirical correlation coefficient r is now easy to derive. We first calculate ζ from (26):

$$\zeta = \frac{w}{r} \sqrt{1-r^2}.$$

Then we form

$$t = \frac{w}{\zeta} \sqrt{n-2} = \frac{r}{\sqrt{1-r^2}} \sqrt{n-2}. \tag{31}$$

Since w has a standard normal distribution and ζ^2 has an independent χ^2-distribution with $n-2$ degrees of freedom, according to § 28 *t has a t-distribution with $n-2$ degrees of freedom.* The cut-off points for t at the usual significance levels (5, 2, and 1 %), together with (31), immediately yield cut-off points for the correlation coefficient r. These are tabulated in Table 13 at the end of the book.

Table 13 is employed in the following manner: *If in a practical problem we observe a value r whose absolute magnitude exceeds the cut-off r_α in the table, we then decide that the variables x and y are dependent.*

We can make the following statements about the error probability of this test.

There exist cases in which x and y are in fact dependent. In these cases we commit no error if we declare x and y dependent on the basis of the test.

Secondly, there exist cases in which x and y are in fact independent and approximately normally distributed. In these cases the error probability of the test is 2α, since all the assumptions under which the test was derived are satisfied.

Thirdly, there exist cases in which x and y are independent, but are not even approximately normally distributed. In these cases the error probability of the test can be somewhat larger than 2α. However, if the standard deviations σ_x and σ_y are assumed to be finite and n is sufficiently large, the deviation from 2α is not very important. The chance fluctuations in r arise chiefly from the numerator

$$\sum (x-\bar{x})(y-\bar{y}). \tag{32}$$

As an approximation if n is large, we can replace $\bar{x}$ and $\bar{y}$ by $\hat{x}$ and $\hat{y}$, in which case we have

$$\sum (x-\hat{x})(y-\hat{y})$$

instead of (32). This is a sum of many independent terms, each of which has a standard deviation $\sigma_x \sigma_y$. Therefore, according to the Central Limit Theorem (§ 24 D), the sum is approximately normally distributed with expectation zero and standard deviation $n^{\frac{1}{2}} \sigma_x \sigma_y$. The denominator of r is $(n-1)\, s_x\, s_y$, and therefore approximately equal to $n\, \sigma_x \sigma_y$. Hence, for $n \to \infty$, r is asymptotically normally distributed with expectation zero and standard deviation $n^{-\frac{1}{2}}$, regardless of whether the x and y are normally distributed or not. The error probability therefore tends toward 2α as $n \to \infty$. If n is finite but not too small, the deviation from 2α is not very large.

Example 47 (from R. A. Fisher, Statistical Methods for Research Workers, 11th Edition, Ex. 27). For the 20 years between 1885 and 1904 the correlation between the wheat crop in eastern England and the rainfall in the autumn was found to be negative, -0.63. From the table we find 0.56 as the cut-off at the 1%-level. The existence of a relationship between rainfall and wheat crop in eastern England therefore appears to be certain.

§ 68. Partial Correlation Coefficients

A. The Concepts of Partial Correlation and Residual Variables

It is possible that the correlation between two variables x and y arises in part, if not entirely, from the fact that both x and y show a considerable correlation with a third variable z. The question which then can be asked is how strongly x and y are still correlated, once the dependence on z is eliminated by replacing x and y by *residual* variables

$$x' = x - \lambda z \quad \text{and} \quad y' = y - \mu z$$

which are no longer correlated with z, in the sense that the correlation coefficient is zero.

We shall first look at the problem with regard to *true* correlation coefficients. If we assume that $\hat{x} = \hat{y} = \hat{z} = 0$, then, for example,

$$\rho_{xy} = \frac{\mathscr{E} x\, y}{\sigma_x \sigma_y}. \tag{1}$$

The factors λ and μ must be chosen so that

$$\mathscr{E} x' z = \mathscr{E}(x z - \lambda z^2) = 0,$$
$$\mathscr{E} y' z = \mathscr{E}(y z - \mu z^2) = 0,$$

which imply that

$$\lambda = \frac{\mathscr{E} x z}{\mathscr{E} z^2} = \frac{\rho_{xz} \sigma_x \sigma_z}{\sigma_z^2} = \rho_{xz} \frac{\sigma_x}{\sigma_z}, \tag{2}$$

$$\mu = \frac{\mathscr{E} y z}{\mathscr{E} z^2} = \frac{\rho_{yz} \sigma_y \sigma_z}{\sigma_z^2} = \rho_{yz} \frac{\sigma_y}{\sigma_z}. \tag{3}$$

The *partial correlation coefficient* $\rho_{xy|z}$ is thus

$$\rho_{xy|z} = \rho_{x'y'} = \frac{\mathscr{E}(x - \lambda z)(y - \mu z)}{\sigma_{x - \lambda z} \sigma_{y - \mu z}}. \tag{4}$$

Its numerator has the value

$$\mathscr{E} x y - \lambda \mathscr{E} z y - \mu \mathscr{E} x z + \lambda \mu \mathscr{E} z^2$$
$$= \rho_{xy} \sigma_x \sigma_y - \lambda \rho_{yz} \sigma_y \sigma_z - \mu \rho_{xz} \sigma_x \sigma_z + \lambda \mu \sigma_z^2.$$

Replacing λ and μ here by their values from (2) and (3), we have

$$\mathscr{E}(x - \lambda z)(y - \mu z) = (\rho_{xy} - \rho_{xz} \rho_{yz}) \sigma_x \sigma_y, \tag{5}$$

and likewise we find

$$\sigma_{x - \lambda z}^2 = \mathscr{E}(x - \lambda z)^2 = (1 - \rho_{xz}^2) \sigma_x^2, \tag{6}$$

$$\sigma_{y - \mu z}^2 = \mathscr{E}(y - \mu z)^2 = (1 - \rho_{yz}^2) \sigma_y^2. \tag{7}$$

Substituting all of this into (4) we get

$$\rho_{xy|z} = \frac{\rho_{xy} - \rho_{xz} \rho_{yz}}{\sqrt{1 - \rho_{xz}^2} \sqrt{1 - \rho_{yz}^2}}. \tag{8}$$

To obtain an estimate of $\rho_{xy|z}$, we replace the true correlation coefficients ρ by the corresponding empirical coefficients r, which gives us

$$r_{xy|z} = \frac{r_{xy} - r_{xz} r_{yz}}{\sqrt{1 - r_{xz}^2} \sqrt{1 - r_{yz}^2}}. \tag{9}$$

We can also derive (9) in a manner similar to (8) by determining linear combinations $x'' = x - a z$ and $y'' = y - b z$ for which the empirical correlation with z is zero. The empirical correlation between x'' and y'' is then found to be just that given by (9).

From this interpretation of expression (9) it follows that $r_{xy|z}$ is not altered if x is replaced by $x-\lambda z$ and y is replaced by $y-\mu z$, where λ and μ are for the moment completely arbitrary. This is true because these linear combinations $x''=x-az$ and $y''=y-bz$, which show no empirical correlation with z, are always the same, regardless of whether we use x and y or $x'=x-\lambda z$ and $y'=y-\mu z$. Hence, in our discussion of the properties of the random variable $r_{xy|z}$ it is always possible to replace x and y by variables x' and y' which exhibit no true correlation with z; i.e., we can assume from the outset that $\rho_{xz}=\rho_{yz}=0$.

B. Distribution Function of $r_{xy|z}$

As in § 67, we shall now take up the problem of how large the empirical partial correlation coefficient $r_{xy|z}$ must be before we can conclude that there exists a real dependence between the residuals $x-\lambda z$ and $y-\mu z$; or, to put it the other way around, what values of $r_{xy|z}$ are to be expected purely by chance, if $x-\lambda z$ and $y-\mu z$ are in fact independent.

From the above remark we can assume that $\lambda=\mu=0$, or $\rho_{xz}=\rho_{yz}=0$. We now make the even stronger assumption that x, y, z are independent and normally distributed. If their standard deviations are normalized to unity, the probability density is

$$f(x,y,z)=(2\pi)^{-\frac{3}{2}}\,e^{-\frac{1}{2}(x^2+y^2+z^2)}. \tag{10}$$

Under these hypotheses we are to determine the distribution function $H(c)$ of the random variable $r_{xy|z}$. It is, of course, equal to the $3n$-fold integral

$$\begin{aligned} H(c)&=\mathscr{P}(r_{xy|z}<c)\\ &=\iiint\cdots\iiint f(x_1,y_1,z_1)\ldots f(x_n,y_n,z_n)\,dx_1\,dy_1\ldots dz_n, \end{aligned} \tag{11}$$

integrated over $r_{xy|z}<c$.

As in § 67, we dispose of the means $\bar{x}, \bar{y}, \bar{z}$ by means of an orthogonal transformation which replaces the variables x_i, y_j, z_k by variables u_i, v_j, w_k such that u_1, v_1, w_1 are proportional to $\bar{x}, \bar{y}, \bar{z}$. In fact, the v_j and w_k are obtained from the y_j and z_k by means of the same transformation as used to go from the x_i to the u_i. If, for brevity, we set

$$[uv]=u_2v_2+\cdots+u_nv_n, \tag{12}$$

with analogous definitions for $[uw]$ and $[vw]$, then

$$r_{xy}=\frac{[uv]}{\sqrt{[uu]\cdot[vv]}},\quad r_{xz}=\frac{[uw]}{\sqrt{[uu]\cdot[ww]}},\quad r_{yz}=\frac{[vw]}{\sqrt{[vv]\cdot[ww]}}. \tag{13}$$

The transformed integral looks exactly like the original integral (11), except that the u, v, w appear in place of the x, y, z. Since u_1, v_1, w_1 do not occur in the definition of the region $r_{xy|z} < c$, we can carry out the integration with respect to u_1, v_1, w_1, which yields

$$H(c) = (2\pi)^{\frac{-3n+3}{2}} \iiint \cdots \iiint e^{-\frac{1}{2}([uu]+[vv]+[ww])}\, du_2\, dv_2 \ldots dw_n, \tag{14}$$

integrated over the region $r_{xy|z} < c$.

We next carry out the integration with respect to $u_2, \ldots, u_n, v_2, \ldots, v_n$ and obtain an inner integral

$$I = \iint \cdots \iint e^{-\frac{1}{2}[uu] - \frac{1}{2}[vv]}\, du_2 \ldots du_n\, dv_2 \ldots dv_n, \tag{15}$$

which we must afterwards integrate with respect to $w_2, \ldots, w_n$.

As in § 67, we now make an orthogonal transformation of the u and v for constant w. We set

$$u'_i = \sum a_{ik} u_k, \qquad v'_i = \sum a_{ik} v_k, \tag{16}$$

and, in particular,

$$u'_2 = \frac{[uw]}{\sqrt{[ww]}}, \qquad v'_2 = \frac{[vw]}{\sqrt{[ww]}}, \tag{17}$$

so that

$$r_{xz} = \frac{u'_2}{\sqrt{[uu]}} = \frac{u'_2}{\sqrt{[u'u']}},$$

$$r_{yz} = \frac{v'_2}{\sqrt{[vv]}} = \frac{v'_2}{\sqrt{[v'v']}},$$

$$r_{xy} = \frac{[uv]}{\sqrt{[uu] \cdot [vv]}} = \frac{[u'v']}{\sqrt{[u'u'] \cdot [v'v']}}.$$

The integral (15) thereby becomes

$$I = \iint \cdots \iint e^{-\frac{1}{2}[u'u'] - \frac{1}{2}[v'v']}\, du'_2 \ldots du'_n\, dv'_2 \ldots dv'_n, \tag{18}$$

integrated as before (but now for constant w) over $r_{xy|z} < c$. Now, however,

$$\begin{aligned} r_{xy|z} &= \frac{[u'v'] - u'_2 v'_2}{\sqrt{[u'u'] - u'^2_2} \cdot \sqrt{[v'v'] - v'^2_2}} \\ &= \frac{u'_3 v'_3 + \cdots + u'_n v'_n}{\sqrt{u'^2_3 + \cdots + u'^2_n} \cdot \sqrt{v'^2_3 + \cdots + v'^2_n}}. \end{aligned} \tag{19}$$

Thus u_2' and v_2' do not occur in the definition of the region of integration and we can carry out the integration with respect to u_2' and v_2', which yields

$$I=(2\pi)\int\cdots\int e^{-\frac{1}{2}(u_3'^2+\cdots+u_n'^2+v_3'^2+\cdots+v_n'^2)}\,du_3'\ldots du_n'\,dv_3'\ldots dv_n', \tag{20}$$

integrated over $r_{xy|z}<c$.

We see that the integral I is independent of $w_2,\ldots,w_n$ and that we can therefore bring I out in front of the integral (14) as a factor and integrate with respect to $w_2,\ldots,w_n$. Thus we obtain, if we write $u_3',\ldots$ simply as $u_3,\ldots,$

$$H(c)=(2\pi)^{-n+2}\int\cdots\int e^{-\frac{1}{2}(u_3^2+\cdots+v_3^2+\cdots+v_n^2)}\,du_3\ldots dv_n, \tag{21}$$

integrated over the region $r<c$, where r is defined according to (19) by

$$r=\frac{u_3 v_3+\cdots+u_n v_n}{\sqrt{u_3^2+\cdots+u_n^2}\cdot\sqrt{v_3^2+\cdots+v_n^2}}. \tag{22}$$

We now compare integral (21) with our previously calculated integral (10) § 67. Instead of the earlier $u_2,\ldots,u_n$ and $v_2,\ldots,v_n$, we now have $u_3,\ldots,u_n$ and $v_3,\ldots,v_n$, and, therefore, $2(n-2)$ instead of $2(n-1)$ variables. Previously the region of integration was defined by the inequalities

$$s_x^2<a, \quad s_y^2<b, \quad r<c.$$

If, however, we let a and b go to ∞, we get simply the region $r<c$. Therefore, the present integral is a special case of the previous one, and we have the following theorem:

The distribution function of the partial correlation coefficient $r_{xy|z}$ is the same as that of the ordinary correlation coefficient r of two independent, normally distributed sequences of variables $x_1,\ldots,x_{n-1}$ and $y_1,\ldots,y_{n-1}$, with $n-1$ instead of n variables. This is the case, to repeat it once more, under the hypothesis that $x-\lambda z$, $y-\mu z$, and z are independent, normally distributed variables. It depends on the independence, but not so much on the normality.

Thus, for judging the dependence of the residual variables $x-\lambda z$ and $y-\mu z$, we can use Table 13 with $n-1$ in place of n.

C. Geometric Representation

The results of this and the preceding paragraph can also be derived geometrically.

Let us take, say, $n=4$ and interpret (u_2,u_3,u_4), (v_2,v_3,v_4), and (w_2,w_3,w_4) as components of three vectors $\mathfrak{u},\mathfrak{v},\mathfrak{w}$ in three-dimensional space. Then $[uu]$ is the square of the length of $\mathfrak{u}$. Likewise, $[uv]$ is the scalar product of $\mathfrak{u}$ and $\mathfrak{v}$, and r_{xy} the cosine of the angle φ between $\mathfrak{u}$ and $\mathfrak{v}$. The probability density of the variables u_i and v_j,

$$(2\pi)^{-(n-1)}\, e^{-\frac{1}{2}[uu]-\frac{1}{2}[vv]},$$

implies that all six vector components u_i and v_j are normally distributed and are independent of one another. This continues to be true when we include the w_k as well.

This distribution law is invariant under orthogonal transformations. Therefore, instead of choosing all six vector components u_i and v_j independently of one another, we can choose first the u_i, then the one component v' of $\mathfrak{v}$ parallel to $\mathfrak{u}$ and the remaining two components of $\mathfrak{v}$ perpendicular to $\mathfrak{u}$, with all components distributed according to the normal distribution law. The parallel component v' has a normal distribution with standard deviation 1, and the sum of the squares of the perpendicular components has a χ^2-distribution with two degrees of freedom. Let us call this sum of squares v''^2. The quotient v'/v'' is the cotangent of the angle φ (Fig. 35), and therefore

$$\frac{v'}{v''}=\frac{\cos\varphi}{\sin\varphi}=\frac{r}{\sqrt{1-r^2}}.$$

This consideration explains the fact that the variable

$$t=\frac{r}{\sqrt{1-r^2}}\sqrt{n-1}$$

has a t-distribution with $n-1$ degrees of freedom.

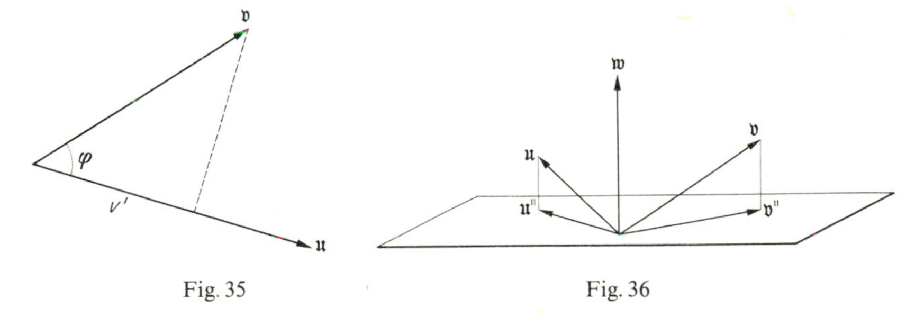

Fig. 35 Fig. 36

If we now include the vector $\mathfrak{w}$ (Fig. 36), then in considering the correlation coefficient between the residuals of x and y, instead of r_{xy}, we replace the vectors $\mathfrak{u}$ and $\mathfrak{v}$ by new vectors $\mathfrak{u}''=\mathfrak{u}-a\mathfrak{w}$ and $\mathfrak{v}''=\mathfrak{v}-b\mathfrak{w}$ which are perpendicular to $\mathfrak{w}$. We can first choose the vector $\mathfrak{w}$ and then choose the components of the vectors $\mathfrak{u}$ and $\mathfrak{v}$ parallel and perpendicular to $\mathfrak{w}$. The perpendicular components define the vectors $\mathfrak{u}''$ and $\mathfrak{v}''$; the cosine of the angle between them is the partial correlation coefficient $r_{xy|z}$. Now, since these vectors in the plane perpendicular to $\mathfrak{w}$ are distributed according to the same probability law as the original vectors $\mathfrak{u}$ and $\mathfrak{v}$, except with $n-2$ instead of $n-1$ dimensions, it is understandable that $r_{xy|z}$ has the same distribution function as an ordinary correlation coefficient r, except with $n-1$ instead of n.

§ 69. Distribution of the Coefficient r for Dependent Variables

A. Normally Distributed Pairs of Variables

Thus far we have always assumed that the variables x and y are in fact independent and that r differs from zero only by chance. The problem becomes significantly more difficult if x and y are in fact dependent. To enable us to assert anything at all about the distribution of r, we must make some assumption about the distributions of x and y.

The simplest assumption is that x is normally distributed and that y is equal to $\lambda x + z$, where z is likewise normally distributed and independent of x. If the variance of x is g^{-1} and that of z is h^{-1}, then the probability density of the pair (x, z) is

$$\frac{\sqrt{gh}}{2\pi} e^{-\frac{1}{2}(gx^2 + hz^2)}, \tag{1}$$

and thus the probability density of the pair (x, y) is

$$\begin{aligned} f(x, y) &= \frac{\sqrt{gh}}{2\pi} e^{-\frac{1}{2}[gx^2 + h(y-\lambda x)^2]} \\ &= \frac{\sqrt{gh}}{2\pi} e^{-\frac{1}{2}[(g+h\lambda^2)x^2 - 2\lambda hxy + hy^2]}. \end{aligned} \tag{2}$$

In place of this we can also write

$$f(x, y) = C\, e^{-\frac{1}{2}(ax^2 - 2bxy + hy^2)}. \tag{3}$$

Simply as a matter of convenience, we have assumed that the expectations of x and of y are equal to zero. For the general case, we need only replace x and y by $x - \hat{x}$ and $y - \hat{y}$.

A distribution of the type (3) with a negative quadratic form in the exponent is called a *bivariate normal distribution of the pair* (x, y). Very frequently such distributions are realized at least approximately in biology, particularly in the study of the hereditary transmission of animal or plant characteristics in non-selected material.

Every distribution of the type (3) can be reduced to form (2); i.e., y can be thought of as the sum of a linear term λx and a variable z independent of x. Since x and y play identical roles in (3), the inverse concept of x as a sum of a linear term μy and a variable z' independent of y is, of course, also possible. It depends on the concrete problem at hand, whether we prefer to think of x or y as the independent variable.

It is always possible to make a change in the scale of x and y so that $a=h=1$. Then $\lambda=b$ and $g=1-b^2$. Hence the probability density (3) takes on the simple form

$$f(x,y)=\frac{1}{2\pi}\sqrt{1-b^2}\,e^{-\frac{1}{2}(x^2-2bxy+y^2)}, \tag{4}$$

which we shall use in the following discussion.

The variance of x is

$$\sigma_x^2=g^{-1}=(1-b^2)^{-1}. \tag{5}$$

The variance of y, of course, has the same value

$$\sigma_y^2=(1-b^2)^{-1}. \tag{6}$$

We calculate the covariance of x and y to be

$$\begin{aligned}\mathscr{E}x\,y&=\mathscr{E}x(\lambda x+z)=\lambda\,\mathscr{E}x^2+\mathscr{E}x\,z\\&=b\,\mathscr{E}x^2+0=b(1-b^2)^{-1}.\end{aligned} \tag{7}$$

Hence the true correlation coefficient is

$$\rho=\frac{\mathscr{E}x\,y}{\sigma_x\,\sigma_y}=b. \tag{8}$$

Thus in (4) we could also have written ρ instead of b:

$$f(x,y)=\frac{1}{2\pi}\sqrt{1-\rho^2}\,e^{-\frac{1}{2}(x^2-2\rho xy+y^2)}. \tag{9}$$

B. Asymptotic Distribution of r for Large n

Let $(x_1,y_1),\ldots,(x_n,y_n)$ be independent bivariate random variables, each of which is distributed according to (9). The probability density of the entire system $(x_1,y_1,\ldots,x_n,y_n)$ is then

$$f(x_1,y_1)\ldots f(x_n,y_n)=(2\pi)^{-n}(1-\rho^2)^{n/2}\,e^{-\frac{1}{2}\Sigma(x_k^2-2\rho x_k y_k+y_k^2)}. \tag{10}$$

Under these assumptions the distribution of the correlation coefficient r is asymptotically a normal distribution as $n\to\infty$. The expectation is asymptotically equal to ρ and the standard deviation is asymptotically *equal to*

$$\sigma = \frac{1-\rho^2}{\sqrt{n-1}}. \tag{11}$$

The proof is very simple. As in § 67, we can first transform the x_k and y_k orthogonally into new variables u_k and v_k, such that $u_1 = \bar{x}\sqrt{n}$, $v_1 = \bar{y}\sqrt{n}$, and

$$r = (u_2 v_2 + \cdots + u_n v_n)(u_2^2 + \cdots + u_n^2)^{-\frac{1}{2}}(v_2^2 + \cdots + v_n^2)^{-\frac{1}{2}}. \tag{12}$$

The probability density of the u_k and v_k looks exactly like that of the x_k and y_k:

$$f(u_1, v_1) \dots f(u_n, v_n) = (2\pi)^{-n}(1-\rho^2)^{n/2}\, e^{-\frac{1}{2}\Sigma(u_k^2 - 2\rho u_k v_k + v_k^2)}. \tag{13}$$

The expectations of u_k^2, v_k^2, and $u_k v_k$ are

$$\mathscr{E} u_k^2 = \sigma_x^2 = g^{-1},$$
$$\mathscr{E} v_k^2 = \sigma_y^2 = g^{-1},$$
$$\mathscr{E} u_k v_k = \rho\, \sigma_x \sigma_y = \rho\, g^{-1}.$$

Therefore we can set

$$\begin{aligned} u_k^2 &= g^{-1}(1+p_k), \\ v_k^2 &= g^{-1}(1+q_k), \\ u_k v_k &= g^{-1}(\rho + r_k). \end{aligned} \tag{14}$$

Substitution into (12) yields

$$r = (m\rho + \sum r_k)(m + \sum p_k)^{-\frac{1}{2}}(m + \sum q_k)^{-\frac{1}{2}}, \tag{15}$$

where $m = n-1$. Here each sum $\sum r_k$, $\sum p_k$, and $\sum q_k$ is composed of m terms. Each individual term has expectation 0 and standard deviation of the order of magnitude 1. Thus in probability the sums are only of the order of magnitude $m^{\frac{1}{2}}$ and therefore small relative to m. Hence we can expand (15) and drop the higher powers:

$$\begin{aligned} r &= (\rho + m^{-1}\sum r_k)(1 + m^{-1}\sum p_k)^{-\frac{1}{2}}(1 + m^{-1}\sum q_k)^{-\frac{1}{2}} \\ &= \rho + m^{-1}\sum r_k - \tfrac{1}{2}\rho\, m^{-1}\sum p_k - \tfrac{1}{2}\rho\, m^{-1}\sum q_k + \cdots, \end{aligned}$$

which gives us the asymptotic formula

$$r - \rho \sim \sum \frac{r_k - \frac{1}{2}\rho\, p_k - \frac{1}{2}\rho\, q_k}{n-1}. \tag{16}$$

On the right-hand side we have a sum of many independent terms, each of which has a bounded standard deviation. Thus, according to the Central Limit Theorem (§ 24D) the sum is asymptotically normally distributed. The expectation is zero and the variance is the sum of the variances of the individual terms. Calculation of this sum gives rise to the standard deviation in formula (11).

To make use of this result, however, presents certain difficulties. In the first place, ρ is unknown and therefore σ can not be calculated. Frequently, ρ is replaced by r in the right-hand side of (11) to give

$$s = \frac{1-r^2}{\sqrt{n-1}} \tag{17}$$

as an estimate of σ.

This estimate is reliable, however, only if n is very large and ρ^2 is not too close to one. If n is only moderately large or if ρ^2 is close to one (in which case r^2 is probably also close to one), then the deviation of s from σ can be considerable. Moreover, in this case the expectation of r deviates from ρ; i.e., the estimate r now has some bias. This we can easily see by calculating a few more terms in the approximation (16), in which case we get

$$\mathscr{E} r = \rho - \frac{\rho}{2} \frac{1-\rho^2}{n-1} + \cdots. \tag{18}$$

Furthermore, if n is only moderate, the exact distribution of r can deviate considerably from the normal distribution, in particular if ρ^2 is close to one. Therefore, as matters stand, we can not use the normal distribution to calculate confidence intervals for ρ based on a given r, but rather we must use the exact distribution function of r, or at least a better approximation.

C. Exact Distribution of s_x^2, s_y^2, and r

To obtain the probability density of the three random variables s_x, s_y, and r we shall first of all introduce the following notation for the estimated variances and covariance:

$$s_{xx} = s_x^2, \qquad s_{yy} = s_y^2, \qquad s_{xy} = r\, s_x\, s_y. \tag{19}$$

Then we shall again set $y = \lambda\, x + z = \rho\, x + z$, where z is independent of x, and likewise form the estimated variances and covariance:

$$s_{xx} = s_x^2, \qquad s_{zz} = s_z^2, \qquad s_{xz} = r'\, s_x\, s_z, \tag{20}$$

where r' is the empirical correlation between x and z. Those variables in (19) can easily be expressed in terms of these in (20):

$$\begin{aligned} s_{xx} &= s_{xx} \\ s_{xy} &= \rho\, s_{xx} + s_{xz} \\ s_{yy} &= \rho^2\, s_{xx} + 2\rho\, s_{xz} + s_{zz}. \end{aligned} \tag{21}$$

Also, as in § 67, we can make yet another change of variables from those in (20) to χ, w, ζ, where

$$\begin{aligned} (n-1)\, s_{xx} &= \sigma_x^2\, \chi^2, \\ (n-1)\, s_{xz} &= \sigma_x\, \sigma_z\, \chi\, w, \\ (n-1)\, s_{zz} &= \sigma_z^2\, (w^2 + \zeta^2). \end{aligned} \tag{22}$$

Here $\sigma_x^2 = g^{-1} = (1-\rho^2)^{-1}$ and $\sigma_z^2 = 1$, as in (5), and the probability density of the triplet χ, w, ζ of independent variables is

$$f_0(\chi, w, \zeta) = \alpha\, e^{-\frac{1}{2}\chi^2 - \frac{1}{2}\zeta^2 - \frac{1}{2}w^2}\, \chi^{n-2}\, \zeta^{n-3}, \tag{23}$$

where α is a constant.

To find the probability density of s_x, s_y, and r we can now make changes of variables in the probability density (23) of χ, w, ζ, first according to (22) to s_{xx}, s_{xz}, and s_{zz}, then according to (21) to s_{xx}, s_{xy}, and s_{yy}, and finally according to (19) to s_x, s_y, and r. *Thus the probability density of the three variables s_x, s_y, and r is*

$$\begin{aligned} f(s_x, s_y, r) = \frac{(n-1)^{n-1}}{\pi\, \Gamma(n-2)} (1-\rho^2)^{\frac{n-1}{2}}\, e^{-\frac{n-1}{2}(s_x^2 - 2\rho r s_x s_y + s_y^2)} \\ \cdot\, s_x^{n-2}\, s_y^{n-2} (1-r^2)^{\frac{n-4}{2}}. \end{aligned} \tag{24}$$

This result is due to R. A. Fisher (Biometrika 10 (1915) p. 507).

D. R. A. Fisher's Auxiliary Variable z

Straight-forward integration with respect to s_x and s_y in (24) yields the probability density of r:

$$f(r) = \frac{1}{\pi\, \Gamma(n-2)} (1-\rho^2)^{\frac{n-1}{2}} (1-r^2)^{\frac{n-4}{4}} \frac{d^{n-2}}{d(\rho\, r)^{n-2}} \frac{\arcsin \rho\, r}{(1-\rho^2\, r^2)^{\frac{1}{2}}}. \tag{25}$$

For the details of the calculation the reader is referred to M. G. Kendall, Advanced Theory of Statistics I, 14.14, or to the above-cited work of R. A. Fisher in Biometrika 10.

The very useful transformation

$$z=\tfrac{1}{2}\ln\frac{1+r}{1-r}, \tag{26}$$

which is due to R. A. Fisher, transforms the distribution (25) into an approximately normal distribution. In this very good approximation, the variable z is normally distributed with standard deviation

$$\sigma_z=(n-3)^{-\frac{1}{2}}, \tag{27}$$

which is independent of ρ, and expectation

$$\mathscr{E}z=\tfrac{1}{2}\ln\frac{1+\rho}{1-\rho}+\frac{\rho}{2(n-1)}. \tag{28}$$

The correction term on the right-hand side in (28) is always small in comparison with the standard deviation σ_z and can therefore be ignored. It plays a role only when we have observed many correlation coefficients and want to consider the average of the z-values.

How very much the normality is improved by the introduction of z in place of r is to be seen from Figs. 37 and 38, which are taken from R. A. Fisher, Mathematical Methods for Research Workers (11th ed., Fig. 8, p. 200).

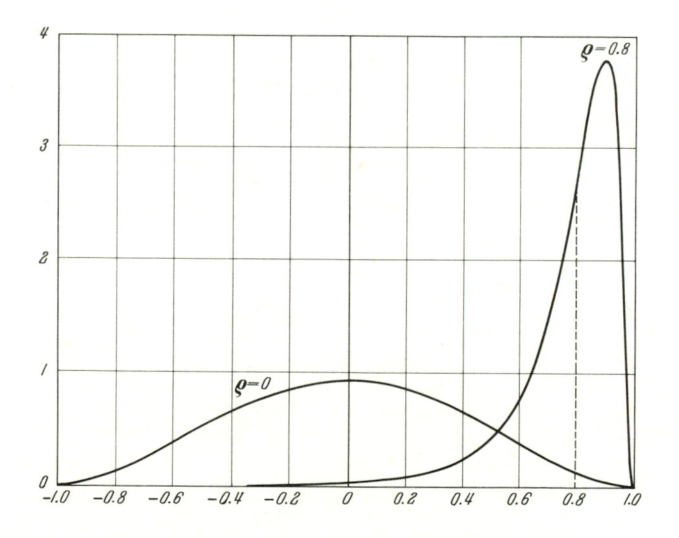

Fig. 37. Probability densities of r for $\rho=0$ and $\rho=0.8$

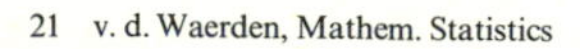

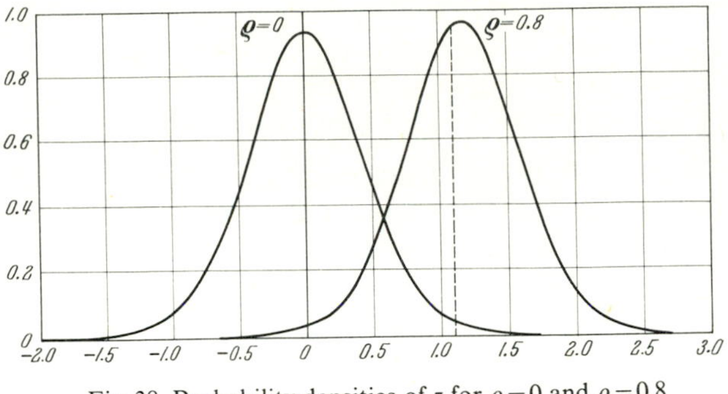

Fig. 38. Probability densities of z for $\rho=0$ and $\rho=0.8$

With the aid of z the following problems can be solved:

1. To test whether an observed correlation agrees with a ρ-value determined theoretically.

2. To determine confidence bounds for ρ once r has been observed.

3. To test whether two observed correlation coefficients r_1 and r_2 can correspond to the same ρ or not.

4. Under the assumption that several observed values $r_1, r_2, \ldots$ correspond to the same ρ, to find the best possible estimate of ρ.

Example 48. In a sample of 25 pairs, the correlation was found to be 0.60. Under the assumption of normality, within what confidence bounds does the true correlation ρ presumably lie?

Formula (26) yields $z=0.693$. From (27) the standard deviation of z is

$$\sigma_z = 22^{-\frac{1}{2}} = 0.2132.$$

The confidence bounds for $\mathscr{E} z$ at the 5%-level are therefore

$$z_1 = 0.693 - 1.96\,\sigma_z = 0.275,$$

$$z_2 = 0.693 + 1.96\,\sigma_z = 1.111.$$

For a given z, r is found by solving (26):

$$r = \frac{e^{2z}-1}{e^{2z}+1}.$$

The confidence bounds for ρ at the 5%-level are thus

$$r_1 = 0.268 \quad \text{and} \quad r_2 = 0.804.$$

Example 49. In a sample of 20 pairs the correlation was found to be $r_1=0.6$ and in one of 25 pairs it was found to be $r_2=0.8$. Is the difference significant?

We find

$$z_1 = 0.693,$$

$$z_2 = 1.099,$$

$$d = z_2 - z_1 = 0.406.$$

The variances are

$$\sigma_1^2 = 17^{-1} = 0.0588,$$

$$\sigma_2^2 = 22^{-1} = 0.0455,$$

$$\sigma_d^2 = \sigma_1^2 + \sigma_2^2 = 0.1043.$$

The ratio of d to σ_d is

$$\frac{d}{\sigma_d} = \frac{0.406}{0.323} = 1.26.$$

The difference d is therefore not significant.

§ 70. Spearman's Rank Correlation R

A. Definition of R

There exist random variables, which we shall call *qualities*, whose values, although not easily measurable, are comparable, so that a rank ordering of the individuals can be established. As examples there come to mind a school child's performance in a particular subject, musical talent, or hair color. If we wish to test the hypothesis of independence between two such qualities, we can use the ranks of a sequence of observations on independent individuals to form a correlation coefficient known as the *rank correlation.*

Customarily we assign the ranks according to descending quality: no. 1 is the best of the class, etc.

We first subtract $(n+1)/2$ from the ranks $1, 2, \ldots, n$ of the n individuals with respect to the two qualities to be compared, so that their arithmetic mean is zero. We then double them so that they are integer-valued and for the one quality denote them by ξ and for the other quality by η. The so-defined rank number ξ, or η, of an individual is equal to $k-l$, if, with respect to the characteristic in question, the individual is exceeded by k other individuals, but itself exceeds l others.

The sum of squares of the ranks ξ, or η, is

$$Q = \sum \xi^2 = \sum \eta^2 = (n-1)^2 + (n-3)^2 + \cdots + (-n+1)^2 = \frac{n(n-1)(n+1)}{3}.$$

The rank correlation R due to Spearman is then defined by

$$R = \frac{\sum \xi \eta}{Q}. \tag{1}$$

It was applied by Spearman in research in psychology[2]. Commonly it is denoted by ρ, but we have already given this letter another meaning.

[2] C. Spearman, The proof and measurement of association between two things. Amer. J. Psychol. 15 (1904) p. 88.

The extreme values of R are again $+1$ and -1, and, to be sure, R takes on the value $+1$ if the two sequences agree completely, and the value -1 if they are just the opposite.

A convenient method of computing R is to assign the usual ranks from 1 through n and form the difference d of the ranks corresponding to each individual. Then

$$R=1-\frac{2\sum d^2}{Q}=1-\frac{6\sum d^2}{n(n-1)(n+1)}. \tag{2}$$

If the two qualities are independent random variables, then the ranks η are likewise independent of the ranks ξ, the expectation of every term in the numerator of (1) is therefore zero, and thus $\mathscr{E}R=0$. Therefore, an observed value of R which is considerably different from zero suggests dependence. In order to make this "considerably" precise, we need to determine how large a deviation of R from zero is still attributable to pure chance in the case of independence; i.e., we must look into the distribution of R in the case of two independent qualities.

B. The Distribution of R for Independent Variables

If the two qualities are independent, then the η-sequence is independent of the ξ-sequence. The η form a permutation of the ξ, and for a given sequence of ξ, all these permutations are equally probable. To each such permutation there is a corresponding value of R. Thus the desired distribution function is a completely determined step function, which is independent of all assumptions about the distribution functions of the variables x and y underlying the qualities, and which, with the necessary patience, can be easily calculated for small n. For $n=5$, for example, it is necessary to calculate the value of R corresponding to each of the 120 permutations of the η. Up to $n=8$ the calculation has actually been carried out; the results are to be found in M. G. Kendall, Rank Correlation Methods (London 1948), Appendix Table 2. For larger n the calculation becomes too tedious.

In the case of independence, the expectation of R is zero. We shall now calculate its variance σ^2. From (1) it follows that

$$Q^2R^2=\left(\sum \xi_i\eta_i\right)\left(\sum \xi_k\eta_k\right)=\sum_i\sum_k \xi_i\xi_k\eta_i\eta_k,$$

$$Q^2\sigma^2=Q^2\mathscr{E}R^2=\sum_i\sum_k \mathscr{E}\xi_i\xi_k\cdot\mathscr{E}\eta_i\eta_k.$$

This sum involves n terms with $i=k$, which, since the index number i or k of an individual does not matter, have the same value, and likewise

$n(n-1)$ equal terms with $i \neq k$. Hence the sum becomes

$$Q^2 \sigma^2 = n \mathscr{E} \xi_1^2 \cdot \mathscr{E} \eta_1^2 + n(n-1) \mathscr{E} \xi_1 \xi_2 \cdot \mathscr{E} \eta_1 \eta_2 . \tag{3}$$

Since $\sum \xi_i = 0$,

$$0 = \mathscr{E}(\xi_1 \sum \xi_i) = \mathscr{E} \xi_1^2 + (n-1) \mathscr{E} \xi_1 \xi_2 ;$$

hence

$$\mathscr{E} \xi_1 \xi_2 = \frac{-1}{n-1} \mathscr{E} \xi_1^2 , \tag{4}$$

and likewise

$$\mathscr{E} \eta_1 \eta_2 = \frac{-1}{n-1} \mathscr{E} \eta_1^2 . \tag{5}$$

Furthermore

$$\sum \xi_i^2 = Q . \tag{6}$$

If we take the expectation on both sides of (6), we thus obtain

$$n \mathscr{E} \xi_1^2 = Q .$$

The same is, of course, true for η_1. Thus we have

$$\mathscr{E} \xi_1^2 = \mathscr{E} \eta_1^2 = Q/n , \tag{7}$$

and, furthermore, from (4) and (5)

$$\mathscr{E} \xi_1 \xi_2 = \mathscr{E} \eta_1 \eta_2 = \frac{-Q}{n(n-1)} . \tag{8}$$

Substitution of (7) and (8) into (3) gives us

$$Q^2 \sigma^2 = \frac{Q^2}{n} + \frac{Q^2}{n(n-1)} = \frac{Q^2}{n-1} ,$$

or, after cancellation of the Q^2,

$$\sigma^2 = \frac{1}{n-1} . \tag{9}$$

The same method also yields the fourth moment of the distribution of R, that is, the expectation of R^4. The result is

$$\mathscr{E} R^4 = \frac{3}{(n-1)^2} \, \frac{25 n^3 - 38 n^2 - 35 n + 72}{25 n (n^2 - 1)} . \tag{10}$$

C. Comparison with the Normal Distribution

If the last two terms in (10), $-35n+72$, are replaced by $-25n+38$, which makes only a slight difference, then (n^2-1) can be cancelled and the result is

$$\mathscr{E} R^4 \sim 3(n-1)^{-2}\left(1-\frac{38}{25n}\right). \tag{11}$$

If we compare this with the fourth moment of a normal distribution with expectation zero and variance $(n-1)^{-1}$,

$$\mu_4 = 3(n-1)^{-2}, \tag{12}$$

we see that (10) is asymptotically equal to (12), although it is somewhat smaller for n finite. This means that those values of R which deviate strongly from zero, and thus make the greatest contribution to the expectation of R^4, are assigned a somewhat smaller probability than they would be assigned by the normal distribution with the same standard deviation.

Calculation of the higher moments $\mathscr{E} R^{2k}$ reveals that after multiplication by $(n-1)^k$ all converge as $n\to\infty$ to the corresponding moments of the normal distribution with expectation zero and standard deviation one[3]; that is, they all converge to

$$\frac{(2k)!}{2^k\, k!}. \tag{13}$$

Hence it follows from the "second limit theorem" (§ 24 F) that the distribution function of $R\sqrt{n-1}$ converges as $n\to\infty$ to the normal distribution with expectation zero and standard deviation one; i.e., R *is asymptotically normally distributed with expectation zero and standard deviation* $\sigma=(n-1)^{-\frac{1}{2}}$.

If, in testing for independence, we act as if R were normally distributed, we remain on the safe side, for, as we have just seen, the extreme values of R actually have a smaller probability than is assigned to them by the normal distribution. We thus get the following simple test for independence:

The hypothesis of independence is to be rejected whenever the rank correlation R *(or, in the case of a two-sided test, the magnitude* $|R|$*) exceeds*

$$R_\alpha = \frac{\Psi(1-\alpha)}{\sqrt{n-1}}. \tag{14}$$

The error probability of the one-sided test is $<\alpha$, that of the two-sided test $<2\alpha$.

[3] Cf. M. G. Kendall, Rank Correlation Methods, p. 61.

D. Comparison with Student's Distribution

Kendall remarked that a distribution with probability density

$$f(R)=B(\tfrac{1}{2},\tfrac{1}{2}n-1)^{-1}(1-R^2)^{\frac{1}{2}n-2}, \tag{15}$$

where $B(p,q)$ is the beta-function discussed in § 12, approximates the distribution of R still somewhat more closely than the normal distribution. The moments of this distribution are

$$\mu_2'=(n-1)^{-1},$$
$$\mu_4'=3(n^2-1)^{-1}.$$

The variance μ_2' is thus exactly correct. The fourth moment μ_4' can also be written as

$$\mu_4'=3(n-1)^{-2}\left(1-\frac{2}{n+1}\right). \tag{16}$$

If we compare this with (11) we see that μ_4' is somewhat smaller. Thus we have an approximation which is not on the safe side.

If in (15) we substitute the new variable

$$t=R\sqrt{\frac{n-2}{1-R^2}}, \tag{17}$$

we get precisely a t-distribution with $n-2$ degrees of freedom. For practical purposes this means that *once we have determined R, we can calculate t from* (17) *and use Student's test.* Of course, in doing so the error probability is somewhat larger than α, or 2α. The use of the bound (14), which is based on the normal distribution, is both simpler and safer.

What we have made plausible here by consideration of the second and fourth moments can be verified directly in the case $n=8$, for which the exact distribution is known. We find the following values for the cut-off bounds:

Level	$2\alpha=1\%$	$2\alpha=2\%$	$2\alpha=5\%$
Exact bounds	0.86 (0.7%)	0.82 (1.5%)	0.72 (4.6%)
Normal distribution	0.97 (0.04%)	0.88 (0.7%)	0.74 (3.6%)
Student's test	0.83 (1.1%)	0.79 (2.2%)	0.71 (5.8%)

The true error probabilities corresponding to these bounds are given in parentheses in each case. As is to be seen, the true error probabilities with Student's test are systematically too large. On the other hand, the true error probabilities with the normal distribution are unnecessarily small. One could, perhaps, choose as a cut-off the arithmetic mean of the two values obtained from the normal distribution and from Student's test; probably, one would still remain on the safe side.

If n is very large, it makes no difference whether one uses the normal distribution or Student's test.

E. The Case of Dependence

We now wish to examine the relationship between the true correlation coefficient ρ and the rank correlation R in the case that the variables are dependent.

We shall assume that underlying the two qualities are normally distributed random variables x and y whose probability density is

$$f(x, y) = \frac{1}{2\pi}(1-\rho^2)^{\frac{1}{2}} e^{-\frac{1}{2}(x^2 - 2\rho x y + y^2)}. \tag{18}$$

For the n independent pairs (x_i, y_i) the probability density is then

$$f(x_1, y_1)\ldots f(x_n, y_n) = (2\pi)^{-n}(1-\rho^2)^{n/2} e^{-\frac{1}{2}\Sigma(x_i^2 - 2\rho x_i y_i + y_i^2)}. \tag{19}$$

From (1) we have

$$QR = \sum \xi_i \eta_i, \tag{20}$$

where ξ_i, for example, is the number of the x_k which are larger than x_i, diminished by the number of the x_k which are smaller than x_i.

We now define a random variable x_{ik} which takes on the value $+1$ if $x_i < x_k$, the value 0 if $x_i = x_k$, and the value -1 if $x_i > x_k$. y_{ik} is defined analogously. Then we have

$$\xi_i = \sum_k x_{ik}, \qquad \eta_i = \sum_k y_{ik}.$$

Substituting this into (20) gives

$$QR = \sum_i \sum_k \sum_l (x_{ik}\, y_{il}). \tag{21}$$

Taking the expectation on both sides yields

$$Q\hat{R} = \sum_i \sum_k \sum_l \mathscr{E}(x_{ik}\, y_{il}). \tag{22}$$

Those terms of this sum with $i=k$ or $i=l$ are zero. Of the remaining terms in the sum, there are $n(n-1)(n-2)$ terms with $k \neq l$, all of which have the same value, and likewise $n(n-1)$ equal terms with $k=l$. Thus

$$Q\hat{R} = n(n-1)(n-2)\,\mathscr{E}(x_{12}\, y_{13}) + n(n-1)\,\mathscr{E}(x_{12}\, y_{12}),$$

or, after division by $Q = \frac{1}{3}n(n-1)(n+1)$,

$$\hat{R} = 3\,\frac{n-2}{n+1}\,\mathscr{E}(x_{12}\, y_{13}) + \frac{3}{n+1}\,\mathscr{E}(x_{12}\, y_{12}). \tag{23}$$

Now the only problem which still remains is that of calculating the expectations of $x_{12}\,y_{12}$ and $x_{12}\,y_{13}$. The variable $x_{12}\,y_{12}$ takes on the value $+1$ if either $x_1<x_2$ and $y_1<y_2$ or $x_1>x_2$ and $y_1>y_2$, and it takes on the value -1 if either $x_1<x_2$ and $y_1>y_2$ or $x_1>x_2$ and $y_1<y_2$. The probability that $x_1<x_2$ and $y_1<y_2$ is equal to the integral of the probability density $f(x_1,y_1)\,f(x_2,y_2)$ over the region $x_1<x_2$, $y_1<y_2$:

$$P_1=(2\pi)^{-2}(1-\rho^2)\iiiint\limits_{\substack{x_1<x_2\\ y_1<y_2}} e^{-\frac{1}{2}(x_1^2-2\rho x_1 y_1+y_1^2)}\, e^{-\frac{1}{2}(x_2^2-2\rho x_2 y_2+y_2^2)} \cdot dx_1\,dy_1\,dx_2\,dy_2.$$

The integral has exactly the same form as the integrals evaluated in general in § 14C:

$$I=\sqrt{2\pi}^{\,-n}\sqrt{g}\int\limits_{\substack{(ux)>0\\ (vx)>0}}\cdots\int e^{-\frac{1}{2}G}\,dx^1\ldots dx^n, \tag{24}$$

where

$$G=\sum g_{ik}\,x^i x^k=x_1^2-2\rho\,x_1\,y_1+y_1^2+x_2^2-2\rho\,x_2\,y_2+y_2^2. \tag{25}$$

In order to employ the notation used at that time, we need to set, say,

$$\begin{array}{llll} x^1=x_1, & x^2=y_1, & x^3=x_2, & x^4=y_2,\\ u_1=-1, & u_2=0, & u_3=+1, & u_4=0,\\ v_1=0, & v_2=-1, & v_3=0, & v_4=+1. \end{array}$$

The form contragredient to G is

$$\sum g^{ik}\,u_i\,u_k=\frac{u_1^2+2\rho\,u_1\,u_2+u_2^2}{1-\rho^2}+\frac{u_3^2+2\rho\,u_3\,u_4+u_4^2}{1-\rho^2}. \tag{26}$$

From this we get the values of the invariants

$$(u\,u)=\sum g^{ik}\,u_i\,u_k=\frac{1}{1-\rho^2}+\frac{1}{1-\rho^2}=\frac{2}{1-\rho^2},$$

$$(u\,v)=\sum g^{ik}\,u_i\,v_k=\frac{\rho}{1-\rho^2}+\frac{\rho}{1-\rho^2}=\frac{2\rho}{1-\rho^2},$$

$$(v\,v)=\sum g^{ik}\,v_i\,v_k=\frac{1}{1-\rho^2}+\frac{1}{1-\rho^2}=\frac{2}{1-\rho^2}.$$

The integral $P_1=I$ thus becomes

$$P_1=\frac{1}{2\pi}\,\text{arc cos}\,\frac{-(u\,v)}{\sqrt{(u\,u)}\,\sqrt{(v\,v)}}=\frac{1}{2\pi}\,\text{arc cos}(-\rho). \tag{27}$$

The probability P_2 that $x_1 > x_2$ and $y_1 > y_2$ has exactly the same value. The probability P_3 that $x_1 > x_2$ and $y_1 < y_2$ is obtained by multiplying the v_i by (-1):

$$P_3 = \frac{1}{2\pi} \operatorname{arc\,cos} \rho. \tag{28}$$

The probability P_4 that $x_1 < x_2$ and $y_1 > y_2$ is the same. Hence the expectation becomes

$$\begin{aligned} \mathscr{E}(x_{12}\,y_{12}) &= (P_1 + P_2)\cdot 1 + (P_3 + P_4)\cdot(-1) = 2P_1 - 2P_3 \\ &= \frac{1}{\pi} \operatorname{arc\,cos}(-\rho) - \frac{1}{\pi} \operatorname{arc\,cos} \rho \\ &= \frac{1}{\pi}\left(\frac{\pi}{2} + \operatorname{arc\,sin} \rho\right) - \frac{1}{\pi}\left(\frac{\pi}{2} - \operatorname{arc\,sin} \rho\right) \\ &= \frac{2}{\pi} \operatorname{arc\,sin} \rho. \end{aligned} \tag{29}$$

The method of calculating the expectation of the variable $x_{12}\,y_{13}$ is exactly the same. The variable has the value $+1$ if either $x_1 < x_2$ and $y_1 < y_3$ or $x_1 > x_2$ and $y_1 > y_3$. The probability of the first event is

$$P_5 = (2\pi)^{-3}(1-\rho^2)^{\frac{3}{2}} \underset{\substack{x_1 < x_2 \\ y_1 < y_3}}{\iiiint\!\!\iint} e^{-\frac{1}{2}G}\, dx_1\, dy_1\, dx_2\, dy_2\, dx_3\, dy_3,$$

where

$$G = (x_1^2 - 2\rho\, x_1\, y_1 + y_1^2) + (x_2^2 - 2\rho\, x_2\, y_2 + y_2^2) + (x_3^2 - 2\rho\, x_3\, y_3 + y_3^2),$$

$$\sum g^{ik}\, u_i\, u_k = \frac{u_1^2 + 2\rho\, u_1\, u_2 + u_2^2}{1-\rho^2} + \frac{u_3^2 + 2\rho\, u_3\, u_4 + u_4^2}{1-\rho^2} + \frac{u_5^2 + 2\rho\, u_5\, u_6 + u_6^2}{1-\rho^2}.$$

The invariants are

$$(u\,u) = \frac{2}{1-\rho^2}, \qquad (u\,v) = \frac{\rho}{1-\rho^2}, \qquad (v\,v) = \frac{2}{1-\rho^2}.$$

Therefore we have

$$P_5 = \frac{1}{2\pi} \operatorname{arc\,cos} \frac{-(u\,v)}{\sqrt{(u\,u)}\sqrt{(v\,v)}} = \frac{1}{2\pi} \operatorname{arc\,cos} \frac{-\rho}{2}.$$

The probability P_6 that $x_1 > x_2$ and $y_1 > y_3$ is the same as P_5. The probabilities P_7 and P_8 corresponding to the two remaining cases possible are

$$P_7 = P_8 = \frac{1}{2\pi} \operatorname{arc\,cos} \frac{\rho}{2}.$$

Thus we have

$$\mathscr{E}(x_{12}\, y_{13}) = (P_5 + P_6) \cdot 1 + (P_7 + P_8) \cdot (-1) = \frac{2}{\pi} \arcsin \frac{\rho}{2}. \tag{30}$$

If we now substitute (29) and (30) into (23), we get

$$\hat{R} = \frac{6}{\pi} \frac{n-2}{n+1} \arcsin \frac{\rho}{2} + \frac{6}{\pi(n+1)} \arcsin \rho. \tag{31}$$

For large n we see from (31) that

$$\hat{R} \sim \frac{6}{\pi} \arcsin \frac{\rho}{2}. \tag{32}$$

For moderate n, $\hat{R}$ is somewhat smaller, since

$$2 \arcsin \frac{\rho}{2} < \arcsin \rho < 3 \arcsin \frac{\rho}{2},$$

and therefore

$$\frac{6}{\pi} \frac{n}{n+1} \arcsin \frac{\rho}{2} < \hat{R} < \frac{6}{\pi} \arcsin \frac{\rho}{2}.$$

The difference between $n/(n+1)$ and 1 is so small that the approximation can also be used for moderately large n. Solving this for ρ we get

$$\rho \sim 2 \sin \frac{\pi}{6} \hat{R}. \tag{33}$$

Thus: *For large n, $2 \sin \frac{\pi}{6} R$ can be used as an estimate of the true correlation coefficient ρ.*

This whole discussion is valid only under the assumption that the x and y are simultaneously normally distributed. If this assumption is not satisfied, it is nevertheless possible to consider R as an estimate of a "true rank correlation" P, which we now define.

Let the distributions of the variables x and y be $F(x)$ and $G(y)$, which are assumed to be continuous, and set[4]

$$\xi = F(x), \qquad \eta = G(y).$$

The variables ξ and η are then uniformly distributed between 0 and 1, and thus their variances are

$$\sigma_\xi^2 = \sigma_\eta^2 = \tfrac{1}{12}.$$

Now P is defined to be the true correlation coefficient between ξ and η:

$$\mathsf{P} = 12\, \mathscr{E}(\xi - \tfrac{1}{2})(\eta - \tfrac{1}{2}). \tag{34}$$

---

[4] See M. G. Kendall, Rank Correlation Methods 9.7 and 10.6. Kendall refers to ξ and η as "grades".

If x and y are simultaneously normally distributed, then between ρ and P there exists a relationship analogous to (33):

$$\rho = 2 \sin \frac{\pi}{6} \mathsf{P}, \tag{35}$$

which was established by Karl Pearson (see Draper's Company Research Memoirs, Biometric Series IV, Cambridge 1907, p. 13).

Example 50 (Karl Pearson, Biometrika 13, p. 304). In an examination of 27 candidates for the news service, the points given in arithmetic ranged from 1 to 300 and those given in four other subjects (orthography, handwriting, geography, and English) ranged from 1 to 200. The points were added and the candidates were ordered according to the total number of points scored. In the following table are given first the ranks and total scores for all subjects taken together and then the ranks and scores for arithmetic.

Total		Arithmetic		Total		Arithmetic	
rank	score	rank	score	rank	score	rank	score
1	907	1	230	15	580	13	131
2	764	9	158	16	561	15	128
3	748	2	228	17	560	18	116
4	746	10	154	18	532	22	82
5	724	8	162	19	529	16	125
6	718	5	182	20	526	17	122
7	710	14	129	21	515	19	114
8	703	7	164	22	484	21	93
9	677	3	187	23	463	25	61
10	665	4	186	24	444	26	38
11	645	11	151	25	386	27	37
12	643	6	167	26	369	23	63
13	634	20	103	27	288	24	62
14	628	12	146				

Pearson found the rank correlation to be

$$R = 0.8834.$$

From formula (33) above this yields as an estimate r' of the true correlation coefficient

$$r' = 2 \sin \frac{\pi}{6} R = 0.893.$$

The correlation coefficient calculated directly from the scores was

$$r = 0.896.$$

Pearson correctly remarks that the agreement between r and r' in this case is excellent.

§ 71. Kendall's Rank Correlation *T*

A rank correlation coefficient τ, or preferably T, which is related to R, was proposed first by Greiner and Esscher and later by M.G. Kendall. Kendall's book, Rank Correlation Methods, which we have already quoted repeatedly, contains a thorough investigation of the characteristics of T. Here we shall touch upon some main points only.

A. Definition of *T*

Let there again be n individuals which are ordered in two ways. For every pair of individuals (i, k) we record a score of $+1$ if they occur in the same order in the two rankings, and -1 otherwise. In the notation of § 70E the contribution of the pair (i, k) is the product $x_{ik} y_{ik}$. The sum S of the contributions of all pairs,

$$S = \sum x_{ik} y_{ik}, \tag{1}$$

can not exceed

$$\binom{n}{2} = \frac{1}{2} n(n-1)$$

in magnitude. If we therefore set

$$T = \frac{S}{\frac{1}{2} n(n-1)}, \tag{2}$$

then T takes on values between -1 and $+1$ only, and, in fact, T is $+1$ only if the two rankings are in perfect agreement and -1 only if they are just the opposite.

If the ranks of the first ordering are arranged according to increasing magnitude from 1 to n and if the ranks of the second ordering are written below each one,

$$\begin{array}{llll} \xi_1 = 1, & \xi_2 = 2, & \ldots, & \xi_n = n, \\ \eta_1, & \eta_2, & \ldots, & \eta_n, \end{array}$$

then the calculation of S is simplified considerably. The procedure is to count how many η_k larger than η_1 stand to the right of η_1, then how many η_k larger than η_2 stand to the right of η_2, etc. If P denotes the sum of all these numbers, then S is the sum of P contributions $+1$ and $\binom{n}{2} - P$ contributions -1. Therefore

$$S = 2P - \tfrac{1}{2} n(n-1), \tag{3}$$

$$T = \frac{2P}{\frac{1}{2} n(n-1)} - 1. \tag{4}$$

B. Distribution of T

The expectation of T is obviously zero if the variables x and y are independent. If they are dependent and normally distributed with correlation ρ, the expectation of S is

$$\begin{aligned}\mathscr{E}S=\mathscr{E}\sum x_{ik}\,y_{ik}&=\frac{n(n-1)}{2}\,\mathscr{E}x_{12}\,y_{12}\\&=\frac{n(n-1)}{2}\,\frac{2}{\pi}\arcsin\rho\,,\end{aligned}$$

according to § 70 formula (29), and therefore

$$\mathscr{E}T=\frac{2}{\pi}\arcsin\rho\,. \tag{5}$$

Therefore, if the joint distribution of x and y is normal, then

$$r''=\sin\left(\frac{\pi}{2}\,T\right) \tag{6}$$

may be used as an estimate of ρ.

We now return to the case of independent variables and use (1) to calculate the variance of S:

$$\sigma_S^2=\mathscr{E}S^2=\mathscr{E}\left(\sum x_{ik}\,y_{ik}\right)^2.$$

Since Kendall carries out the calculation completely in his book, we record only the result:

$$\sigma_S^2=\frac{n(n-1)(2n+5)}{18}\,. \tag{7}$$

This, together with (2), yields

$$\sigma_T^2=\frac{2(2n+5)}{9n(n-1)}\,. \tag{8}$$

If we were to calculate likewise the higher moments $\mathscr{E}T^4$, $\mathscr{E}T^6, \ldots$ (the odd moments are zero, since the values T and $-T$ always have the same probability), we would see that asymptotically, as $n\to\infty$, they are equal to the moments of the normal distribution with standard deviation σ_T:

$$\mu_{2r}=\frac{(2r)!}{2^r\,r!}\,\sigma_T^{2r}\,. \tag{9}$$

Hence T *is asymptotically normally distributed with expectation zero and variance* (8).

This asymptotic normality is valid even for dependent variables with an arbitrary distribution function, so long as $\mathscr{E}T$ does not lie too near $+1$ or -1. For the proof see Kendall, Rank Correlation Methods 5.21.

For dependent variables the theorem on the asymptotic normality is not very useful for practical purposes, because even for n moderately large the distribution can still be considerably skew and, moreover, its standard deviation is unknown. For independent variables, however, the standard deviation is known from (8), and the normal curve already gives a good approximation for n as small as 8. Therefore T can well be used as a test statistic for testing the hypothesis of independence. For $n \leqq 10$ the exact distribution of T is known (Kendall, Appendix Table 1), and for $n > 10$ the normal approximation can be used; i.e., the hypothesis of independence can be rejected whenever T itself or, for the two-sided test, the absolute value $|T|$, exceeds the bound

$$T_\alpha = \sigma_T \, \Psi(1-\alpha). \tag{10}$$

The error probability is then α for the one-sided test and 2α for the two-sided.

C. Comparison of R and T

Power series expansion of the expected values $\hat{R}$ and $\hat{T}$ calculated under the normality assumption yields

$$\hat{R} = \frac{3}{\pi}\left(\frac{n}{n+1}\rho + \frac{1}{24}\,\frac{n+6}{n+1}\rho^3 + \cdots\right), \tag{11}$$

$$\hat{T} = \frac{2}{\pi}\left(\rho + \frac{1}{6}\rho^3 + \cdots\right). \tag{12}$$

From this we see that for large n and small ρ the ratio of the expected values is approximately 3:2. If, on the other hand, we compare the standard deviations for $\rho = 0$, namely,

$$\sigma_R = (n-1)^{-\frac{1}{2}}, \tag{13}$$

$$\sigma_T = \left[\frac{2(2n+5)}{9n(n-1)}\right]^{\frac{1}{2}} = \frac{2}{3}\left(1 + \frac{5}{2n}\right)^{\frac{1}{2}}(n-1)^{-\frac{1}{2}}, \tag{14}$$

we find that for large n the same ratio 3:2 holds again. We can thus come to the tentative conclusion that R and T maintain approximately the ratio 3:2, so long as both are sufficiently smaller than 1 in magnitude. This conclusion is consistent with a result of H. E. Daniels, who has shown that the correlation coefficient between R and T tends to 1 as $n \to \infty$, and that even for moderately large n it is very close to 1 (Kendall, Rank Correlation Methods 5.14).

Which test statistic, then, yields the better test for independence, R or T? The one-sided R-test rejects the hypothesis of independence whenever $R > R_\alpha$, just as the T-test does whenever $T > T_\alpha$. The problem, then, is to attempt to determine which of these two tests has the greater power (in the sense of § 59). By the power of a test we mean, in this case, the probability that the test leads to the rejection of the hypothesis of independence, if in fact the variables x and y are dependent; i.e., the probability that it leads to a detection of the deviation from the independence hypothesis.

In order to answer the question we must first make an assumption about the distribution of the variables x and y. Once more we shall assume that they are simultaneously normally distributed. Their joint probability density can then be written (as in § 69) as

$$f(x, y) = \frac{1}{2\pi}(1-\rho^2)^{\frac{1}{2}} e^{-\frac{1}{2}(x^2 - 2\rho x y + y^2)}. \tag{15}$$

The power of any test is a function of ρ.

We shall assume for the moment that n is so large that the two variables R and T are nearly normally distributed. The cut-off points R_α and T_α are then found from the normal distribution to be

$$R_\alpha = \sigma_R \cdot \Psi(1-\alpha) = (n-1)^{-\frac{1}{2}} \Psi(1-\alpha), \tag{16}$$

$$T_\alpha = \sigma_T \cdot \Psi(1-\alpha) = \frac{2}{3}\left(1 + \frac{5}{2n}\right)^{\frac{1}{2}} (n-1)^{-\frac{1}{2}} \Psi(1-\alpha). \tag{17}$$

In (16) and (17) σ_R and σ_T have, of course, been calculated for $\rho = 0$. The deviation of ρ from zero effects a change in σ_R and σ_T as well, but this change is only of the order of magnitude ρ^2 and will be ignored for the moment. Hence we shall make our calculations both for R and for T from normal distributions with expected values given by (11) and (12) and standard deviations given by (13) and (14). The power of the R-test, or the probability of the event $R > R_\alpha$, is then

$$M_R(\rho) = 1 - \Phi\left(\frac{R_\alpha - \hat{R}}{\sigma_R}\right) = \Phi\left(\frac{\hat{R} - R_\alpha}{\sigma_R}\right), \tag{18}$$

and the power of the T-test is likewise

$$M_T(\rho) = \Phi\left(\frac{\hat{T} - T_\alpha}{\sigma_T}\right). \tag{19}$$

Whenever ρ is large relative to $(n-1)^{-\frac{1}{2}}$, the two expectations $\hat{R}$ and $\hat{T}$ are large relative to σ_R and σ_T, respectively, and the power functions expressed by (18) and (19) both take on values very close to one. Therefore we can assume that ρ^2 is at most of the order of magnitude n^{-1}. If we now drop all terms in (18) and (19) of the order of magnitude n^{-1} and smaller, we get for $M_R(\rho)$ and $M_T(\rho)$ exactly the same asymptotic function $M(\rho)$, namely,

$$M(\rho)=\Phi\left\{\frac{3}{\pi}\rho(n-1)^{\frac{1}{2}}-\Psi(1-\alpha)\right\}. \tag{20}$$

Hence in this approximation both tests have the same power function, which is represented in Fig. 39 by the continuous-line curve denoted by $M(\rho)$.

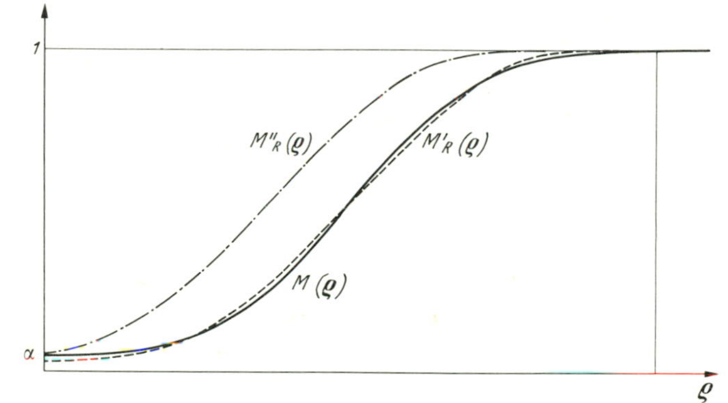

Fig. 39. Power functions for the R- and T-tests. The continuous-line curve holds for both R and T in the first approximation. The dashed-line curve and the dot-dash curve represent improved approximations to the power function for R

We shall now refine the approximation step by step. For the moment we shall retain the normality assumption, but include the terms of order of magnitude ρ^3 in (11) and (12) and the term $5/2n$ in (17). Calculation yields practically the same result: as before, the two curves corresponding to $M_R(\rho)$ and $M_T(\rho)$ very nearly coincide.

As we have seen, the normal distribution is a good approximation to the distribution of T, but not to that of R. For positive R-values the true distribution function of R tends more quickly toward 1, and for negative R-values more quickly towards 0, than does the normal distribution. If we take this into consideration, but leave the cut-off value R_α unchanged for the time-being, the result for $M_R(\rho)$ is a curve of the form shown in Fig. 39 as a dashed-line curve and denoted by M'_R. At first, for small ρ, this curve lies below the continuous-line curve, then above it until $M=\frac{1}{2}$, then again below it, and finally, for large ρ, again above it.

The figure is only qualitatively correct: the differences between the curves are somewhat exaggerated.

For large ρ the skewness of the density functions of R and T would have to be taken into consideration. This skewness effects a decrease in the values of the power functions $M_R(\rho)$ and $M_T(\rho)$ in the vicinity of $\rho=1$, and, in fact, a somewhat greater decrease for M_R than for M_T. This reduction in power is, however, fairly immaterial and is not indicated in Fig. 39.

The last correction which can be taken into consideration, that of the cut-off R_α, is, however, decisive. As we have seen, the cut-off values calculated from the normal distribution are approximately correct for T, since T is approximately normally distributed, but considerably too large for R. For $n=8$, at the 1 %-level we obtain the cut-off point $R_\alpha=0.97$, whereas the exact cut-off is only 0.86. Thus we are faced with a choice. If, in order to be safe, we leave the cut-off R_α unchanged, then the power of the R-test also remains unchanged, but the significance level of the test is considerably smaller than α. If, on the other hand, we take a smaller value for R_α, so that the significance level remains exactly $\leqq \alpha$, then the dashed-line curve is shifted a considerable distance to the left and we get a larger power function $M_R''(\rho)$, which is represented in the figure by the dot-dash curve. Therefore:

Compared with the T-test, the R-test with approximately the same power has a smaller significance level, and that with the same significance level is more powerful.

In addition, the calculation of R demands less computation than that of T. Therefore, it would seem that from both theoretical and practical considerations the older Spearman rank correlation R is preferable to its newer competitor T.

Chapter Fourteen

Tables

Table 1. *Normal distribution function* $\Phi(t)=\frac{1}{\sqrt{2\pi}}\int_{-\infty}^{t} e^{-\frac{1}{2}x^2}\,dx$

t	0	1	2	3	4	5	6	7	8	9
−0.0	.5000	.4960	.4920	.4880	.4840	.4801	.4761	.4721	.4681	.4641
−0.1	.4602	.4562	.4522	.4483	.4443	.4404	.4364	.4325	.4286	.4247
−0.2	.4207	.4168	.4129	.4090	.4052	.4013	.3974	.3936	.3897	.3859
−0.3	.3821	.3783	.3745	.3707	.3669	.3632	.3594	.3557	.3520	.3483
−0.4	.3446	.3409	.3372	.3336	.3300	.3264	.3228	.3192	.3156	.3121
−0.5	.3085	.3050	.3015	.2981	.2946	.2912	.2877	.2843	.2810	.2776
−0.6	.2743	.2709	.2676	.2643	.2611	.2578	.2546	.2514	.2483	.2451
−0.7	.2420	.2389	.2358	.2327	.2297	.2266	.2236	.2206	.2177	.2148
−0.8	.2119	.2090	.2061	.2033	.2005	.1977	.1949	.1922	.1894	.1867
−0.9	.1841	.1814	.1788	.1762	.1736	.1711	.1685	.1660	.1635	.1611
−1.0	.1587	.1562	.1539	.1515	.1492	.1469	.1446	.1423	.1401	.1379
−1.1	.1357	.1335	.1314	.1292	.1271	.1251	.1230	.1210	.1190	.1170
−1.2	.1151	.1131	.1112	.1093	.1075	.1056	.1038	.1020	.1003	.0985
−1.3	.0968	.0951	.0934	.0918	.0901	.0885	.0869	.0853	.0838	.0823
−1.4	.0808	.0793	.0778	.0764	.0749	.0735	.0721	.0708	.0694	.0681
−1.5	.0668	.0655	.0643	.0630	.0618	.0606	.0594	.0582	.0571	.0559
−1.6	.0548	.0537	.0526	.0516	.0505	.0495	.0485	.0475	.0465	.0455
−1.7	.0446	.0436	.0427	.0418	.0409	.0401	.0392	.0384	.0375	.0367
−1.8	.0359	.0351	.0344	.0336	.0329	.0322	.0314	.0307	.0301	.0294
−1.9	.0288	.0281	.0274	.0268	.0262	.0256	.0250	.0244	.0239	.0233
−2.0	.0228	.0222	.0217	.0212	.0207	.0202	.0197	.0192	.0188	.0183
−2.1	.0179	.0174	.0170	.0166	.0162	.0158	.0154	.0150	.0146	.0143
−2.2	.0139	.0136	.0132	.0129	.0125	.0122	.0119	.0116	.0113	.0110
−2.3	.0107	.0104	.0102	.0099	.0096	.0094	.0091	.0089	.0087	.0084
−2.4	.0082	.0080	.0078	.0075	.0073	.0071	.0069	.0068	.0066	.0064
−2.5	.0062	.0060	.0059	.0057	.0055	.0054	.0052	.0051	.0049	.0048
−2.6	.0047	.0045	.0044	.0043	.0041	.0040	.0039	.0038	.0037	.0036
−2.7	.0035	.0034	.0033	.0032	.0031	.0030	.0029	.0028	.0027	.0026
−2.8	.0026	.0025	.0024	.0023	.0023	.0022	.0021	.0021	.0020	.0019
−2.9	.0019	.0018	.0018	.0017	.0016	.0016	.0015	.0015	.0014	.0014

$t=$ −3.0	−3.1	−3.2	−3.3	−3.4	−3.5	−3.6	−3.7	−3.8	−3.9
$\Phi(t)=$ .0013	.0010	.0007	.0005	.0003	.0002	.0002	.0001	.0001	.0000

Table 1 (Continuation). $\Phi(t)$

t	0	1	2	3	4	5	6	7	8	9
0.0	.5000	.5040	.5080	.5120	.5160	.5199	.5239	.5279	.5319	.5359
0.1	.5398	.5438	.5478	.5517	.5557	.5596	.5636	.5675	.5714	.5753
0.2	.5793	.5832	.5871	.5910	.5948	.5987	.6026	.6064	.6103	.6141
0.3	.6179	.6217	.6255	.6293	.6331	.6368	.6406	.6443	.6480	.6517
0.4	.6554	.6591	.6628	.6664	.6700	.6736	.6772	.6808	.6844	.6879
0.5	.6915	.6950	.6985	.7019	.7054	.7088	.7123	.7157	.7190	.7224
0.6	.7257	.7291	.7324	.7357	.7389	.7422	.7454	.7486	.7517	.7549
0.7	.7580	.7611	.7642	.7673	.7703	.7734	.7764	.7794	.7823	.7852
0.8	.7881	.7910	.7939	.7967	.7995	.8023	.8051	.8078	.8106	.8133
0.9	.8159	.8186	.8212	.8238	.8264	.8289	.8315	.8340	.8365	.8389
1.0	.8413	.8438	.8461	.8485	.8508	.8531	.8554	.8577	.8599	.8621
1.1	.8643	.8665	.8686	.8708	.8729	.8749	.8770	.8790	.8810	.8830
1.2	.8849	.8869	.8888	.8907	.8925	.8944	.8962	.8980	.8997	.9015
1.3	.9032	.9049	.9066	.9082	.9099	.9115	.9131	.9147	.9162	.9177
1.4	.9192	.9207	.9222	.9236	.9251	.9265	.9279	.9292	.9306	.9319
1.5	.9332	.9345	.9357	.9370	.9382	.9394	.9406	.9418	.9429	.9441
1.6	.9452	.9463	.9474	.9484	.9495	.9505	.9515	.9525	.9535	.9545
1.7	.9554	.9564	.9573	.9582	.9591	.9599	.9608	.9616	.9625	.9633
1.8	.9641	.9649	.9656	.9664	.9671	.9678	.9686	.9693	.9699	.9706
1.9	.9713	.9719	.9726	.9732	.9738	.9744	.9750	.9756	.9761	.9767
2.0	.9772	.9778	.9783	.9788	.9793	.9798	.9803	.9808	.9812	.9817
2.1	.9821	.9826	.9830	.9834	.9838	.9842	.9846	.9850	.9854	.9857
2.2	.9861	.9864	.9868	.9871	.9875	.9878	.9881	.9884	.9887	.9890
2.3	.9893	.9896	.9898	.9901	.9904	.9906	.9909	.9911	.9913	.9916
2.4	.9918	.9920	.9922	.9925	.9927	.9929	.9931	.9932	.9934	.9936
2.5	.9938	.9940	.9941	.9943	.9945	.9946	.9948	.9949	.9951	.9952
2.6	.9953	.9955	.9956	.9957	.9959	.9960	.9961	.9962	.9963	.9964
2.7	.9965	.9966	.9967	.9968	.9969	.9970	.9971	.9972	.9973	.9974
2.8	.9974	.9975	.9976	.9977	.9977	.9978	.9979	.9979	.9980	.9981
2.9	.9981	.9982	.9982	.9983	.9984	.9984	.9985	.9985	.9986	.9986

$t=$3.0	3.1	3.2	3.3	3.4	3.5	3.6	3.7	3.8	3.9
$\Phi(t)=$.9987	.9990	.9993	.9995	.9997	.9998	.9998	.9999	.9999	1.0000

Table 2. *Inverse function* $\Psi(x)$

x	0	1	2	3	4	5	6	7	8	9
0.00	$-\infty$	-3.09	-2.88	-2.75	-2.65	-2.58	-2.51	-2.46	-2.41	-2.37
0.01	-2.33	-2.29	-2.26	-2.23	-2.20	-2.17	-2.14	-2.12	-2.10	-2.07
0.02	-2.05	-2.03	-2.01	-2.00	-1.98	-1.96	-1.94	-1.93	-1.91	-1.90
0.03	-1.88	-1.87	-1.85	-1.84	-1.83	-1.81	-1.80	-1.79	-1.77	-1.76
0.04	-1.75	-1.74	-1.73	-1.72	-1.71	-1.70	-1.68	-1.67	-1.66	-1.65
0.05	-1.64	-1.64	-1.63	-1.62	-1.61	-1.60	-1.59	-1.58	-1.57	-1.56
0.06	-1.55	-1.55	-1.54	-1.53	-1.52	-1.51	-1.51	-1.50	-1.49	-1.48
0.07	-1.48	-1.47	-1.46	-1.45	-1.45	-1.44	-1.43	-1.43	-1.42	-1.41
0.08	-1.41	-1.40	-1.39	-1.39	-1.38	-1.37	-1.37	-1.36	-1.35	-1.35
0.09	-1.34	-1.33	-1.33	-1.32	-1.32	-1.31	-1.30	-1.30	-1.29	-1.29
0.10	-1.28	-1.28	-1.27	-1.26	-1.26	-1.25	-1.25	-1.24	-1.24	-1.23
0.11	-1.23	-1.22	-1.22	-1.21	-1.21	-1.20	-1.20	-1.19	-1.19	-1.18
0.12	-1.18	-1.17	-1.17	-1.16	-1.16	-1.15	-1.15	-1.14	-1.14	-1.13
0.13	-1.13	-1.12	-1.12	-1.11	-1.11	-1.10	-1.10	-1.09	-1.09	-1.09
0.14	-1.08	-1.08	-1.07	-1.07	-1.06	-1.06	-1.05	-1.05	-1.05	-1.04
0.15	-1.04	-1.03	-1.03	-1.02	-1.02	-1.02	-1.01	-1.01	-1.00	-1.00
0.16	-0.99	-0.99	-0.99	-0.98	-0.98	-0.97	-0.97	-0.97	-0.96	-0.96
0.17	-0.95	-0.95	-0.95	-0.94	-0.94	-0.93	-0.93	-0.93	-0.92	-0.92
0.18	-0.92	-0.91	-0.91	-0.90	-0.90	-0.90	-0.89	-0.89	-0.89	-0.88
0.19	-0.88	-0.87	-0.87	-0.87	-0.86	-0.86	-0.86	-0.85	-0.85	-0.85
0.20	-0.84	-0.84	-0.83	-0.83	-0.83	-0.82	-0.82	-0.82	-0.81	-0.81
0.21	-0.81	-0.80	-0.80	-0.80	-0.79	-0.79	-0.79	-0.78	-0.78	-0.78
0.22	-0.77	-0.77	-0.77	-0.76	-0.76	-0.76	-0.75	-0.75	-0.75	-0.74
0.23	-0.74	-0.74	-0.73	-0.73	-0.73	-0.72	-0.72	-0.72	-0.71	-0.71
0.24	-0.71	-0.70	-0.70	-0.70	-0.69	-0.69	-0.69	-0.68	-0.68	-0.68
0.25	-0.67	-0.67	-0.67	-0.67	-0.66	-0.66	-0.66	-0.65	-0.65	-0.65
0.26	-0.64	-0.64	-0.64	-0.63	-0.63	-0.63	-0.63	-0.62	-0.62	-0.62
0.27	-0.61	-0.61	-0.61	-0.60	-0.60	-0.60	-0.59	-0.59	-0.59	-0.59
0.28	-0.58	-0.58	-0.58	-0.57	-0.57	-0.57	-0.57	-0.56	-0.56	-0.56
0.29	-0.55	-0.55	-0.55	-0.54	-0.54	-0.54	-0.54	-0.53	-0.53	-0.53
0.30	-0.52	-0.52	-0.52	-0.52	-0.51	-0.51	-0.51	-0.50	-0.50	-0.50
0.31	-0.50	-0.49	-0.49	-0.49	-0.48	-0.48	-0.48	-0.48	-0.47	-0.47
0.32	-0.47	-0.46	-0.46	-0.46	-0.46	-0.45	-0.45	-0.45	-0.45	-0.44
0.33	-0.44	-0.44	-0.43	-0.43	-0.43	-0.43	-0.42	-0.42	-0.42	-0.42
0.34	-0.41	-0.41	-0.41	-0.40	-0.40	-0.40	-0.40	-0.39	-0.39	-0.39
0.35	-0.39	-0.38	-0.38	-0.38	-0.37	-0.37	-0.37	-0.37	-0.36	-0.36
0.36	-0.36	-0.36	-0.35	-0.35	-0.35	-0.35	-0.34	-0.34	-0.34	-0.33
0.37	-0.33	-0.33	-0.33	-0.32	-0.32	-0.32	-0.32	-0.31	-0.31	-0.31
0.38	-0.31	-0.30	-0.30	-0.30	-0.30	-0.29	-0.29	-0.29	-0.28	-0.28
0.39	-0.28	-0.28	-0.27	-0.27	-0.27	-0.27	-0.26	-0.26	-0.26	-0.26
0.40	-0.25	-0.25	-0.25	-0.25	-0.24	-0.24	-0.24	-0.24	-0.23	-0.23
0.41	-0.23	-0.23	-0.22	-0.22	-0.22	-0.21	-0.21	-0.21	-0.21	-0.20
0.42	-0.20	-0.20	-0.20	-0.19	-0.19	-0.19	-0.19	-0.18	-0.18	-0.18
0.43	-0.18	-0.17	-0.17	-0.17	-0.17	-0.16	-0.16	-0.16	-0.16	-0.15
0.44	-0.15	-0.15	-0.15	-0.14	-0.14	-0.14	-0.14	-0.13	-0.13	-0.13
0.45	-0.13	-0.12	-0.12	-0.12	-0.12	-0.11	-0.11	-0.11	-0.11	-0.10
0.46	-0.10	-0.10	-0.10	-0.09	-0.09	-0.09	-0.09	-0.08	-0.08	-0.08
0.47	-0.08	-0.07	-0.07	-0.07	-0.07	-0.06	-0.06	-0.06	-0.06	-0.05
0.48	-0.05	-0.05	-0.05	-0.04	-0.04	-0.04	-0.04	-0.03	-0.03	-0.03
0.49	-0.03	-0.02	-0.02	-0.02	-0.02	-0.01	-0.01	-0.01	-0.01	-0.00

Table 2 (Continuation). $\Psi(x)$

x	0	1	2	3	4	5	6	7	8	9
0.50	0.00	0.00	0.01	0.01	0.01	0.01	0.02	0.02	0.02	0.02
0.51	0.03	0.03	0.03	0.03	0.04	0.04	0.04	0.04	0.05	0.05
0.52	0.05	0.05	0.06	0.06	0.06	0.06	0.07	0.07	0.07	0.07
0.53	0.08	0.08	0.08	0.08	0.09	0.09	0.09	0.09	0.10	0.10
0.54	0.10	0.10	0.11	0.11	0.11	0.11	0.12	0.12	0.12	0.12
0.55	0.13	0.13	0.13	0.13	0.14	0.14	0.14	0.14	0.15	0.15
0.56	0.15	0.15	0.16	0.16	0.16	0.16	0.17	0.17	0.17	0.17
0.57	0.18	0.18	0.18	0.18	0.19	0.19	0.19	0.19	0.20	0.20
0.58	0.20	0.20	0.21	0.21	0.21	0.21	0.22	0.22	0.22	0.23
0.59	0.23	0.23	0.23	0.24	0.24	0.24	0.24	0.25	0.25	0.25
0.60	0.25	0.26	0.26	0.26	0.26	0.27	0.27	0.27	0.27	0.28
0.61	0.28	0.28	0.28	0.29	0.29	0.29	0.30	0.30	0.30	0.30
0.62	0.31	0.31	0.31	0.31	0.32	0.32	0.32	0.32	0.33	0.33
0.63	0.33	0.33	0.34	0.34	0.34	0.35	0.35	0.35	0.35	0.36
0.64	0.36	0.36	0.36	0.37	0.37	0.37	0.37	0.38	0.38	0.38
0.65	0.39	0.39	0.39	0.39	0.40	0.40	0.40	0.40	0.41	0.41
0.66	0.41	0.42	0.42	0.42	0.42	0.43	0.43	0.43	0.43	0.44
0.67	0.44	0.44	0.45	0.45	0.45	0.45	0.46	0.46	0.46	0.46
0.68	0.47	0.47	0.47	0.48	0.48	0.48	0.48	0.49	0.49	0.49
0.69	0.50	0.50	0.50	0.50	0.51	0.51	0.51	0.52	0.52	0.52
0.70	0.52	0.53	0.53	0.53	0.54	0.54	0.54	0.54	0.55	0.55
0.71	0.55	0.56	0.56	0.56	0.57	0.57	0.57	0.57	0.58	0.58
0.72	0.58	0.59	0.59	0.59	0.59	0.60	0.60	0.60	0.61	0.61
0.73	0.61	0.62	0.62	0.62	0.63	0.63	0.63	0.63	0.64	0.64
0.74	0.64	0.65	0.65	0.65	0.66	0.66	0.66	0.67	0.67	0.67
0.75	0.67	0.68	0.68	0.68	0.69	0.69	0.69	0.70	0.70	0.70
0.76	0.71	0.71	0.71	0.72	0.72	0.72	0.73	0.73	0.73	0.74
0.77	0.74	0.74	0.75	0.75	0.75	0.76	0.76	0.76	0.77	0.77
0.78	0.77	0.78	0.78	0.78	0.79	0.79	0.79	0.80	0.80	0.80
0.79	0.81	0.81	0.81	0.82	0.82	0.82	0.83	0.83	0.83	0.84
0.80	0.84	0.85	0.85	0.85	0.86	0.86	0.86	0.87	0.87	0.87
0.81	0.88	0.88	0.89	0.89	0.89	0.90	0.90	0.90	0.91	0.91
0.82	0.92	0.92	0.92	0.93	0.93	0.93	0.94	0.94	0.95	0.95
0.83	0.95	0.96	0.96	0.97	0.97	0.97	0.98	0.98	0.99	0.99
0.84	0.99	1.00	1.00	1.01	1.01	1.02	1.02	1.02	1.03	1.03
0.85	1.04	1.04	1.05	1.05	1.05	1.06	1.06	1.07	1.07	1.08
0.86	1.08	1.09	1.09	1.09	1.10	1.10	1.11	1.11	1.12	1.12
0.87	1.13	1.13	1.14	1.14	1.15	1.15	1.16	1.16	1.17	1.17
0.88	1.18	1.18	1.19	1.19	1.20	1.20	1.21	1.21	1.22	1.22
0.89	1.23	1.23	1.24	1.24	1.25	1.25	1.26	1.26	1.27	1.28
0.90	1.28	1.29	1.29	1.30	1.30	1.31	1.32	1.32	1.33	1.33
0.91	1.34	1.35	1.35	1.36	1.37	1.37	1.38	1.39	1.39	1.40
0.92	1.41	1.41	1.42	1.43	1.43	1.44	1.45	1.45	1.46	1.47
0.93	1.48	1.48	1.49	1.50	1.51	1.51	1.52	1.53	1.54	1.55
0.94	1.55	1.56	1.57	1.58	1.59	1.60	1.61	1.62	1.63	1.64
0.95	1.64	1.65	1.66	1.67	1.68	1.70	1.71	1.72	1.73	1.74
0.96	1.75	1.76	1.77	1.79	1.80	1.81	1.83	1.84	1.85	1.87
0.97	1.88	1.90	1.91	1.93	1.94	1.96	1.98	2.00	2.01	2.03
0.98	2.05	2.07	2.10	2.12	2.14	2.17	2.20	2.23	2.26	2.29
0.99	2.33	2.37	2.41	2.46	2.51	2.58	2.65	2.75	2.88	3.09

Table 3. *The factor g from the normal distribution and the upper bound g^2 for χ^2 with a single degree of freedom*

Error probability		g	g^2
one-sided (%)	two-sided (%)	normal distribution	bound for χ^2
5	10	1.64	2.71
2.5	5	1.96	3.84
1	2	2.33	5.02
0.5	1	2.58	6.63
0.1	0.2	3.09	9.55
0.05	0.1	3.29	10.83

Table 4. *Δ-Test. Exact and asymptotic one-sided bounds for the maximal difference between the true and the empirical distribution*

n	5% Error probability			1% Error probability		
	exact	asymptotic	ratio	exact	asymptotic	ratio
5	0.5094	0.5473	1.074	0.6271	0.6786	1.082
8	0.4096	0.4327	1.056	0.5065	0.5365	1.059
10	0.3687	0.3870	1.050	0.4566	0.4799	1.051
20	0.2647	0.2737	1.034	0.3285	0.3393	1.033
40	0.1891	0.1935	1.023	0.2350	0.2399	1.021
50	0.1696	0.1731	1.021	0.2107	0.2146	1.019

For larger n we may use the asymptotic bound

$$\bar{\varepsilon}_\alpha = \sqrt{\frac{-\ln\alpha}{2n}} \qquad \text{(error probability } \alpha\text{)},$$

which is always conservative. From the table we see that we can reduce the asymptotic bound by as much as $1/6n$ and yet always remain on the safe side.

Table 4 has been extracted from Z. W. Birnbaum and F. H. Tingey, One-sided confidence contours for probability distribution functions. Ann. Math. Stat. 22 (1951) p. 595.

Table 5. *Kolmogorov's test. Exact and asymptotic two-sided bounds for the maximal difference between the true and the empirical distribution*

n	5% Error probability			1% Error probability		
	exact	asymptotic	ratio	exact	asymptotic	ratio
5	0.5633	0.6074	1.078	0.6685	0.7279	1.089
10	0.4087	0.4295	1.051	0.4864	0.5147	1.058
15	0.3375	0.3507	1.039	0.4042	0.4202	1.040
20	0.2939	0.3037	1.033	0.3524	0.3639	1.033
25	0.2639	0.2716	1.029	0.3165	0.3255	1.028
30	0.2417	0.2480	1.026	0.2898	0.2972	1.025
40	0.2101	0.2147	1.022	0.2521	0.2574	1.021
50	0.1884	0.1921	1.019	0.2260	0.2302	1.018
60	0.1723	0.1753	1.018	0.2067	0.2101	1.016
70	0.1597	0.1623	1.016	0.1917	0.1945	1.015
80	0.1496	0.1518	1.015	0.1795	0.1820	1.014
90	0.1412	0.1432	1.014			
100	0.1340	0.1358	1.013			

For larger n we may use the asymptotic bounds

$$\bar{\varepsilon}_{0.05} = 1.36 n^{-\frac{1}{2}} \quad \text{and} \quad \bar{\varepsilon}_{0.01} = 1.63 n^{-\frac{1}{2}},$$

which are always conservative. From the table we see that we can reduce the asymptotic bound by as much as $1/6n$ and yet always remain on the safe side.

Table 5 has been extracted from Z.W. Birnbaum, Numerical tabulation of the distribution of Kolmogorov's statistic. J. Amer. Statist. Assoc. 47 (1952) p. 431.

Table 6. *Bounds for χ^2 with f degrees of freedom*

f	5%	1%	0.1%	f	5%	1%	0.1%
1	3.84	6.63	10.8	41	56.9	65.0	74.7
2	5.99	9.21	13.8	42	58.1	66.2	76.1
3	7.81	11.3	16.3	43	59.3	67.5	77.4
4	9.49	13.3	18.5	44	60.5	68.7	78.7
5	11.1	15.1	20.5	45	61.7	70.0	80.1
6	12.6	16.8	22.5	46	62.8	71.2	81.4
7	14.1	18.5	24.3	47	64.0	72.4	82.7
8	15.5	20.1	26.1	48	65.2	73.7	84.0
9	16.9	21.7	27.9	49	66.3	74.9	85.4
10	18.3	23.2	29.6	50	67.5	76.2	86.7
11	19.7	24.7	31.3	51	68.7	77.4	88.0
12	21.0	26.2	32.9	52	69.8	78.6	89.3
13	22.4	27.7	34.5	53	71.0	79.8	90.6
14	23.7	29.1	36.1	54	72.2	81.1	91.9
15	25.0	30.6	37.7	55	73.3	82.3	93.2
16	26.3	32.0	39.3	56	74.5	83.5	94.5
17	27.6	33.4	40.8	57	75.6	84.7	95.8
18	28.9	34.8	42.3	58	76.8	86.0	97.0
19	30.1	36.2	43.8	59	77.9	87.2	98.3
20	31.4	37.6	45.3	60	79.1	88.4	99.6
21	32.7	38.9	46.8	61	80.2	89.6	100.9
22	33.9	40.3	48.3	62	81.4	90.8	102.2
23	35.2	41.6	49.7	63	82.5	92.0	103.4
24	36.4	43.0	51.2	64	83.7	93.2	104.7
25	37.7	44.3	52.6	65	84.8	94.4	106.0
26	38.9	45.6	54.1	66	86.0	95.6	107.3
27	40.1	47.0	55.5	67	87.1	96.8	108.5
28	41.3	48.3	56.9	68	88.3	98.0	109.8
29	42.6	49.6	58.3	69	89.4	99.2	111.1
30	43.8	50.9	59.7	70	90.5	100.4	112.3
31	45.0	52.2	61.1	71	91.7	101.6	113.6
32	46.2	53.5	62.5	72	92.8	102.8	114.8
33	47.4	54.8	63.9	73	93.9	104.0	116.1
34	48.6	56.1	65.2	74	95.1	105.2	117.3
35	49.8	57.3	66.6	75	96.2	106.4	118.6
36	51.0	58.6	68.0	76	97.4	107.6	119.9
37	52.2	59.9	69.3	77	98.5	108.8	121.1
38	53.4	61.2	70.7	78	99.6	110.0	122.3
39	54.6	62.4	72.1	79	100.7	111.1	123.6
40	55.8	63.7	73.4	80	101.9	112.3	124.8

Tables 6 and 7 have been extracted from A. Hald, Statistical Tables and Formulas. New York: John Wiley and Sons 1952. In the last column of Table 7, three entries have been corrected in accordance with Table 12 of E. S. Pearson and H. O. Hartley, Biometrika Tables for Statisticians I.

Table 6 (Continuation).
Bounds for χ^2

f	5%	1%	0.1%
81	103.0	113.5	126.1
82	104.1	114.7	127.3
83	105.3	115.9	128.6
84	106.4	117.1	129.8
85	107.5	118.2	131.0
86	108.6	119.4	132.3
87	109.8	120.6	133.5
88	110.9	121.8	134.7
89	112.0	122.9	136.0
90	113.1	124.1	137.2
91	114.3	125.3	138.4
92	115.4	126.5	139.7
93	116.5	127.6	140.9
94	117.6	128.8	142.1
95	118.8	130.0	143.3
96	119.9	131.1	144.6
97	121.0	132.3	145.8
98	122.1	133.5	147.0
99	123.2	134.6	148.2
100	124.3	135.8	149.4

For larger values of f, the upper cut-off point for χ^2 with error probability α is

$$\chi_\alpha^2 = \tfrac{1}{2}[\sqrt{2f-1} + \Psi(1-\alpha)]^2.$$

Table 7. *Student's test. Bounds for t with f degrees of freedom*

f ↓	Two-sided			
	5%	2%	1%	0.1%
1	12.71	31.82	63.66	636.6
2	4.303	6.965	9.925	31.60
3	3.182	4.541	5.841	12.92
4	2.776	3.747	4.604	8.610
5	2.571	3.365	4.032	6.869
6	2.447	3.143	3.707	5.959
7	2.365	2.998	3.499	5.408
8	2.306	2.896	3.355	5.041
9	2.262	2.821	3.250	4.781
10	2.228	2.764	3.169	4.587
11	2.201	2.718	3.106	4.437
12	2.179	2.681	3.055	4.318
13	2.160	2.650	3.012	4.221
14	2.145	2.624	2.977	4.140
15	2.131	2.602	2.947	4.073
16	2.120	2.583	2.921	4.015
17	2.110	2.567	2.898	3.965
18	2.101	2.552	2.878	3.922
19	2.093	2.539	2.861	3.883
20	2.086	2.528	2.845	3.850
21	2.080	2.518	2.831	3.819
22	2.074	2.508	2.819	3.792
23	2.069	2.500	2.807	3.767
24	2.064	2.492	2.797	3.745
25	2.060	2.485	2.787	3.725
26	2.056	2.479	2.779	3.707
27	2.052	2.473	2.771	3.690
28	2.048	2.467	2.763	3.674
29	2.045	2.462	2.756	3.659
30	2.042	2.457	2.750	3.646
40	2.021	2.423	2.704	3.551
50	2.009	2.403	2.678	3.495
60	2.000	2.390	2.660	3.460
80	1.990	2.374	2.639	3.415
100	1.984	2.365	2.626	3.389
200	1.972	2.345	2.601	3.339
500	1.965	2.334	2.586	3.310
∞	1.960	2.326	2.576	3.291
↑	2.5%	1%	0.5%	0.05%
f	One-sided			

Linear interpolation in Table 7 is reliable only to two decimal places.

Table 8 A. *Bounds for $F = s_1^2/s_2^2$ with error probability 5%. The degrees of freedom are f_1 in the numerator and f_2 in the denominator*

f_2 ↓	Degrees of freedom for the numerator f_1														
	1	2	3	4	5	6	7	8	9	10	11	12	13	14	15
1	161	200	216	225	230	234	237	239	241	242	243	244	245	245	246
2	18.5	19.0	19.2	19.2	19.3	19.3	19.4	19.4	19.4	19.4	19.4	19.4	19.4	19.4	19.4
3	10.1	9.55	9.28	9.12	9.01	8.94	8.89	8.85	8.81	8.79	8.76	8.74	8.73	8.71	8.70
4	7.71	6.94	6.59	6.39	6.26	6.16	6.09	6.04	6.00	5.96	5.94	5.91	5.89	5.87	5.86
5	6.61	5.79	5.41	5.19	5.05	4.95	4.88	4.82	4.77	4.74	4.70	4.68	4.66	4.64	4.62
6	5.99	5.14	4.76	4.53	4.39	4.28	4.21	4.15	4.10	4.06	4.03	4.00	3.98	3.96	3.94
7	5.59	4.74	4.35	4.12	3.97	3.87	3.79	3.73	3.68	3.64	3.60	3.57	3.55	3.53	3.51
8	5.32	4.46	4.07	3.84	3.69	3.58	3.50	3.44	3.39	3.35	3.31	3.28	3.26	3.24	3.22
9	5.12	4.26	3.86	3.63	3.48	3.37	3.29	3.23	3.18	3.14	3.10	3.07	3.05	3.03	3.01
10	4.96	4.10	3.71	3.48	3.33	3.22	3.14	3.07	3.02	2.98	2.94	2.91	2.89	2.86	2.85
11	4.84	3.98	3.59	3.36	3.20	3.09	3.01	2.95	2.90	2.85	2.82	2.79	2.76	2.74	2.72
12	4.75	3.89	3.49	3.26	3.11	3.00	2.91	2.85	2.80	2.75	2.72	2.69	2.66	2.64	2.62
13	4.67	3.81	3.41	3.18	3.03	2.92	2.83	2.77	2.71	2.67	2.63	2.60	2.58	2.55	2.53
14	4.60	3.74	3.34	3.11	2.96	2.85	2.76	2.70	2.65	2.60	2.57	2.53	2.51	2.48	2.46
15	4.54	3.68	3.29	3.06	2.90	2.79	2.71	2.64	2.59	2.54	2.51	2.48	2.45	2.42	2.40
16	4.49	3.63	3.24	3.01	2.85	2.74	2.66	2.59	2.54	2.49	2.46	2.42	2.40	2.37	2.35
17	4.45	3.59	3.20	2.96	2.81	2.70	2.61	2.55	2.49	2.45	2.41	2.38	2.35	2.33	2.31
18	4.41	3.55	3.16	2.93	2.77	2.66	2.58	2.51	2.46	2.41	2.37	2.34	2.31	2.29	2.27
19	4.38	3.52	3.13	2.90	2.74	2.63	2.54	2.48	2.42	2.38	2.34	2.31	2.28	2.26	2.23
20	4.35	3.49	3.10	2.87	2.71	2.60	2.51	2.45	2.39	2.35	2.31	2.28	2.25	2.22	2.20
21	4.32	3.47	3.07	2.84	2.68	2.57	2.49	2.42	2.37	2.32	2.28	2.25	2.22	2.20	2.18
22	4.30	3.44	3.05	2.82	2.66	2.55	2.46	2.40	2.34	2.30	2.26	2.23	2.20	2.17	2.15
23	4.28	3.42	3.03	2.80	2.64	2.53	2.44	2.37	2.32	2.27	2.23	2.20	2.18	2.15	2.13
24	4.26	3.40	3.01	2.78	2.62	2.51	2.42	2.36	2.30	2.25	2.21	2.18	2.15	2.13	2.11
25	4.24	3.39	2.99	2.76	2.60	2.49	2.40	2.34	2.28	2.24	2.20	2.16	2.14	2.11	2.09
26	4.23	3.37	2.98	2.74	2.59	2.47	2.39	2.32	2.27	2.22	2.18	2.15	2.12	2.09	2.07
27	4.21	3.35	2.96	2.73	2.57	2.46	2.37	2.31	2.25	2.20	2.17	2.13	2.10	2.08	2.06
28	4.20	3.34	2.95	2.71	2.56	2.45	2.36	2.29	2.24	2.19	2.15	2.12	2.09	2.06	2.04
29	4.18	3.33	2.93	2.70	2.55	2.43	2.35	2.28	2.22	2.18	2.14	2.10	2.08	2.05	2.03
30	4.17	3.32	2.92	2.69	2.53	2.42	2.33	2.27	2.21	2.16	2.13	2.09	2.06	2.04	2.01
32	4.15	3.29	2.90	2.67	2.51	2.40	2.31	2.24	2.19	2.14	2.10	2.07	2.04	2.01	1.99
34	4.13	3.28	2.88	2.65	2.49	2.38	2.29	2.23	2.17	2.12	2.08	2.05	2.02	1.99	1.97
36	4.11	3.26	2.87	2.63	2.48	2.36	2.28	2.21	2.15	2.11	2.07	2.03	2.00	1.98	1.95
38	4.10	3.24	2.85	2.62	2.46	2.35	2.26	2.19	2.14	2.09	2.05	2.02	1.99	1.96	1.94
40	4.08	3.23	2.84	2.61	2.45	2.34	2.25	2.18	2.12	2.08	2.04	2.00	1.97	1.95	1.92
42	4.07	3.22	2.83	2.59	2.44	2.32	2.24	2.17	2.11	2.06	2.03	1.99	1.96	1.93	1.91
44	4.06	3.21	2.82	2.58	2.43	2.31	2.23	2.16	2.10	2.05	2.01	1.98	1.95	1.92	1.90
46	4.05	3.20	2.81	2.57	2.42	2.30	2.22	2.15	2.09	2.04	2.00	1.97	1.94	1.91	1.89
48	4.04	3.19	2.80	2.57	2.41	2.29	2.21	2.14	2.08	2.03	1.99	1.96	1.93	1.90	1.88
50	4.03	3.18	2.79	2.56	2.40	2.29	2.20	2.13	2.07	2.03	1.99	1.95	1.92	1.89	1.87
60	4.00	3.15	2.76	2.53	2.37	2.25	2.17	2.10	2.04	1.99	1.95	1.92	1.89	1.86	1.84
70	3.98	3.13	2.74	2.50	2.35	2.23	2.14	2.07	2.02	1.97	1.93	1.89	1.86	1.84	1.81
80	3.96	3.11	2.72	2.49	2.33	2.21	2.13	2.06	2.00	1.95	1.91	1.88	1.84	1.82	1.79
90	3.95	3.10	2.71	2.47	2.32	2.20	2.11	2.04	1.99	1.94	1.90	1.86	1.83	1.80	1.78
100	3.94	3.09	2.70	2.46	2.31	2.19	2.10	2.03	1.97	1.93	1.89	1.85	1.82	1.79	1.77
125	3.92	3.07	2.68	2.44	2.29	2.17	2.08	2.01	1.96	1.91	1.87	1.83	1.80	1.77	1.75
150	3.90	3.06	2.66	2.43	2.27	2.16	2.07	2.00	1.94	1.89	1.85	1.82	1.79	1.76	1.73
200	3.89	3.04	2.65	2.42	2.26	2.14	2.06	1.98	1.93	1.88	1.84	1.80	1.77	1.74	1.72
300	3.87	3.03	2.63	2.40	2.24	2.13	2.04	1.97	1.91	1.86	1.82	1.78	1.75	1.72	1.70
500	3.86	3.01	2.62	2.39	2.23	2.12	2.03	1.96	1.90	1.85	1.81	1.77	1.74	1.71	1.69
1,000	3.85	3.00	2.61	2.38	2.22	2.11	2.02	1.95	1.89	1.84	1.80	1.76	1.73	1.70	1.68

Whenever f_2 exceeds 1,000, the bound corresponding to $f_2 = 1{,}000$ is to be used.

Table 8 A (Continuation). *Bounds for* $F = s_1^2/s_2^2$ *with error probability 5%. The degrees of freedom are* f_1 *in the numerator and* f_2 *in the denominator*

f_2 ↓	Degrees of freedom for the numerator f_1														
	16	17	18	19	20	22	24	26	28	30	40	50	60	80	100
1	246	247	247	248	248	249	249	249	250	250	251	252	252	252	253
2	19.4	19.4	19.4	19.4	19.4	19.5	19.5	19.5	19.5	19.5	19.5	19.5	19.5	19.5	19.5
3	8.69	8.68	8.67	8.67	8.66	8.65	8.64	8.63	8.62	8.62	8.59	8.58	8.57	8.56	8.55
4	5.84	5.83	5.82	5.81	5.80	5.79	5.77	5.76	5.75	5.75	5.72	5.70	5.69	5.67	5.66
5	4.60	4.59	4.58	4.57	4.56	4.54	4.53	4.52	4.50	4.50	4.46	4.44	4.43	4.41	4.41
6	3.92	3.91	3.90	3.88	3.87	3.86	3.84	3.83	3.82	3.81	3.77	3.75	3.74	3.72	3.71
7	3.49	3.48	3.47	3.46	3.44	3.43	3.41	3.40	3.39	3.38	3.34	3.32	3.30	3.29	3.27
8	3.20	3.19	3.17	3.16	3.15	3.13	3.12	3.10	3.09	3.08	3.04	3.02	3.01	2.99	2.97
9	2.99	2.97	2.96	2.95	2.94	2.92	2.90	2.89	2.87	2.86	2.83	2.80	2.79	2.77	2.76
10	2.83	2.81	2.80	2.78	2.77	2.75	2.74	2.72	2.71	2.70	2.66	2.64	2.62	2.60	2.59
11	2.70	2.69	2.67	2.66	2.65	2.63	2.61	2.59	2.58	2.57	2.53	2.51	2.49	2.47	2.46
12	2.60	2.58	2.57	2.56	2.54	2.52	2.51	2.49	2.48	2.47	2.43	2.40	2.38	2.36	2.35
13	2.51	2.50	2.48	2.47	2.46	2.44	2.42	2.41	2.39	2.38	2.34	2.31	2.30	2.27	2.26
14	2.44	2.43	2.41	2.40	2.39	2.37	2.35	2.33	2.32	2.31	2.27	2.24	2.22	2.20	2.19
15	2.38	2.37	2.35	2.34	2.33	2.31	2.29	2.27	2.26	2.25	2.20	2.18	2.16	2.14	2.12
16	2.33	2.32	2.30	2.29	2.28	2.25	2.24	2.22	2.21	2.19	2.15	2.12	2.11	2.08	2.07
17	2.29	2.27	2.26	2.24	2.23	2.21	2.19	2.17	2.16	2.15	2.10	2.08	2.06	2.03	2.02
18	2.25	2.23	2.22	2.20	2.19	2.17	2.15	2.13	2.12	2.11	2.06	2.04	2.02	1.99	1.98
19	2.21	2.20	2.18	2.17	2.16	2.13	2.11	2.10	2.08	2.07	2.03	2.00	1.98	1.96	1.94
20	2.18	2.17	2.15	2.14	2.12	2.10	2.08	2.07	2.05	2.04	1.99	1.97	1.95	1.92	1.91
21	2.16	2.14	2.12	2.11	2.10	2.07	2.05	2.04	2.02	2.01	1.96	1.94	1.92	1.89	1.88
22	2.13	2.11	2.10	2.08	2.07	2.05	2.03	2.01	2.00	1.98	1.94	1.91	1.89	1.86	1.85
23	2.11	2.09	2.07	2.06	2.05	2.02	2.00	1.99	1.97	1.96	1.91	1.88	1.86	1.84	1.82
24	2.09	2.07	2.05	2.04	2.03	2.00	1.98	1.97	1.95	1.94	1.89	1.86	1.84	1.82	1.80
25	2.07	2.05	2.04	2.02	2.01	1.98	1.96	1.95	1.93	1.92	1.87	1.84	1.82	1.80	1.78
26	2.05	2.03	2.02	2.00	1.99	1.97	1.95	1.93	1.91	1.90	1.85	1.82	1.80	1.78	1.76
27	2.04	2.02	2.00	1.99	1.97	1.95	1.93	1.91	1.90	1.88	1.84	1.81	1.79	1.76	1.74
28	2.02	2.00	1.99	1.97	1.96	1.93	1.91	1.90	1.88	1.87	1.82	1.79	1.77	1.74	1.73
29	2.01	1.99	1.97	1.96	1.94	1.92	1.90	1.88	1.87	1.85	1.81	1.77	1.75	1.73	1.71
30	1.99	1.98	1.96	1.95	1.93	1.91	1.89	1.87	1.85	1.84	1.79	1.76	1.74	1.71	1.70
32	1.97	1.95	1.94	1.92	1.91	1.88	1.86	1.85	1.83	1.82	1.77	1.74	1.71	1.69	1.67
34	1.95	1.93	1.92	1.90	1.89	1.86	1.84	1.82	1.80	1.80	1.75	1.71	1.69	1.66	1.65
36	1.93	1.92	1.90	1.88	1.87	1.85	1.82	1.81	1.79	1.78	1.73	1.69	1.67	1.64	1.62
38	1.92	1.90	1.88	1.87	1.85	1.83	1.81	1.79	1.77	1.76	1.71	1.68	1.65	1.62	1.61
40	1.90	1.89	1.87	1.85	1.84	1.81	1.79	1.77	1.76	1.74	1.69	1.66	1.64	1.61	1.59
42	1.89	1.87	1.86	1.84	1.83	1.80	1.78	1.76	1.74	1.73	1.68	1.65	1.62	1.59	1.57
44	1.88	1.86	1.84	1.83	1.81	1.79	1.77	1.75	1.73	1.72	1.67	1.63	1.61	1.58	1.56
46	1.87	1.85	1.83	1.82	1.80	1.78	1.76	1.74	1.72	1.71	1.65	1.62	1.60	1.57	1.55
48	1.86	1.84	1.82	1.81	1.79	1.77	1.75	1.73	1.71	1.70	1.64	1.61	1.59	1.56	1.54
50	1.85	1.83	1.81	1.80	1.78	1.76	1.74	1.72	1.70	1.69	1.63	1.60	1.58	1.54	1.52
60	1.82	1.80	1.78	1.76	1.75	1.72	1.70	1.68	1.66	1.65	1.59	1.56	1.53	1.50	1.48
70	1.79	1.77	1.75	1.74	1.72	1.70	1.67	1.65	1.64	1.62	1.57	1.53	1.50	1.47	1.45
80	1.77	1.75	1.73	1.72	1.70	1.68	1.65	1.63	1.62	1.60	1.54	1.51	1.48	1.45	1.43
90	1.76	1.74	1.72	1.70	1.69	1.66	1.64	1.62	1.60	1.59	1.53	1.49	1.46	1.43	1.41
100	1.75	1.73	1.71	1.69	1.68	1.65	1.63	1.61	1.59	1.57	1.52	1.48	1.45	1.41	1.39
125	1.72	1.70	1.69	1.67	1.65	1.63	1.60	1.58	1.57	1.55	1.49	1.45	1.42	1.39	1.36
150	1.71	1.69	1.67	1.66	1.64	1.61	1.59	1.57	1.55	1.53	1.48	1.44	1.41	1.37	1.34
200	1.69	1.67	1.66	1.64	1.62	1.60	1.57	1.55	1.53	1.52	1.46	1.41	1.39	1.35	1.32
300	1.68	1.66	1.64	1.62	1.61	1.58	1.55	1.53	1.51	1.50	1.43	1.39	1.36	1.32	1.30
500	1.66	1.64	1.62	1.61	1.59	1.56	1.54	1.52	1.50	1.48	1.42	1.38	1.34	1.30	1.28
1,000	1.65	1.63	1.61	1.60	1.58	1.55	1.53	1.51	1.49	1.47	1.41	1.36	1.33	1.29	1.26

Whenever f_2 exceeds 1,000, the bound corresponding to $f_2 = 1{,}000$ is to be used.

Table 8 B. *Bounds for* $F = s_1^2/s_2^2$ *with error probability 1%. The degrees of freedom are* f_1 *in the numerator and* f_2 *in the denominator*

f_2 ↓	Degrees of freedom for the numerator f_1														
	1	2	3	4	5	6	7	8	9	10	11	12	13	14	15
2	98.5	99.0	99.2	99.2	99.3	99.3	99.4	99.4	99.4	99.4	99.4	99.4	99.4	99.4	99.4
3	34.1	30.8	29.5	28.7	28.2	27.9	27.7	27.5	27.3	27.2	27.1	27.1	27.0	26.9	26.9
4	21.2	18.0	16.7	16.0	15.5	15.2	15.0	14.8	14.7	14.5	14.4	14.4	14.3	14.2	14.2
5	16.3	13.3	12.1	11.4	11.0	10.7	10.5	10.3	10.2	10.1	9.96	9.89	9.82	9.77	9.72
6	13.7	10.9	9.78	9.15	8.75	8.47	8.26	8.10	7.98	7.87	7.79	7.72	7.66	7.60	7.56
7	12.2	9.55	8.45	7.85	7.46	7.19	6.99	6.84	6.72	6.62	6.54	6.47	6.41	6.36	6.31
8	11.3	8.65	7.59	7.01	6.63	6.37	6.18	6.03	5.91	5.81	5.73	5.67	5.61	5.56	5.52
9	10.6	8.02	6.99	6.42	6.06	5.80	5.61	5.47	5.35	5.26	5.18	5.11	5.05	5.00	4.96
10	10.0	7.56	6.55	5.99	5.64	5.39	5.20	5.06	4.94	4.85	4.77	4.71	4.65	4.60	4.56
11	9.65	7.21	6.22	5.67	5.32	5.07	4.89	4.74	4.63	4.54	4.46	4.40	4.34	4.29	4.25
12	9.33	6.93	5.95	5.41	5.06	4.82	4.64	4.50	4.39	4.30	4.22	4.16	4.10	4.05	4.01
13	9.07	6.70	5.74	5.21	4.86	4.62	4.44	4.30	4.19	4.10	4.02	3.96	3.91	3.86	3.82
14	8.86	6.51	5.56	5.04	4.69	4.46	4.28	4.14	4.03	3.94	3.86	3.80	3.75	3.70	3.66
15	8.68	6.36	5.42	4.89	4.56	4.32	4.14	4.00	3.89	3.80	3.73	3.67	3.61	3.56	3.52
16	8.53	6.23	5.29	4.77	4.44	4.20	4.03	3.89	3.78	3.69	3.62	3.55	3.50	3.45	3.41
17	8.40	6.11	5.18	4.67	4.34	4.10	3.93	3.79	3.68	3.59	3.52	3.46	3.40	3.35	3.31
18	8.29	6.01	5.09	4.58	4.25	4.01	3.84	3.71	3.60	3.51	3.43	3.37	3.32	3.27	3.23
19	8.18	5.93	5.01	4.50	4.17	3.94	3.77	3.63	3.52	3.43	3.36	3.30	3.24	3.19	3.15
20	8.10	5.85	4.94	4.43	4.10	3.87	3.70	3.56	3.46	3.37	3.29	3.23	3.18	3.13	3.09
21	8.02	5.78	4.87	4.37	4.04	3.81	3.64	3.51	3.40	3.31	3.24	3.17	3.12	3.07	3.03
22	7.95	5.72	4.82	4.31	3.99	3.76	3.59	3.45	3.35	3.26	3.18	3.12	3.07	3.02	2.98
23	7.88	5.66	4.76	4.26	3.94	3.71	3.54	3.41	3.30	3.21	3.14	3.07	3.02	2.97	2.93
24	7.82	5.61	4.72	4.22	3.90	3.67	3.50	3.36	3.26	3.17	3.09	3.03	2.98	2.93	2.89
25	7.77	5.57	4.68	4.18	3.86	3.63	3.46	3.32	3.22	3.13	3.06	2.99	2.94	2.89	2.85
26	7.72	5.53	4.64	4.14	3.82	3.59	3.42	3.29	3.18	3.09	3.02	2.96	2.90	2.86	2.82
27	7.68	5.49	4.60	4.11	3.78	3.56	3.39	3.26	3.15	3.06	2.99	2.93	2.87	2.82	2.78
28	7.64	5.45	4.57	4.07	3.75	3.53	3.36	3.23	3.12	3.03	2.96	2.90	2.84	2.79	2.75
29	7.60	5.42	4.54	4.04	3.73	3.50	3.33	3.20	3.09	3.00	2.93	2.87	2.81	2.77	2.73
30	7.56	5.39	4.51	4.02	3.70	3.47	3.30	3.17	3.07	2.98	2.91	2.84	2.79	2.74	2.70
32	7.50	5.34	4.46	3.97	3.65	3.43	3.26	3.13	3.02	2.93	2.86	2.80	2.74	2.70	2.66
34	7.44	5.29	4.42	3.93	3.61	3.39	3.22	3.09	2.98	2.89	2.82	2.76	2.70	2.66	2.62
36	7.40	5.25	4.38	3.89	3.57	3.35	3.18	3.05	2.95	2.86	2.79	2.72	2.67	2.62	2.58
38	7.35	5.21	4.34	3.86	3.54	3.32	3.15	3.02	2.92	2.83	2.75	2.69	2.64	2.59	2.55
40	7.31	5.18	4.31	3.83	3.51	3.29	3.12	2.99	2.89	2.80	2.73	2.66	2.61	2.56	2.52
42	7.28	5.15	4.29	3.80	3.49	3.27	3.10	2.97	2.86	2.78	2.70	2.64	2.59	2.54	2.50
44	7.25	5.12	4.26	3.78	3.47	3.24	3.08	2.95	2.84	2.75	2.68	2.62	2.56	2.52	2.47
46	7.22	5.10	4.24	3.76	3.44	3.22	3.06	2.93	2.82	2.73	2.66	2.60	2.54	2.50	2.45
48	7.19	5.08	4.22	3.74	3.43	3.20	3.04	2.91	2.80	2.72	2.64	2.58	2.53	2.48	2.44
50	7.17	5.06	4.20	3.72	3.41	3.19	3.02	2.89	2.79	2.70	2.63	2.56	2.51	2.46	2.42
55	7.12	5.01	4.16	3.68	3.37	3.15	2.98	2.85	2.75	2.66	2.59	2.53	2.47	2.42	2.38
60	7.08	4.98	4.13	3.65	3.34	3.12	2.95	2.82	2.72	2.63	2.56	2.50	2.44	2.39	2.35
70	7.01	4.92	4.08	3.60	3.29	3.07	2.91	2.78	2.67	2.59	2.51	2.45	2.40	2.35	2.31
80	6.96	4.88	4.04	3.56	3.26	3.04	2.87	2.74	2.64	2.55	2.48	2.42	2.36	2.31	2.27
90	6.93	4.85	4.01	3.54	3.23	3.01	2.84	2.72	2.61	2.52	2.45	2.39	2.33	2.29	2.24
100	6.90	4.82	3.98	3.51	3.21	2.99	2.82	2.69	2.59	2.50	2.43	2.37	2.31	2.26	2.22
125	6.84	4.78	3.94	3.47	3.17	2.95	2.79	2.66	2.55	2.47	2.39	2.33	2.28	2.23	2.19
150	6.81	4.75	3.92	3.45	3.14	2.92	2.76	2.63	2.53	2.44	2.37	2.31	2.25	2.20	2.16
200	6.76	4.71	3.88	3.41	3.11	2.89	2.73	2.60	2.50	2.41	2.34	2.27	2.22	2.17	2.13
300	6.72	4.68	3.85	3.38	3.08	2.86	2.70	2.57	2.47	2.38	2.31	2.24	2.19	2.14	2.10
500	6.69	4.65	3.82	3.36	3.05	2.84	2.68	2.55	2.44	2.36	2.28	2.22	2.17	2.12	2.07
1,000	6.66	4.63	3.80	3.34	3.04	2.82	2.66	2.53	2.43	2.34	2.27	2.20	2.15	2.10	2.06

Whenever $f_2 = 1$, the bound for F is the square of the two-sided bound for t with f_1 degrees of freedom (Table 7).

Table 8 B (Continuation). *Bounds for* $F = s_1^2/s_2^2$ *with error probability 1%. The degrees of freedom are* f_1 *in the numerator and* f_2 *in the denominator*

f_2 ↓	Degrees of freedom for the numerator f_1														
	16	17	18	19	20	22	24	26	28	30	40	50	60	80	100
2	99.4	99.4	99.4	99.4	99.4	99.5	99.5	99.5	99.5	99.5	99.5	99.5	99.5	99.5	99.5
3	26.8	26.8	26.8	26.7	26.7	26.6	26.6	26.6	26.5	26.5	26.4	26.4	26.3	26.3	26.2
4	14.2	14.1	14.1	14.0	14.0	14.0	13.9	13.9	13.9	13.8	13.7	13.7	13.7	13.6	13.6
5	9.68	9.64	9.61	9.58	9.55	9.51	9.47	9.43	9.40	9.38	9.29	9.24	9.20	9.16	9.13
6	7.52	7.48	7.45	7.42	7.40	7.35	7.31	7.28	7.25	7.23	7.14	7.09	7.06	7.01	6.99
7	6.27	6.24	6.21	6.18	6.16	6.11	6.07	6.04	6.02	5.99	5.91	5.86	5.82	5.78	5.75
8	5.48	5.44	5.41	5.38	5.36	5.32	5.28	5.25	5.22	5.20	5.12	5.07	5.03	4.99	4.96
9	4.92	4.89	4.86	4.83	4.81	4.77	4.73	4.70	4.67	4.65	4.57	4.52	4.48	4.44	4.42
10	4.52	4.49	4.46	4.43	4.41	4.36	4.33	4.30	4.27	4.25	4.17	4.12	4.08	4.04	4.01
11	4.21	4.18	4.15	4.12	4.10	4.06	4.02	3.99	3.96	3.94	3.86	3.81	3.78	3.73	3.71
12	3.97	3.94	3.91	3.88	3.86	3.82	3.78	3.75	3.72	3.70	3.62	3.57	3.54	3.49	3.47
13	3.78	3.75	3.72	3.69	3.66	3.62	3.59	3.56	3.53	3.51	3.43	3.38	3.34	3.30	3.27
14	3.62	3.59	3.56	3.53	3.51	3.46	3.43	3.40	3.37	3.35	3.27	3.22	3.18	3.14	3.11
15	3.49	3.45	3.42	3.40	3.37	3.33	3.29	3.26	3.24	3.21	3.13	3.08	3.05	3.00	2.98
16	3.37	3.34	3.31	3.28	3.26	3.22	3.18	3.15	3.12	3.10	3.02	2.97	2.93	2.89	2.86
17	3.27	3.24	3.21	3.18	3.16	3.12	3.08	3.05	3.03	3.00	2.92	2.87	2.83	2.79	2.76
18	3.19	3.16	3.13	3.10	3.08	3.03	3.00	2.97	2.94	2.92	2.84	2.78	2.75	2.70	2.68
19	3.12	3.08	3.05	3.03	3.00	2.96	2.92	2.89	2.87	2.84	2.76	2.71	2.67	2.63	2.60
20	3.05	3.02	2.99	2.96	2.94	2.90	2.86	2.83	2.80	2.78	2.69	2.64	2.61	2.56	2.54
21	2.99	2.96	2.93	2.90	2.88	2.84	2.80	2.77	2.74	2.72	2.64	2.58	2.55	2.50	2.48
22	2.94	2.91	2.88	2.85	2.83	2.78	2.75	2.72	2.69	2.67	2.58	2.53	2.50	2.45	2.42
23	2.89	2.86	2.83	2.80	2.78	2.74	2.70	2.67	2.64	2.62	2.54	2.48	2.45	2.40	2.37
24	2.85	2.82	2.79	2.76	2.74	2.70	2.66	2.63	2.60	2.58	2.49	2.44	2.40	2.36	2.33
25	2.81	2.78	2.75	2.72	2.70	2.66	2.62	2.59	2.56	2.54	2.45	2.40	2.36	2.32	2.29
26	2.78	2.74	2.72	2.69	2.66	2.62	2.58	2.55	2.53	2.50	2.42	2.36	2.33	2.28	2.25
27	2.75	2.71	2.68	2.66	2.63	2.59	2.55	2.52	2.49	2.47	2.38	2.33	2.29	2.25	2.22
28	2.72	2.68	2.65	2.63	2.60	2.56	2.52	2.49	2.46	2.44	2.35	2.30	2.26	2.22	2.19
29	2.69	2.66	2.63	2.60	2.57	2.53	2.49	2.46	2.44	2.41	2.33	2.27	2.23	2.19	2.16
30	2.66	2.63	2.60	2.57	2.55	2.51	2.47	2.44	2.41	2.39	2.30	2.25	2.21	2.16	2.13
32	2.62	2.58	2.55	2.53	2.50	2.46	2.42	2.39	2.36	2.34	2.25	2.20	2.16	2.11	2.08
34	2.58	2.55	2.51	2.49	2.46	2.42	2.38	2.35	2.32	2.30	2.21	2.16	2.12	2.07	2.04
36	2.54	2.51	2.48	2.45	2.43	2.38	2.35	2.32	2.29	2.26	2.17	2.12	2.08	2.03	2.00
38	2.51	2.48	2.45	2.42	2.40	2.35	2.32	2.28	2.26	2.23	2.14	2.09	2.05	2.00	1.97
40	2.48	2.45	2.42	2.39	2.37	2.33	2.29	2.26	2.23	2.20	2.11	2.06	2.02	1.97	1.94
42	2.46	2.43	2.40	2.37	2.34	2.30	2.26	2.23	2.20	2.18	2.09	2.03	1.99	1.94	1.91
44	2.44	2.40	2.37	2.35	2.32	2.28	2.24	2.21	2.18	2.15	2.06	2.01	1.97	1.92	1.89
46	2.42	2.38	2.35	2.33	2.30	2.26	2.22	2.19	2.16	2.13	2.04	1.99	1.95	1.90	1.86
48	2.40	2.37	2.33	2.31	2.28	2.24	2.20	2.17	2.14	2.12	2.02	1.97	1.93	1.88	1.84
50	2.38	2.35	2.32	2.29	2.27	2.22	2.18	2.15	2.12	2.10	2.01	1.95	1.91	1.86	1.82
55	2.34	2.31	2.28	2.25	2.23	2.18	2.15	2.11	2.08	2.06	1.97	1.91	1.87	1.81	1.78
60	2.31	2.28	2.25	2.22	2.20	2.15	2.12	2.08	2.05	2.03	1.94	1.88	1.84	1.78	1.75
70	2.27	2.23	2.20	2.18	2.15	2.11	2.07	2.03	2.01	1.98	1.89	1.83	1.78	1.73	1.70
80	2.23	2.20	2.17	2.14	2.12	2.07	2.03	2.00	1.97	1.94	1.85	1.79	1.75	1.69	1.66
90	2.21	2.17	2.14	2.11	2.09	2.04	2.00	1.97	1.94	1.92	1.82	1.76	1.72	1.66	1.62
100	2.19	2.15	2.12	2.09	2.07	2.02	1.98	1.94	1.92	1.89	1.80	1.73	1.69	1.63	1.60
125	2.15	2.11	2.08	2.05	2.03	1.98	1.94	1.91	1.88	1.85	1.76	1.69	1.65	1.59	1.55
150	2.12	2.09	2.06	2.03	2.00	1.96	1.92	1.88	1.85	1.83	1.73	1.66	1.62	1.56	1.52
200	2.09	2.06	2.02	2.00	1.97	1.93	1.89	1.85	1.82	1.79	1.69	1.63	1.58	1.52	1.48
300	2.06	2.03	1.99	1.97	1.94	1.89	1.85	1.82	1.79	1.76	1.66	1.59	1.55	1.48	1.44
500	2.04	2.00	1.97	1.94	1.92	1.87	1.83	1.79	1.76	1.74	1.63	1.56	1.52	1.45	1.41
1,000	2.02	1.98	1.95	1.92	1.90	1.85	1.81	1.77	1.74	1.72	1.61	1.54	1.50	1.43	1.38

Whenever f_2 exceeds 1,000, the bound corresponding to $f_2 = 1{,}000$ is to be used.

Table 8C. *Bounds for $F = s_1^2/s_2^2$ with error probability 0.1 %. The degrees of freedom are f_1 in the numerator and f_2 in the denominator*

f_2 ↓	Degrees of freedom for the numerator f_1														
	1	2	3	4	5	6	7	8	9	10	15	20	30	50	100
2	998	999	999	999	999	999	999	999	999	999	999	999	999	999	999
3	168	148	141	137	135	133	132	131	130	129	127	126	125	125	124
4	74.1	61.2	56.2	53.4	51.7	50.5	49.7	49.0	48.5	48.0	46.8	46.1	45.4	44.9	44.5
5	47.0	36.6	33.2	31.1	29.8	28.8	28.2	27.6	27.2	26.9	25.9	25.4	24.9	24.4	24.1
6	35.5	27.0	23.7	21.9	20.8	20.0	19.5	19.0	18.7	18.4	17.6	17.1	16.7	16.3	16.0
7	29.2	21.7	18.8	17.2	16.2	15.5	15.0	14.6	14.3	14.1	13.3	12.9	12.5	12.2	11.9
8	25.4	18.5	15.8	14.4	13.5	12.9	12.4	12.0	11.8	11.5	10.8	10.5	10.1	9.80	9.57
9	22.9	16.4	13.9	12.6	11.7	11.1	10.7	10.4	10.1	9.89	9.24	8.90	8.55	8.26	8.04
10	21.0	14.9	12.6	11.3	10.5	9.92	9.52	9.20	8.96	8.75	8.13	7.80	7.47	7.19	6.98
11	19.7	13.8	11.6	10.4	9.58	9.05	8.66	8.35	8.12	7.92	7.32	7.01	6.68	6.41	6.21
12	18.6	13.0	10.8	9.63	8.89	8.38	8.00	7.71	7.48	7.29	6.71	6.40	6.09	5.83	5.63
13	17.8	12.3	10.2	9.07	8.35	7.86	7.49	7.21	6.98	6.80	6.23	5.93	5.62	5.37	5.17
14	17.1	11.8	9.73	8.62	7.92	7.43	7.08	6.80	6.58	6.40	5.85	5.56	5.25	5.00	4.80
15	16.6	11.3	9.34	8.25	7.57	7.09	6.74	6.47	6.26	6.08	5.53	5.25	4.95	4.70	4.51
16	16.1	11.0	9.00	7.94	7.27	6.81	6.46	6.19	5.98	5.81	5.27	4.99	4.70	4.45	4.26
17	15.7	10.7	8.73	7.68	7.02	6.56	6.22	5.96	5.75	5.58	5.05	4.78	4.48	4.24	4.05
18	15.4	10.4	8.49	7.46	6.81	6.35	6.02	5.76	5.56	5.39	4.87	4.59	4.30	4.06	3.87
19	15.1	10.2	8.28	7.26	6.61	6.18	5.84	5.59	5.39	5.22	4.70	4.43	4.14	3.90	3.71
20	14.8	9.95	8.10	7.10	6.46	6.02	5.69	5.44	5.24	5.08	4.56	4.29	4.01	3.77	3.58
22	14.4	9.61	7.80	6.81	6.19	5.76	5.44	5.19	4.99	4.83	4.32	4.06	3.77	3.53	3.34
24	14.0	9.34	7.55	6.59	5.98	5.55	5.23	4.99	4.80	4.64	4.14	3.87	3.59	3.35	3.16
26	13.7	9.12	7.36	6.41	5.80	5.38	5.07	4.83	4.64	4.48	3.99	3.72	3.45	3.20	3.01
28	13.5	8.93	7.19	6.25	5.66	5.24	4.93	4.69	4.50	4.35	3.86	3.60	3.32	3.08	2.89
30	13.3	8.77	7.05	6.12	5.53	5.12	4.82	4.58	4.39	4.24	3.75	3.49	3.22	2.98	2.79
40	12.6	8.25	6.60	5.70	5.13	4.73	4.43	4.21	4.02	3.87	3.40	3.15	2.87	2.64	2.44
50	12.2	7.95	6.34	5.46	4.90	4.51	4.22	4.00	3.82	3.67	3.20	2.95	2.68	2.44	2.24
60	12.0	7.76	6.17	5.31	4.76	4.37	4.09	3.87	3.69	3.54	3.08	2.83	2.56	2.31	2.11
80	11.7	7.54	5.97	5.13	4.58	4.21	3.92	3.70	3.53	3.39	2.93	2.68	2.40	2.16	1.95
100	11.5	7.41	5.85	5.01	4.48	4.11	3.83	3.61	3.44	3.30	2.84	2.59	2.32	2.07	1.87
200	11.2	7.15	5.64	4.81	4.29	3.92	3.65	3.43	3.26	3.12	2.67	2.42	2.15	1.90	1.68
500	11.0	7.01	5.51	4.69	4.18	3.82	3.54	3.33	3.16	3.02	2.58	2.33	2.05	1.80	1.57
∞	10.8	6.91	5.42	4.62	4.10	3.74	3.47	3.27	3.10	2.96	2.51	2.27	1.99	1.73	1.49

Linear interpolation yields bounds which are somewhat too large. The bounds for $f_2 = 1{,}000$ lie approximately midway between the bounds for 500 and ∞.

Tables 8A, B, and C have been extracted from A. Hald, Statistical Tables and Formulas. New York: John Wiley and Sons 1952.

Table 9. *Bounds for the sign test*

One-sided	2.5%		1%		0.5%		One-sided	2.5%		1%		0.5%	
$n=5$	0	5	0	5	0	5	$n=53$	19	34	18	35	17	36
6	1	5	0	6	0	6	54	20	34	19	35	18	36
7	1	6	1	6	0	7	55	20	35	19	36	18	37
8	1	7	1	7	1	7	56	21	35	19	37	18	38
9	2	7	1	8	1	8	57	21	36	20	37	19	38
10	2	8	1	9	1	9	58	22	36	20	38	19	39
11	2	9	2	9	1	10	59	22	37	21	38	20	39
12	3	9	2	10	2	10	60	22	38	21	39	20	40
13	3	10	2	11	2	11	61	23	38	21	40	21	40
14	3	11	3	11	2	12	62	23	39	22	40	21	41
15	4	11	3	12	3	12	63	24	39	22	41	21	42
16	4	12	3	13	3	13	64	24	40	23	41	22	42
17	5	12	4	13	3	14	65	25	40	23	42	22	43
18	5	13	4	14	4	14	66	25	41	24	42	23	43
19	5	14	5	14	4	15	67	26	41	24	43	23	44
20	6	14	5	15	4	16	68	26	42	24	44	23	45
21	6	15	5	16	5	16	69	26	43	25	44	24	45
22	6	16	6	16	5	17	70	27	43	25	45	24	46
23	7	16	6	17	5	18	71	27	44	26	45	25	46
24	7	17	6	18	6	18	72	28	44	26	46	25	47
25	8	17	7	18	6	19	73	28	45	27	46	26	47
26	8	18	7	19	7	19	74	29	45	27	47	26	48
27	8	19	8	19	7	20	75	29	46	27	48	26	49
28	9	19	8	20	7	21	76	29	47	28	48	27	49
29	9	20	8	21	8	21	77	30	47	28	49	27	50
30	10	20	9	21	8	22	78	30	48	29	49	28	50
31	10	21	9	22	8	23	79	31	48	29	50	28	51
32	10	22	9	23	9	23	80	31	49	30	50	29	51
33	11	22	10	23	9	24	81	32	49	30	51	29	52
34	11	23	10	24	10	24	82	32	50	31	51	29	53
35	12	23	11	24	10	25	83	33	50	31	52	30	53
36	12	24	11	25	10	26	84	33	51	31	53	30	54
37	13	24	11	26	11	26	85	33	52	32	53	31	54
38	13	25	12	26	11	27	86	34	52	32	54	31	55
39	13	26	12	27	12	27	87	34	53	33	54	32	55
40	14	26	13	27	12	28	88	35	53	33	55	32	56
41	14	27	13	28	12	29	89	35	54	34	55	32	57
42	15	27	14	28	13	29	90	36	54	34	56	33	57
43	15	28	14	29	13	30	91	36	55	34	57	33	58
44	16	28	14	30	14	30	92	37	55	35	57	34	58
45	16	29	15	30	14	31	93	37	56	35	58	34	59
46	16	30	15	31	14	32	94	38	56	36	58	35	59
47	17	30	16	31	15	32	95	38	57	36	59	35	60
48	17	31	16	32	15	33	96	38	58	37	59	35	61
49	18	31	16	33	16	33	97	39	58	37	60	36	61
50	18	32	17	33	16	34	98	39	59	38	60	36	62
51	19	32	17	34	16	35	99	40	59	38	61	37	62
52	19	33	18	34	17	35	100	40	60	38	62	37	63
Two-sided	5%		2%		1%		Two-sided	5%		2%		1%	

An observation lying outside these bounds may be safely assumed to indicate the presence of an effect.

Table 10. *Significance probabilities for the Wilcoxon test*

The number of inversions (x following y in the sequence of the x and the y arranged according to increasing magnitude) is a random variable U, which takes on in particular cases the values u. If $u > \frac{1}{2} g h$, the roles of x and y are to be interchanged. The *significance probability* is the probability of the event $U \leqq u$ under the "null hypothesis". *If this probability is* $\leqq \alpha$, *the null hypothesis is to be rejected.* The significance level is α for the one-sided test, 2α for the two-sided test. The table gives, in percentages, those significance probabilities which do not exceed 5%.

u ↓	Numbers g; h or h; g of observations x and y													
	2; 5	2; 6	2; 7	2; 8	2; 9	2; 10	3; 3	3; 4	3; 5	3; 6	3; 7	3; 8	3; 9	3; 10
0	4.76	3.57	2.78	2.22	1.82	1.52	5.00	2.86	1.79	1.19	0.83	0.61	0.45	0.35
1				4.44	3.64	3.03			3.57	2.38	1.67	1.21	0.91	0.70
2										4.76	3.33	2.42	1.82	1.40
3												4.24	3.18	2.45
4													5.00	3.85

u	4; 4	4; 5	4; 6	4; 7	4; 8	4; 9	4; 10	5; 5	5; 6	5; 7	5; 8	5; 9	5; 10	6; 6
0	1.43	0.79	0.48	0.30	0.20	0.14	0.10	0.40	0.22	0.13	0.08	0.05	0.03	0.11
1	2.86	1.59	0.95	0.61	0.40	0.28	0.20	0.79	0.43	0.25	0.16	0.10	0.07	0.22
2		3.17	1.90	1.21	0.81	0.56	0.40	1.59	0.87	0.51	0.31	0.20	0.13	0.43
3			3.33	2.12	1.41	0.98	0.70	2.78	1.52	0.88	0.54	0.35	0.23	0.76
4				3.64	2.42	1.68	1.20	4.76	2.60	1.52	0.93	0.60	0.40	1.30
5					3.64	2.52	1.80		4.11	2.40	1.48	0.95	0.63	2.06
6						3.78	2.70			3.66	2.25	1.45	0.97	3.25
7							3.80				3.26	2.10	1.40	4.65
8											4.66	3.00	2.00	
9												4.15	2.76	
10													3.76	
11													4.96	

Table 10 (Continuation). *Significance probabilities for the Wilcoxon test*

u ↓	Numbers g; h or h; g of observations x and y													
	6; 7	6; 8	6; 9	6; 10	7; 7	7; 8	7; 9	7; 10	8; 8	8; 9	8; 10	9; 9	9; 10	10;10
0	0.06	0.03	0.02	0.01	0.03	0.02	0.01	0.01	0.01	0.00				
1	0.12	0.07	0.04	0.02	0.06	0.03	0.02	0.01	0.02	0.01	0.00	0.00		
2	0.23	0.13	0.08	0.05	0.12	0.06	0.03	0.02	0.03	0.02	0.01	0.01	0.00	
3	0.41	0.23	0.14	0.09	0.20	0.11	0.06	0.04	0.05	0.03	0.02	0.01	0.01	0.00
4	0.70	0.40	0.24	0.15	0.35	0.19	0.10	0.06	0.09	0.05	0.03	0.02	0.01	0.01
5	1.11	0.63	0.38	0.24	0.55	0.30	0.17	0.10	0.15	0.08	0.04	0.04	0.02	0.01
6	1.75	1.00	0.60	0.37	0.87	0.47	0.26	0.15	0.23	0.12	0.07	0.06	0.03	0.02
7	2.56	1.47	0.88	0.55	1.31	0.70	0.39	0.23	0.35	0.19	0.10	0.09	0.05	0.02
8	3.67	2.13	1.28	0.80	1.89	1.03	0.58	0.34	0.52	0.28	0.15	0.14	0.07	0.04
9		2.96	1.80	1.12	2.65	1.45	0.82	0.48	0.74	0.39	0.22	0.20	0.10	0.05
10		4.06	2.48	1.56	3.64	2.00	1.15	0.68	1.03	0.56	0.31	0.28	0.15	0.08
11			3.32	2.10	4.87	2.70	1.56	0.93	1.41	0.76	0.43	0.39	0.21	0.10
12			4.40	2.80		3.61	2.09	1.25	1.90	1.03	0.58	0.53	0.28	0.14
13				3.63		4.69	2.74	1.65	2.49	1.37	0.78	0.71	0.38	0.19
14				4.67			3.56	2.15	3.25	1.80	1.03	0.94	0.51	0.26
15							4.54	2.77	4.15	2.32	1.33	1.22	0.66	0.34
16								3.51		2.96	1.71	1.57	0.86	0.45
17								4.39		3.72	2.17	2.00	1.10	0.57
18										4.64	2.73	2.52	1.40	0.73
19											3.38	3.13	1.75	0.93
20											4.16	3.85	2.17	1.16
21												4.70	2.67	1.44
22													3.26	1.77
23													3.94	2.16
24													4.74	2.62
25														3.15
26														3.76
27														4.46

For larger g and h, the significance probabilities P given by

$$P=\Phi\left(\frac{u+\frac{1}{2}-\frac{1}{2}gh}{\sqrt{\frac{1}{12}gh(g+h+1)}}\right)$$

are sufficiently exact.

Table 10 has been abridged from H. R. van der Vaart, Gebruiksaanwijzing voor de toets van Wilcoxon. Mathematisch Centrum Amsterdam Rapport S 32 (1952).

Table 11. *Bounds on X for the X-test*

One-sided 2.5%				One-sided 1%				One-sided 0.5%			
n	$g-h=$ 0 or 1	$g-h=$ 2 or 3	$g-h=$ 4 or 5	n	$g-h=$ 0 or 1	$g-h=$ 2 or 3	$g-h=$ 4 or 5	n	$g-h=$ 0 or 1	$g-h=$ 2 or 3	$g-h=$ 4 or 5
6	∞	∞	∞	6	∞	∞	∞	6	∞	∞	∞
7	∞	∞	∞	7	∞	∞	∞	7	∞	∞	∞
8	2.40	2.30	∞	8	∞	∞	∞	8	∞	∞	∞
9	2.38	2.20	∞	9	2.80	∞	∞	9	∞	∞	∞
10	2.60	2.49	2.30	10	3.00	2.90	2.80	10	3.20	3.10	∞
11	2.72	2.58	2.40	11	3.20	3.00	2.90	11	3.40	3.40	∞
12	2.86	2.79	2.68	12	3.29	3.30	3.20	12	3.60	3.58	3.40
13	2.96	2.91	2.78	13	3.50	3.36	3.18	13	3.71	3.68	3.50
14	3.11	3.06	3.00	14	3.62	3.55	3.46	14	3.94	3.88	3.76
15	3.24	3.19	3.06	15	3.74	3.68	3.57	15	4.07	4.05	3.88
16	3.39	3.36	3.28	16	3.92	3.90	3.80	16	4.26	4.25	4.12
17	3.49	3.44	3.36	17	4.06	4.01	3.90	17	4.44	4.37	4.23
18	3.63	3.60	3.53	18	4.23	4.21	4.14	18	4.60	4.58	4.50
19	3.73	3.69	3.61	19	4.37	4.32	4.23	19	4.77	4.71	4.62
20	3.86	3.84	3.78	20	4.52	4.50	4.44	20	4.94	4.92	4.85
21	3.96	3.92	3.85	21	4.66	4.62	4.53	21	5.10	5.05	4.96
22	4.08	4.06	4.01	22	4.80	4.78	4.72	22	5.26	5.24	5.17
23	4.18	4.15	4.08	23	4.92	4.89	4.81	23	5.40	5.36	5.27
24	4.29	4.27	4.23	24	5.06	5.04	4.99	24	5.55	5.53	5.48
25	4.39	4.36	4.30	25	5.18	5.14	5.08	25	5.68	5.65	5.58
26	4.50	4.48	4.44	26	5.30	5.29	5.24	26	5.83	5.81	5.76
27	4.59	4.56	4.51	27	5.42	5.39	5.33	27	5.95	5.92	5.85
28	4.69	4.68	4.64	28	5.54	5.52	5.48	28	6.09	6.07	6.03
29	4.78	4.76	4.72	29	5.65	5.62	5.57	29	6.22	6.19	6.13
30	4.88	4.87	4.84	30	5.77	5.75	5.72	30	6.35	6.34	6.30
31	4.97	4.95	4.91	31	5.87	5.85	5.80	31	6.47	6.44	6.39
32	5.07	5.06	5.03	32	5.99	5.97	5.94	32	6.60	6.58	6.55
33	5.15	5.13	5.10	33	6.09	6.07	6.02	33	6.71	6.69	6.64
34	5.25	5.24	5.21	34	6.20	6.19	6.16	34	6.84	6.82	6.79
35	5.33	5.31	5.28	35	6.30	6.28	6.24	35	6.95	6.92	6.88
36	5.42	5.41	5.38	36	6.40	6.39	6.37	36	7.06	7.05	7.02
37	5.50	5.48	5.45	37	6.50	6.48	6.45	37	7.17	7.15	7.11
38	5.59	5.58	5.55	38	6.60	6.59	6.57	38	7.28	7.27	7.25
39	5.67	5.65	5.62	39	6.70	6.68	6.65	39	7.39	7.37	7.33
40	5.75	5.74	5.72	40	6.80	6.79	6.77	40	7.50	7.49	7.47
41	5.83	5.81	5.79	41	6.89	6.88	6.85	41	7.62	7.60	7.56
42	5.91	5.90	5.88	42	6.99	6.98	6.96	42	7.72	7.71	7.69
43	5.99	5.97	5.95	43	7.08	7.07	7.04	43	7.82	7.81	7.77
44	6.06	6.06	6.04	44	7.17	7.17	7.14	44	7.93	7.92	7.90
45	6.14	6.12	6.10	45	7.26	7.25	7.22	45	8.02	8.01	7.98
46	6.21	6.21	6.19	46	7.35	7.35	7.32	46	8.13	8.12	8.10
47	6.29	6.27	6.25	47	7.44	7.43	7.40	47	8.22	8.21	8.18
48	6.36	6.35	6.34	48	7.53	7.52	7.50	48	8.32	8.31	8.29
49	6.43	6.42	6.39	49	7.61	7.60	7.57	49	8.41	8.40	8.37
50	6.50	6.50	6.48	50	7.70	7.69	7.68	50	8.51	8.50	8.48
Two-sided 5%				Two-sided 2%				Two-sided 1%			

Outside these bounds an effect may be safely assumed to be present.

Table 12. *Auxiliary table for the X-test*

$$Q=\frac{1}{n}\sum_{1}^{n}\Psi^2\left(\frac{r}{n+1}\right)$$

n	Q	n	Q	n	Q
1	0.000	51	0.872	101	0.923
2	0.186	52	0.874	102	0.924
3	0.303	53	0.876	103	0.924
4	0.386	54	0.877	104	0.925
5	0.449	55	0.879	105	0.926
6	0.497	56	0.880	106	0.926
7	0.537	57	0.882	107	0.927
8	0.570	58	0.884	108	0.927
9	0.598	59	0.885	109	0.928
10	0.622	60	0.887	110	0.928
11	0.642	61	0.888	111	0.929
12	0.661	62	0.889	112	0.929
13	0.677	63	0.891	113	0.930
14	0.692	64	0.892	114	0.930
15	0.705	65	0.893	115	0.931
16	0.716	66	0.894	116	0.931
17	0.727	67	0.895	117	0.932
18	0.737	68	0.897	118	0.932
19	0.746	69	0.898	119	0.932
20	0.755	70	0.899	120	0.933
21	0.763	71	0.900	121	0.933
22	0.770	72	0.901	122	0.934
23	0.777	73	0.902	123	0.934
24	0.783	74	0.903	124	0.935
25	0.789	75	0.904	125	0.935
26	0.794	76	0.905	126	0.935
27	0.799	77	0.906	127	0.936
28	0.804	78	0.907	128	0.936
29	0.809	79	0.908	129	0.937
30	0.813	80	0.908	130	0.937
31	0.817	81	0.909	131	0.937
32	0.821	82	0.910	132	0.938
33	0.825	83	0.911	133	0.938
34	0.829	84	0.912	134	0.938
35	0.833	85	0.913	135	0.939
36	0.836	86	0.913	136	0.939
37	0.839	87	0.914	137	0.939
38	0.842	88	0.915	138	0.940
39	0.845	89	0.916	139	0.940
40	0.848	90	0.916	140	0.940
41	0.850	91	0.917	141	0.941
42	0.853	92	0.918	142	0.941
43	0.855	93	0.918	143	0.941
44	0.858	94	0.919	144	0.942
45	0.860	95	0.920	145	0.942
46	0.862	96	0.920	146	0.942
47	0.864	97	0.921	147	0.943
48	0.866	98	0.922	148	0.943
49	0.868	99	0.922	149	0.943
50	0.870	100	0.923	150	0.944

Table 13. *Bounds for the correlation coefficient r*

$f=n-2$ for total correlation; $f=n-k-2$ for partial correlation, where k is the number of variables eliminated.

f ↓	Two-sided				f ↓	Two-sided			
	5%	2%	1%	0.1%		5%	2%	1%	0.1%
1	0.997	1.000	1.000	1.000	16	0.468	0.543	0.590	0.708
2	0.950	0.980	0.990	0.999	17	0.456	0.529	0.575	0.693
3	0.878	0.934	0.959	0.991	18	0.444	0.516	0.561	0.679
4	0.811	0.882	0.917	0.974	19	0.433	0.503	0.549	0.665
5	0.754	0.833	0.875	0.951	20	0.423	0.492	0.537	0.652
6	0.707	0.789	0.834	0.925	25	0.381	0.445	0.487	0.597
7	0.666	0.750	0.798	0.898	30	0.349	0.409	0.449	0.554
8	0.632	0.715	0.765	0.872	35	0.325	0.381	0.418	0.519
9	0.602	0.685	0.735	0.847	40	0.304	0.358	0.393	0.490
10	0.576	0.658	0.708	0.823	45	0.288	0.338	0.372	0.465
11	0.553	0.634	0.684	0.801	50	0.273	0.322	0.354	0.443
12	0.532	0.612	0.661	0.780	60	0.250	0.295	0.325	0.408
13	0.514	0.592	0.641	0.760	70	0.232	0.274	0.302	0.380
14	0.497	0.574	0.623	0.742	80	0.217	0.257	0.283	0.357
15	0.482	0.558	0.606	0.725	90	0.205	0.242	0.267	0.338
					100	0.195	0.230	0.254	0.321
↑ f	2.5%	1%	0.5%	0.05%	↑ f	2.5%	1%	0.5%	0.05%
	One-sided					One-sided			

For larger values of f we may calculate

$$t=f^{\frac{1}{2}}(1-r^2)^{-\frac{1}{2}}r$$

and refer to Table 7 with f degrees of freedom.

Table 13 has been extracted from E.S. Pearson and H.O. Hartley, Biometrika Tables for Statisticians I, Table 13 p. 138. Cambridge University Press 1954.

Examples Grouped According to Subject Matter

The number behind the key-word indicates the page number

Probability Theory

1. Probabilities with dice 4
3. Drawing from an urn without replacement 6
21. Estimation of an unknown probability 153
28. Estimation of a probability 167
43. Testing a hypothesis specifying a probability 262

Error Theory, Normal Distributions, and Related Topics

4. Expectation and standard deviation of the Gaussian error curve 15
5. Confidence limits for the expectation of a normal distribution 22
13. Rounding observations 85
22. Estimation of the mean and standard deviation 153
25. Estimation of the mean 165
26. Estimation of the variance 166
27. Method of least squares 166
29. χ^2-distribution with an unknown factor 175
30. Rectangular distribution 176
42. Most powerful test for expectation zero 261

Physics and Chemistry

10. Cosmic radiation intensities 50
16. Period of oscillation of a pendulum 112
23. Repeated measurements of concentrations 154
24. Estimation of the location of a radioactive source 158
39. Analyses of natural gas 243

Astronomy, Geodesy, and Meteorology

12. Rain in Rothamsted 82
14. Rounding errors in Hill's Saturn tables 107
15. Latitude of Cape Town 109
18. Triangulation 132
19. Byzantine table of the sun's movements 141

Biology and Physiology

11. Johannsen's selection experiment 68
32. Linked hereditary factors 187
34. Human blood types, estimating gene frequencies 206
35. Human blood types, testing Bernstein's hypothesis 210
36. Reaction of rabbits to gelsemicine 217
37. Deviations from Mendelian law 239
38. Deviations from normality in beans 241
40. Bacteria in soil samples 251

41. Implantation of partial genital-imaginal disks 254
44. Radiation of *Drosophila* eggs 269
46. Correlation between pollen-grain size and the number of places of exit 304

Medicine and Hygiene

7. Deaths from bronchiecstasis 35
9. Treating thrombosis with anti-coagulants 45
17. Weight increases in children 126

Demography and Econometrics

6. Birth frequencies of boys and girls 30
8. Sampling procedures 39
20. Pig iron production 146
47. Relation between rainfall and wheat harvest 310

Industrial Applications

31. Quality control 178
45. Reduction of waiting times 299

Statistics of Target-Shooting

2. Target-shooting 4
33. Aiming at a point 188

Correlation and Experimental Psychology

48. Confidence bounds for the true correlation coefficient 322
49. Difference between two correlation coefficients 322
50. Correlation between test scores 332

Author- and Subject Index

acceptance of a false hypothesis 257
Addition Rule 5
analysis of variance 246
— — table 250
Anderson 217
anomaly 141
— mean 141
— true 141
anticoagulants 45
approximation 156
— successive 156
— stochastic 223
arithmetic mean 153
asymptotically efficient 184
— equivalent 206
— normal 99, 100
— normally distributed 277
— unbiased 184, 203
asymptotic distribution of u 276
— efficiency 286
— expansion 11, 77
— mean 183
— normality 277
— power function 285
— variance 183, 203
axioms of Kolmogorov 5

bacteria counts 251
beans 68, 241
Behrens 214
Berkson, J. 221
Bernoulli's formula 24
Bernstein, F. 206
Bertani, G. 254
beta function 58
— incomplete 36
bias 30, 160, 163
binomial distribution 24, 93
bio-assay 212
Birkhoff, G. 4
Birnbaum, Z. W. 73, 75, 345
— and Tingey 344
birth of boys 30
birth of girls 30
bivariate normal distribution 316
Bliss, C. I. 217
Boolean algebra 4
bounds for χ^2 344, 346
— for F 243, 348
— for R 327, 338
— for r 309, 358
— for s^2 117
— for t 121, 347
— for U 274
— for W 274
— for x 293, 356
—, one-sided 31
—, two-sided 31
bronchiecstasis 35
Brownlee 222
bunch graphs 150

Carathéodory, C. 4, 8, 13
card 289
Carver 187
Cassel, G. 146
central moments 234
characteristic function 89
Chen 217
Chi-Square distribution 95, 102, 175, 197
— — with factor 175
— method of 235
— test 41, 51, 209, 225
Chung, K. L. 223
class interval 83
— midpoint 83
Clopper 35, 263
— and Pearson 263
Cochran, W. G. 238
coefficients of skewness 234
— of excess 234
combustion 244
comparison of probabilities 40, 228, 231
— of two frequencies 50
complete family 175
— boundedly 265
composite hypothesis 263

conditional probability 5
conditional expectation 167, 168
confidence bounds 32, 271
— — for the median 271
— —, two-sided 271
confidence limits for the median 270
— —, exact 35
consistent 99, 182
contragredient 62
converge in probability 99
correction, estimated 131
— Sheppard's 85, 236
correlation 253, 301
— coefficients 301
— —, empirical 302
— —, partial 310
— —, true 301, 310
—, partial 310
Courant-Hilbert 176
covariance 301
—, empirical 303
Cramér 92, 97, 160, 234
critical region 256, 272
cure 40
curve of Quetelet 67
cut-of point for R 338
— — for r 309
— — for t 321, 347
— — for x 293
Czuber 109

Daniels, H. E. 335
definite-positive 63
— -negative 63
degree of freedom 95, 209
demography 39
De Moivre 27
dependence 305, 328
deviation, mean 88
—, probable 87
die 4
distribution, binomial 24, 93
—, Cauchy- 86, 158
—, hypergeometric 38
—, normal 12, 94, 112, 234
— of χ^2 95, 102, 175, 197
— of R 324
— of r 316
— of $r_{xy|z}$ 312
— of s^2 113
distribution of T 334
— of t 309
— of U 276
— of x 295
—, Poisson- 95, 232
—, Students- 113
—, triangular 106
distribution-free 267
distribution function 8, 227
— —, empirical 69
— —, normal 340
Doob 183
dose 212
— response diagram 212
Drosophila melanogaster 254
— eggs 265
Dugué 97

eccentricity 141
economics 39
economic variable 149
efficient 184, 206
—, asymptotically 184
efficiency 203, 206
ejaculatory duct 254
ellipse 33
empirical distribution function 69
equation of center 141
error function 10
— —, Gaussian 10
— integral 11
—, mean 108
— probability 32, 229
—, random 108
—, systematic 108. 160
—, Type I 256
—, Type II 256
estimable 130
estimate 133, 153
—, consistent 99, 182
—, efficient 206
—, regular 204
—, sufficient 163
—, super-efficient 184
—, unbiased 134, 160
estimated corrections 131
estimation 29, 151, 185
— of a probability 167
— of the mean 165
— of the variance 139, 166

Euler's beta function 58
— gamma function 55
event 3
— rare 47
exact confidence limits 35
expectation 1, 12, 79, 89
— conditional 167, 168
— of u 276
excess 234
experiment 7
experimental animals 267

F-test 242, 249
Feller, W. 27, 75, 101
field of sets 4
Finney, D. J. 219
Fisher, R. A. 2, 45, 78, 103, 151, 163, 230, 251, 310, 320
—'s auxiliary variable 320
fourfold table 46
Fourier 90
— analysis 144
Fréchet 2, 13, 78, 160, 162
frequency 3, 26, 185, 297
Frisch, R. 150
Fritz-Niggli, H. 269
Fuchsia Globosa 304
functional equation 55
fundamental argument 107

Gaddum 216
Galera, J. 254
Galton 68
game of chance 261
gamma function 55
—, incomplete 55, 116
Gauss 1, 101, 108, 127, 137
—' theory of errors 108
Gaussian error function 10
genes 206
genital-imaginal disk 254
Gosset 120
graphical assay 218
Greiner and Esscher 333
grouping 68
group means 248

Hadorn, E. 254
Hald, A. 346, 352
Haldane, J. B. S. 238
half-space 201
Hanau, A. 149
height of a shot 188
Helmert, F. R. 95
Hemelrijk, J. 270
hereditary factors 185, 239
heuristic 290
Hill 107
Hipparchus 144
Hodges, J. L. 160, 184, 222
homogeneous 206
Hood, W. C. 150
Hotelling 183
human blood types 206
hypergeometric distribution 38
hyperplane 199
hypothesis, alternative 257
—, composite 263
— of elementary errors 112
—, simple 263

improvement of an estimate 173
independence 229
independent 7, 14
— variable 14
inequality 160
— Cramér-Rao 160
— Fréchet 162
— information 160, 162
— Schwarz 160, 162
information 157, 183
— inequality 160
— in the sample 157
inference 185
infrequent events 232
interpolation linear 107
— quadratic 107
integral equation 174
—, Fourier- 90
—, improper 54
—, Lebesgue- 13
—, Riemann- 18
—, Stiltjes- 13
intraclass correlation 253
invariant 63
inverse function 2, 342
inverse matrix 62
inversion 274
— formulae 92
— — of Cramér 92
— — of Levi 92

Johannsen, W. 68, 241

Kappos, D. 4
Kärber, G. 214
Kempthorne 256
Kendall, M. G. 126, 320, 324, 327
—'s rank correlation 333
Khinchin 99
Klein, L. R. 150
Kollektivmaßlehre 1
Koller, D. 236
Kolmogorov 1, 4, 5, 75, 169, 234
— axioms of 5
—'s test 74, 272, 355

Lagrange 127
Laplace 27, 101
latitude 109
lattice theory 4
— point 199
law of large numbers 27, 78, 99
least squares 127, 166
Lebesgue 90
Le Cam, L. 184
Lehmann, E. L. 2, 160, 167, 179, 284
— and Scheffé 2, 167, 179, 264
— and Stein 266
lethality 269
level of significance 121, 256, 264
Lévy 97, 101
— -Cramér 97
Liapounov 101
life insurance 32
likelihood equation 155, 186
— function 152
— method 151, 185
— ratio 259
— — test 259
limit theorems 97
— central 100, 251, 288
— elementary 103
— of Lévy-Cramér 97
— second 103, 277
Lindeberg 101
linked 187
logarithmic response curve 212
logistic curve 221
logits 221
loss 128, 178

Mann and Wald 236
— and Whitney 274, 280

manufacturer 178
maximum likelihood 151, 228
mean deviation 88
— error 108
— lethal dose 214
— of the integral 84
— of the population 80
measurable function 8
— set 7, 17
measure 17
measurement of concentration 154
median 75, 85, 270
Mendelean law 239
method, Newton's 156
— of χ^2 235
— of least squares 127, 166
— of maximum likelihood 151
— of moments 234, 277, 295
— of repetitions 132
— of successive approximation 156
metric 193
middle value 75
minimum χ^2 192
— χ_0^2 192
— χ_x^2 195
— variance 162
moment 91
Monro, S. 223
Mood, A. M. 221
Morrison, N. 169
mortality 212, 213
— rate 45
most likely value 152
most powerful 256, 257
— — uniformly 262
— — unbiased test 264

Newton's method 156
Neyman, J. 2, 206, 225
— and Pearson, E. S. 231, 256, 264
Nievergelt, E. 295, 299
Nikodym 169
Noether, G. E. 295
normal distribution 12, 94, 234
— — bivariate 316
— equation 129
— response curve 213, 236
null hypothesis 271, 274, 292

number of hits 50
— of inversions 274

observed point 135
one-point method 220
one-sided 31, 44
open set 17
order of magnitude 27, 284
— statistic 75, 270
orthogonal 60
orthogonalization 145

$\mathscr{P}$ 4
Pearson, E. S. 2, 35, 68, 225, 346
— and Hartley 346, 358
Pearson, K. 199, 200, 230, 332
period of oscillation of a pendulum 112
piecewise continuous 18
place of exit 304
Plackett, R. L. 137
Poisson distribution 95, 232
—'s formula 47
polar coordinates 53
— form 62
pollen 304
population 1
— mean 1, 79
Power 257
— function 285
prices of pigs 149
Prigge, R. 218
Primula sinensis 239
probable deviation 87
— error 82
probability 1, 3
— conditional 5
— total 7
probability density 9
— bivariate 18
— unknown 153
probit method 213
probits 218
Product Rule 5
production of pig iron 146
pure lines 69
pythagorean theorem 137

qualities 323, 328
quartiles 75, 87
Quetelet 67
— curve 67

radiation counter 50
rainfall 82, 310
random error 108
— sample 1
— variable 8
range 79
rank correlation R 323
— — T 323
— number 273
— test 298
Rao, C. R. 2, 130, 160, 167, 179
recombination rate 187
rectangle 17
rectangular 105
— distribution 70, 104, 176
red rabbits 217
region 52
regression 144
— coefficient 146, 301
— — empirical 304
— — sample 146
— line 112
— —, empirical 304
— —, sample 146
regular 204
Reichenbach, H. 4
rejection of a true hypothesis 256
— region 256, 261
residual 310
response curve logarithmic 212
— —, normal 213, 236
Robbins 217, 223
Rosenblatt 222
rounding 81
— errors 104

sample 68
— mean 80
— median 75, 85
— quartiles 75, 87
— sextiles 75
— size 185
— standard deviation 81
— variance 81
sampling 37
Sankhyā, I. 167
scalar product 63
Schäfer, W. 218
Scheffé 2, 167, 179

Schwarz inequality 160, 162
Schwerd 132
scintillation 158
scores 156
self-fertilization 68
Shepherd, M. 243
Sheppard 82
— correction 82, 236, 241
Shohat, J. 103
significance level 32, 256, 264
— probability 281, 354
similar to the sample space 264
simple hypothesis 263
skewness 234
Smirnov 74, 75
—'s test 272
smoothing 127
Spearman 323
—'s rank correlation 323
standard deviation 14, 79
— error 108
— preparation 213
starchy 187
statistical concept 3
step function 9
stochastic 8
— approximation 223
— variable 8
Stocker, D. J. 295
Stone, M. H. 4
Student 120
— and Fisher 123
—'s test 108, 118, 266, 284, 347
successive approximation 156
sufficiency 162
sufficient 163
— statistics 170
sugary 187
super-efficient 184
surface area of the unit sphere 57
surgeon 24, 35, 40
symmetric about zero 270
systematic error 30, 108, 160

table of the sun 141
Tammes, P. M. L. 304
Tchebychev's inequality 16
tend to 98
tensor 61
test χ^2 41, 51, 209, 225
— Δ 74, 344
— F 242, 249
test likelihood ratio 259
— most powerful 256
— non parametric 267
— of Kolmogorov 74, 234, 345
— of Smirnov 272
— of symmetry 270
— order 267
— rank 298
— sign 268
— Student's 118, 121, 266, 284, 347
— t 252
— variance ratio 242, 271
— Wilcoxon's 273, 284, 354
— X 290, 292
testing for normality 234
— of hypothesis 225
— the independence 229
theory of errors 108
thrombosis 45
ties 269
Tinbergen, J. 150
Tingey, F. H. 73
Tintner, G. 150
Tippett 78
transformation of Fisher 321
— of integrals 52
— of Rao 30
— orthogonal 60
trend 146
trial 7
true anomaly 141
true point 136
t-distribution 309
— Test 252
— — one-sided 266
two-point method 220
two-sample problem 270
two-sided 31, 44
2×2 table 229
Type-I-error 256
— II-error 256

unbiased 134, 263
— estimation 174
— asymptotically 184, 203
uniformly distributed 104
— most powerful 262
unknown probability 153
"up and down" method 221
urn 6, 37

van der Vaart, H. R. 282, 355
variance 14, 79
— analysis of 246
— asymptotic 183, 203
— empirical 303
— minimum 162
— of U 276
— ratio 242
— repeatability 245
variation between classes 247
— — groups 246
— within classes 247
— — groups 246
vector 61
— contravariant 61
— covariant 61

waiting times 299
Wald, A. 182, 189, 295
Wegmann, F. 35
wheat crop 310
Wilcoxon's test 273, 284, 354
Wilks, S. S. 78
willow leaves 67
Wise, M. E. 28
Wolfowitz, J. 160, 182, 189, 295

X-test 290, 292
— one-sided 293
— two-sided 293

Yule and Brownlee 230

Druck der Universitätsdruckerei H. Stürtz AG, Würzburg

Die Grundlehren der mathematischen Wissenschaften in Einzeldarstellungen mit besonderer Berücksichtigung der Anwendungsgebiete

2. Knopp: Theorie und Anwendung der unendlichen Reihen. DM 48,– ; US $ 12.00
3. Hurwitz: Vorlesungen über allgemeine Funktionentheorie und elliptische Funktionen. DM 49,– ; US $ 12.25
4. Madelung: Die mathematischen Hilfsmittel des Physikers. DM 49,70; US $ 12.45
10. Schouten: Ricci-Calculus. DM 58,60; US $ 14.65
14. Klein: Elementarmathematik vom höheren Standpunkt aus. 1. Band: Arithmetik. Algebra. Analysis. DM 24,– ; US $ 6.00
15. Klein: Elementarmathematik vom höheren Standpunkt aus. 2. Band: Geometrie. DM 24,– ; US $ 6.00
16. Klein: Elementarmathematik vom höheren Standpunkt aus. 3. Band: Präzisions- und Approximationsmathematik. DM 19,80; US $ 4.95
19. Pólya/Szegö: Aufgaben und Lehrsätze aus der Analysis I: Reihen, Integralrechnung, Funktionentheorie. DM 34,– ; US $ 8.50
20. Pólya/Szegö: Aufgaben und Lehrsätze aus der Analysis II: Funktionentheorie, Nullstellen, Polynome, Determinanten, Zahlentheorie. DM 38,– ; US $ 9.50
22. Klein: Vorlesungen über höhere Geometrie. DM 28,– ; US $ 7.00
26. Klein: Vorlesungen über nicht-euklidische Geometrie. DM 24,– ; US $ 6.00
27. Hilbert/Ackermann: Grundzüge der theoretischen Logik. DM 38,– ; US $ 9.50
30. Lichtenstein: Grundlagen der Hydromechanik. DM 38,– ; US $ 9.50
31. Kellogg: Foundations of Potential Theory. DM 32,– ; US $ 8.00
32. Reidemeister: Vorlesungen über Grundlagen der Geometrie. DM 18,– ; US $ 4.50
38. Neumann: Mathematische Grundlagen der Quantenmechanik. DM 28,– ; US $ 7.00
40. Hilbert/Bernays: Grundlagen der Mathematik I. DM 68,– ; US $ 17.00
43. Neugebauer: Vorlesungen über Geschichte der antiken mathematischen Wissenschaften. 1. Band: Vorgriechische Mathematik. DM 48,– ; US $ 12.00
50. Hilbert/Bernays: Grundlagen der Mathematik II. DM 68,– ; US $ 17.00
52. Magnus/Oberhettinger/Soni: Formulas and Theorems for the Special Functions of Mathematical Physics. DM 66,– ; US $ 16.50
57. Hamel: Theoretische Mechanik. DM 84,– ; US $ 21.00
58. Blaschke/Reichardt: Einführung in die Differentialgeometrie. DM 24,– ; US $ 6.00
59. Hasse: Vorlesungen über Zahlentheorie. DM 69,– ; US $ 17.25
60. Collatz: The Numerical Treatment of Differential Equations. DM 78,– ; US $ 19.50
61. Maak: Fastperiodische Funktionen. DM 38,– ; US $ 9.50
62. Sauer: Anfangswertprobleme bei partiellen Differentialgleichungen. DM 41,– ; US $ 10.25
64. Nevanlinna: Uniformisierung. DM 49,50; US $ 12.40
66. Bieberbach: Theorie der gewöhnlichen Differentialgleichungen. DM 58,50; US $ 14.65
68. Aumann: Reelle Funktionen. DM 59,60; US $ 14.90
69. Schmidt: Mathematische Gesetze der Logik I. DM 79,– ; US $ 19.75
71. Meixner/Schäfke: Mathieusche Funktionen und Sphäroidfunktionen mit Anwendungen auf physikalische und technische Probleme. DM 52,60; US $ 13.15
73. Hermes: Einführung in die Verbandstheorie. DM 46,– ; US $ 11.50
74. Boerner: Darstellungen von Gruppen. DM 58,– ; US $ 14.50
75. Rado/Reichelderfer: Continuous Transformations in Analysis, with an Introduction to Algebraic Topology. DM 59,60; US $ 14.90
76. Tricomi: Vorlesungen über Orthogonalreihen. DM 37,60; US $ 9.40

77. Behnke/Sommer: Theorie der analytischen Funktionen einer komplexen Veränderlichen. DM 79,– ; US $ 19.75
78. Lorenzen: Einführung in die operative Logik und Mathematik. DM 54,– ; US $ 13.50
80. Pickert: Projektive Ebenen. DM 48,60; US $ 12.15
81. Schneider: Einführung in die transzendenten Zahlen. DM 24,80; US $ 6.20
82. Specht: Gruppentheorie. DM 69,60; US $ 17.40
84. Conforto: Abelsche Funktionen und algebraische Geometrie. DM 41,80; US $ 10.45
86. Richter: Wahrscheinlichkeitstheorie. DM 68,– ; US $ 17.00
87. van der Waerden: Mathematische Statistik. DM 49,60; US $ 12.40
88. Müller: Grundprobleme der mathematischen Theorie elektromagnetischer Schwingungen. DM 52,80; US $ 13.20
89. Pfluger: Theorie der Riemannschen Flächen. DM 39,20; US $ 9.80
90. Oberhettinger: Tabellen zur Fourier Transformation. DM 39,50; US $ 9.90
91. Prachar: Primzahlverteilung. DM 58,– ; US $ 14.50
93. Hadwiger: Vorlesungen über Inhalt, Oberfläche und Isoperimetrie. DM 49,80; US $ 12.45
94. Funk: Variationsrechnung und ihre Anwendung in Physik und Technik. DM 98,– ; US $ 24.50
95. Maeda: Kontinuierliche Geometrien. DM 39,– ; US $ 9.75
97. Greub: Lineare Algebra. DM 39,20; US $ 9.80
98. Saxer: Versicherungsmathematik. 2. Teil. DM 48,60; US $ 12.15
99. Cassels: An Introduction to the Geometry of Numbers. DM 69,– ; US $ 17.25
100. Koppenfels/Stallmann: Praxis der konformen Abbildung. DM 69,– ; US $ 17.25
101. Rund: The Differential Geometry of Finsler Spaces. DM 59,60; US $ 14.90
103. Schütte: Beweistheorie. DM 48,– ; US $ 12.00
104. Chung: Markov Chains with Stationary Transition Probabilities. DM 56,– ; US $ 14.00
105. Rinow: Die innere Geometrie der metrischen Räume. DM 83,– ; US $ 20.75
106. Scholz/Hasenjaeger: Grundzüge der mathematischen Logik. DM 98,– ; US $ 24.50
107. Köthe: Topologische Lineare Räume I. DM 78,– ; US $ 19.50
108. Dynkin: Die Grundlagen der Theorie der Markoffschen Prozesse. DM 33,80; US $ 8.45
110. Dinghas: Vorlesungen über Funktionentheorie. DM 69,– ; US $ 17.25
111. Lions: Equations différentielles opérationnelles et problèmes aux limites. DM 64,– ; US $ 16.00
112. Morgenstern/Szabó: Vorlesungen über theoretische Mechanik. DM 69,– ; US $ 17.25
113. Meschkowski: Hilbertsche Räume mit Kernfunktion. DM 58,– ; US $ 14.50
114. MacLane: Homology. DM 62,– ; US $ 15.50
115. Hewitt/Ross: Abstract Harmonic Analysis. Vol. 1: Structure of Topological Groups. Integration Theory. Group Representations. DM 76,– ; US $ 19.00
116. Hörmander: Linear Partial Differential Operators. DM 42,– ; US $ 10.50
117. O'Meara: Introduction to Quadratic Forms. DM 48,– ; US $ 12.00
118. Schäfke: Einführung in die Theorie der speziellen Funktionen der mathematischen Physik. DM 49,40; US $ 12.35
119. Harris: The Theory of Branching Processes. DM 36,– ; US $ 9.00
120. Collatz: Funktionalanalysis und numerische Mathematik. DM 58,– ; US $ 14.50
121./122. Dynkin: Markov Processes. DM 96,– ; US $ 24.00
123. Yosida: Functional Analysis. DM 66,– ; US $ 16.50
124. Morgenstern: Einführung in die Wahrscheinlichkeitsrechnung und mathematische Statistik. DM 38,– ; US $ 9.50
125. Itô/McKean: Diffusion Processes and Their Sample Paths. DM 58,– ; US $ 14.50

126. Letho/Virtanen: Quasikonforme Abbildungen. DM 38,– ; US $ 9.50
127. Hermes: Enumerability, Decidability, Computability. DM 39,– ; US $ 9.75
128. Braun/Koecher: Jordan-Algebren. DM 48,– ; US $ 12.00
129. Nikodým: The Mathematical Apparatus for Quantum-Theories. DM 144,– ; US $ 36.00
130. Morrey: Multiple Integrals in the Calculus of Variations. DM 78,– ; US $ 19.50
131. Hirzebruch: Topological Methods in Algebraic Geometry. DM 38,– ; US $ 9.50
132. Kato: Perturbation Theory for Linear Operators. DM 79,20; US $ 19.80
133. Haupt/Künneth: Geometrische Ordnungen. DM 68,– ; US $ 17.00
134. Huppert: Endliche Gruppen I. DM 156,– ; US $ 39.00
135. Handbook for Automatic Computation. Vol. 1/Part a: Rutishauser: Description of ALGOL 60. DM 58,– ; US $ 14.50
136. Greub: Multilinear Algebra. DM 32,– ; US $ 8.00
137. Handbook for Automatic Computation. Vol. 1/Part b: Grau/Hill/Langmaack: Translation of ALGOL 60. DM 64,– ; US $ 16.00
138. Hahn: Stability of Motion. DM 72,– ; US $ 18.00
139. Mathematische Hilfsmittel des Ingenieurs. Herausgeber: Sauer/Szabó. 1. Teil. DM 88,– ; US $ 22.00
140. Mathematische Hilfsmittel des Ingenieurs. Herausgeber: Sauer/Szabó. 2. Teil. DM 136,– ; US $ 34.00
141. Mathematische Hilfsmittel des Ingenieurs. Herausgeber: Sauer/Szabó. 3. Teil. DM 98,– ; US $ 24.50
142. Mathematische Hilfsmittel des Ingenieurs. Herausgeber: Sauer/Szabó. 4. Teil. In Vorbereitung
143. Schur/Grunsky: Vorlesungen über Invariantentheorie. DM 32,– ; US $ 8.00
144. Weil: Basic Number Theory. DM 48,– ; US $ 12.00
145. Butzer/Berens: Semi-Groups of Operators and Approximation. DM 56,– ; US $ 14.00
146. Treves: Locally Convex Spaces and Linear Partial Differential Equations. DM 36,– ; US $ 9.00
147. Lamotke: Semisimpliziale algebraische Topologie. DM 48,– ; US $ 12.00
148. Chandrasekharan: Introduction to Analytic Number Theory. DM 28,– ; US $ 7.00
149. Sario/Oikawa: Capacity Functions. DM 96,– ; US $ 24.00
150. Iosifescu/Theodorescu: Random Processes and Learning. DM 68,– ; US $ 17.00
151. Mandl: Analytical Treatment of One-dimensional Markov Processes. DM 36,– ; US $ 9.00
152. Hewitt/Ross: Abstract Harmonic Analysis. Vol. II. In preparation
153. Federer: Geometric Measure Theory. DM 118,– ; US $ 29.50
154. Singer: Bases in Banach Spaces I. In preparation
155. Müller: Foundations of the Mathematical Theory of Electromagnetic Waves. DM 58,– ; US $ 14.50
156. van der Waerden: Mathematical Statistics. DM 68,– ; US $ 17.00
157. Prohorov/Rozanov: Probability Theory. DM 68,– ; US $ 17.00
159. Köthe: Topological Vector Spaces I. DM 78,– ; US $ 19.50
160. Agrest/Maksimov: Theory of Incomplete Cylindrical Functions. In preparation
161. Bhatia/Szegö: Stability Theory of Dynamical Systems. In preparation
162. Nevanlinna: Analytic Functions. Approx. DM 68,– ; approx. US $ 17.00
163. Stoer/Witzgall: Convexity and Optimization in Finite Dimensions I. DM 54,– ; US $ 13.50
164. Sario/Nakai: Classification Theory of Riemann Surfaces. In preparation
165. Mitrinović: Analytic Inequalities. In preparation
166. Grothendieck/Dieudonné: Eléments de Géometrie Algébrique. In Vorbereitung
167. Chandrasekharan: Arithmetical Functions. In preparation
168. Palamodov: Linear Differential Operators. In preparation